Jones & Bergman's
JCT Intermediate
Form of Contract

Jones & Bergman's
JCT Intermediate Form of Contract

Third Edition

Neil F. Jones
& Simon E. Baylis

**Blackwell
Science**

© Neil F. Jones & David Bergman 1985, 1990; Neil F. Jones and Simon E. Baylis 1999

The JCT Intermediate Form is the copyright of the Joint Contracts Tribunal Limited and clauses are reproduced by kind permission.

Blackwell Science Ltd
Editorial Offices:
Osney Mead, Oxford OX2 0EL
25 John Street, London WC1N 2BL
23 Ainslie Place, Edinburgh EH3 6AJ
350 Main Street, Malden
    MA 02148 5018, USA
54 University Street, Carlton
    Victoria 3053, Australia
10, rue Casimir Delavigne
    75006 Paris, France

Other Editorial Offices:

Blackwell Wissenschafts-Verlag GmbH
Kurfürstendamm 57
10707 Berlin, Germany

Blackwell Science KK
MG Kodenmacho Building
7–10 Kodenmacho Nihombashi
Chuo-ku, Tokyo 104, Japan

The right of the Author to be identified as the Author of this Work has been asserted in accordance with the Copyright, Designs and Patents Act 1988.

First edition published as a Commentary on the JCT Intermediate Form of Building Contract by Collins Professional and Technical Books 1985; second edition published by BSP Professional Books 1990; third edition published as Jones & Bergman's JCT Intermediate Form of Contract by Blackwell Science Ltd, 1999.

Set in 10/12pt Palatino
by DP Photosetting, Aylesbury, Bucks
Printed and bound in Great Britain by
MPG Books Ltd, Bodmin, Cornwall

The Blackwell Science logo is a trade mark of Blackwell Science Ltd, registered at the United Kingdom Trade Marks Registry

DISTRIBUTORS

Marston Book Services Ltd
PO Box 269
Abingdon
Oxon OX14 4YN
(*Orders:* Tel: 01235 465500
          Fax: 01235 465555)

USA
Blackwell Science, Inc.
Commerce Place
350 Main Street
Malden, MA 02148 5018
(*Orders:* Tel: 800 759 6102
                 781 388 8250
           Fax: 781 388 8255)

Canada
Login Brothers Book Company
324 Saulteaux Crescent
Winnipeg, Manitoba R3J 3T2
(*Orders:* Tel: 204 837-2987
           Fax: 204 837-3116)

Australia
Blackwell Science Pty Ltd
54 University Street
Carlton, Victoria 3053
(*Orders:* Tel: 03 9347 0300
           Fax: 03 9347 5001)

A catalogue record for this title
is available from the British Library

ISBN 0-632-04257-5

Library of Congress
Cataloging-in-Publication Data
Jones, Neil F.
    [JCT intermediate form of contract]
    Jones & Bergman's JCT intermediate form of contract/Neil F. Jones & Simon E. Baylis. — 3rd ed.
       p. cm.
    Rev. ed. of: A commentary on the JCT intermediate form of building contract/Neil F. Jones, David Bergman.
    Includes index.
    ISBN 0-632-04257-5 (hb)
    1. Construction contracts—Great Britain.
I. Baylis, Simon E.   II. Jones, Neil F. Commentary on the JCT intermediate form of building contract.
III. Joint Contracts Tribunal.   IV. Title.   V. Title:
Jones and Bergman's Joint Contracts Tribunal intermediate form of contract.
KD1641.J64   1999
343.73'078624—dc21   99-28281
                                                  CIP

For further information on Blackwell Science, visit our website: www.blackwell-science.com

Contents

Preface

In 1984, the Joint Contracts Tribunal produced the Intermediate Form of Building Contract to fill a perceived gap between the JCT 80 Standard Form of Building Contract which was devised for sizeable and complex contract works on the one hand and the Agreement for Minor Building Works which was clearly unsuitable for contract works other than for those of a simple nature and generally of low value. Since then the Intermediate Form has become popular and well used.

This third edition deals with all of the important changes which have taken place in the last nine years. These changes cover amendments to the contract form itself, developments in case law and the introduction of relevant legislation.

Among the major amendments to the contract itself are those dealing with determination (Amendment 7, April 1994) which introduced a completely new set of provisions on determination of the contractor's employment; amendments to deal with the coming into force of the Construction (Design and Management) Regulations 1994 (Amendment 8, March 1995); an amendment to reword the conclusive nature of the final certificate (Amendment 9, July 1995) following some important cases one of which was a case on the Intermediate Form itself (see below); and the highly significant amendment (Amendment 12) catering for the coming into force of the Housing Grants, Construction and Regeneration Act 1996 (see below) together with the implementation of some, at least, of the recommendations for reform contained in the 1994 Latham Review. Finally, following the coming into force of the Arbitration Act 1996 on 31 January 1997, a new set of arbitration rules governing all JCT contracts was introduced, namely the JCT 1998 Edition of the 'Construction Industry Model Arbitration Rules'.

In November 1998, the Joint Contracts Tribunal Limited published a 1998 Edition of the Intermediate Form of Building Contract (known as IFC 98). This is for the most part a consolidation of IFC 84 with its 12 amendments (although the Tribunal took the opportunity of introducing with the 1998 Edition an option for disputes to be taken to litigation rather than arbitration). Accordingly, this third edition will be of value to anyone concerned with either the IFC 84 or IFC 98 edition of the contract.

Many important cases are covered in this new edition including *Emson Eastern Ltd* v. *E.M.E. Developments* (1991) (what amounts to practical completion); *Southway Group Ltd* v. *Wolff* (1991) (limits to the ability to sub-let work); *Colbart* v. *Kumar* (1992) (a case on IFC 84) and *Crown Estate Commissioners* v. *Mowlem* (1994) in the Court of Appeal (both dealing with the conclusiveness of the final certificate); *J. & J. Fee Ltd* v. *The Express Lift Co.* (1993) and *Allridge* v. *Grandactual* (1996) (dealing with implied terms as to co-operation in construction contracts); *Philips (Hong Kong) Ltd* v. *Attorney General of Hong Kong* (1993) (a Privy Council case which appears to mark a judicial change supporting liquidated damages provisions when they are attacked as being a penalty clause); *Linden Gardens Trust Ltd* v.

Lenesta Sludge Disposals Ltd (1994) (a House of Lords' decision dealing with both the question of consent to assignment under construction contracts and also the issue of whether substantial damages can be claimed for defective work even where no loss is actually suffered); *West Faulkner Associates* v. *London Borough of Newham* (1994) (a Court of Appeal case dealing with the very important question of what amounts to a failure to progress 'regularly and diligently with the works'); *Beaufort Developments (NI) Ltd* v. *Gilbert-Ash NI Ltd* (1998) (a House of Lords' case finally overruling the Court of Appeal decision in 1984 in the case of *North West Regional Health Authority* v. *Derek Crouch Construction Ltd*; and finally *Macob Civil Engineering Ltd* v. *Morrison Construction Ltd* (1999), the first decided case on the enforcement of adjudicators' decisions under the statutory Adjudication Scheme introduced pursuant to the Housing Grants, Construction and Regeneration Act 1996 – see below).

1994 saw the coming into force of the Construction (Design and Management) Regulations which has had a significant impact on the IFC Contract and which is dealt with in this edition. Of fundamental importance has been the coming into force, on 1 May 1998, of Part 2 of the Housing Grants, Construction and Regeneration Act 1996 which, when it came into force on 1 May 1998, marked an unprecedented statutory excursion, some might say incursion, into the field of construction contracts. Its effects in the areas of dispute resolution cannot be over-emphasised, particularly adjudication, and payment matters, including the right to interim payment; the provision of an adequate payment mechanism; notice of proposed payment; notice of intention to withhold payment and a partial prohibition on the inclusion of pay when paid clauses.

Looking ahead, there is presently passing through Parliament the Contracts (Rights of Third Parties) Bill. It is likely in its current or amended form to become law before the end of 1999. It will have significant implications in the sphere of contracts generally, including construction contracts. It will create a significant exception to the rule relating to privity of contract by enabling parties to a contract to expressly grant benefits to a third party with that third party being able to enforce its entitlement. It has particular application to the construction industry including the position of potential tenants and purchasers etc. of buildings who are not themselves parties to a construction contract, but who can be expected to have benefits conferred on them in such contracts.

The citing of legal cases has been used principally where matters of a controversial nature are dealt with or where the decision is particularly germane to some aspects of the contract. Wherever possible, recent cases have been referred to.*

For this third edition, David Bergman steps aside and is replaced by Simon Baylis. David is now enjoying a richly deserved retirement at the end of a long and

---

*Since writing the book, two cases have been reported which bear on the text. First, in the case of *British Fermentation Products Ltd* v. *Compair Reavell Ltd* (1999), Judge Bowsher entertained the possibility, without deciding the point, that a JCT standard form could in appropriate circumstances be regarded as the standard terms of business of one of the parties (whether employer or contractor) for the purposes of Section 3 Unfair Contract Terms Act 1977. See page 43 of the book.

Secondly, in the case of *How Engineering Services Ltd* v. *Lindner Ceilings Floors Partitions Plc* (1999), Mr Justice Dyson in referring to the words of Judge Lloyd the McAlpine case (quoted in the text on page 251 of the book) made it clear that there is room for the exercise of judgement in the process of ascertainment and that the general principles applying to the quantification of damages for breach of contract are relevant.

successful career in the construction industry. It is probably true to say that without his sowing the seeds for a book such as this back in 1984 (and earlier) it might never have been written. We both wish him well and thank him for his past contribution.

Finally, we are both enormously indebted to Mary, wife of the first named co-author, for undertaking the task of typing the script and helping in the process of checking and cross-checking.

We have tried to state the law as at 1 February 1999.

We take this opportunity to thank the Joint Contracts Tribunal Limited for kindly giving us permission to reproduce the IFC contract and some linked documentation.

Neil Jones
Simon Baylis

List of abbreviations

ACOP	Health and Safety Commission approved code of practice
BEC	Building Employers Confederation (now Construction Confederation)
CC	Construction Confederation (formerly Building Employers Confederation)
CDM Regulations	Construction (Design and Management) Regulations 1994
CIMAR	Construction Industry Model Arbitration Rules
DOM/1	Standard Form of Subcontract for Domestic Subcontractors where main contractor under JCT 98
DOM/2	Standard Form of Subcontract for Domestic Subcontractors where main contractor under JCT 98 with Contractors Design
EDI	electronic document interchange
ESA/1	Employer/Subcontractor Agreement
FASS	Federation of Association of Specialists and Subcontractors
IFC 84	JCT Intermediate Form of Building Contract 1984
IFC 98	1998 Edition of the JCT Intermediate Form of Building Contract
IN/SC	Articles of Agreement and Sub-Contract Conditions between main contractor and domestic sub-contractors where main contractor under IFC
JCT	Joint Contracts Tribunal Ltd
JCT 63	Joint Contracts Tribunal Standard Form of Building Contract 1963 Edition
JCT 80	Joint Contracts Tribunal Standard Form of Building Contract 1980 Edition
JCT 98	Joint Contracts Tribunal Ltd Standard Form of Building Contract 1998 Edition
NAM/SC	JCT Standard Form of Sub-Contract Conditions for Sub-Contractors where main contractor under IFC
NAM/T	Form of Sub-Contract Tender and Agreement for use with NAM/SC

NFBTE	National Federation of Building Trades Employers
NSCC	Nominated Sub-Contract Conditions where main contractor under JCT 98
NSC/W	JCT Standard Form of Employer/Nominated Sub-Contractor Agreement where main contractor under JCT 98
RIBA	Royal Institute of British Architects
SECG	Specialist Engineering Contractors' Group
SMM 7	Standard Method of Measurement of Building Works, 7th edition
the Tribunal	Joint Contracts Tribunal Ltd
UNCITRAL	United Nations Commission on International Trade Law

Chapter 1
Background to the Intermediate Form

The Joint Contracts Tribunal, as it then was, decided in 1984 to produce the Intermediate Form of Building Contract to fill a perceived gap between the JCT 80 Standard Form of Building Contract which was designed for sizeable and complex contract works on the one hand and the Agreement for Minor Building Works (Minor Works Form) which was clearly unsuitable for contract works other than those of a simple nature and of low value. It was published in September 1984 and was called the JCT Intermediate Form of Building Contract 1984 (IFC 84).

Following the formation of The Joint Contracts Tribunal Limited (the Tribunal) which commenced trading on 1 May 1998, the JCT Council was formed, and this Council now performs the functions of the former Tribunal under the direction of the new Tribunal. The Council is made up of representatives formed into five colleges representing particular sectors of the industry:

The Employer's/Client's College (including local authorities)
The Contractor's College
Specialists and subcontractors' College
The Consultants' College
The Scottish building industry interests

Between original publication in 1984 and April 1998, twelve amendments were issued, many of which themselves were made up of a considerable number of items covering a wide variety of matters. For example, Amendment 12 issued in April 1998 covered 15 separate items. The current Tribunal decided to consolidate these amendments into a new edition known as the 1998 Edition of the Intermediate Form of Building Contract (IFC 98). Though essentially a consolidating edition of IFC 84 with its amendments, it nevertheless introduced two new provisions: firstly, the inclusion of a provision enabling electronic data interchange in respect of the communication of contractual and other documentation (clause 1.16); and secondly, the introduction of an option for the resolution of disputes by litigation as well as arbitration (article 9B and clause 9C). Additionally, provisions for partial possession which were previously contained in an additional clause 2.11 in the IFC Practice Note IN/1 have now been brought into the contract as a standard provision. Finally, the opportunity was taken to incorporate a considerable number of corrections, some of which may be substantive. It is therefore important to realise that IFC 98 is not precisely the same as a consolidation of IFC 84 with Amendments 1 to 12.

The form is intended for use in England and Wales. It is not used in Scotland. For works which are to be carried out in Northern Ireland, an Adaption Schedule for use with the form is available from the Royal Society of Ulster Architects in Belfast. The previous Tribunal also issued the following documents for use with IFC 84:

- IFC 84 Supplemental Conditions C and D – Fluctuations Clauses (1995)
- JCT Consolidated Main Contract Formula Rules October 1987
- Form of Sub-Contract Tender and Agreement for persons named under clause 3.3 (NAM/T)
- Sub-Contract Conditions (including fluctuations clauses), referred to in NAM/T (NAM/SC)
- Sub-Contract Formula Rules, referred to in NAM/T and NAM/SC
- Clause for insertion in IFC 84 as '2.11 – Partial Possession by Employer' (to be found in Appendix to Practice Note IN/1), if required. This clause has now been incorporated into clause 2 of the conditions in IFC 98
- IFC 84 Sectional Completion Supplement 1985 edition revised January 1996.

All of the above have been or are being reissued for use with IFC 98, amended as appropriate in the form of a 1998 edition.

In addition, a very useful Practice Note IN/1 has been issued. Furthermore, although not a Tribunal document, a separate agreement to be entered into between employer and a named person as sub-contractor dealing with the matter of design has been prepared by the RIBA and SECG for use with IFC 84. It is known as ESA/1. ESA/1 is now in need of an overhaul. Hopefully this will happen soon. Alternatively, the Tribunal may decide to produce its own form of warranty for use between employer and a named person as sub-contractor.

IFC 98 is designed to be suitable for building works:

(1) Of a simple content involving the normally recognised basic trades and skills of the industry; *and*
(2) Without any building service installations of a complex nature, or other specialist work of a similar nature; *and*
(3) Adequately specified, or specified and billed, as appropriate prior to the invitation of tenders.

Finally, though again not a Tribunal document, the CC, SECG and NSCC have produced standard Articles of Agreement and Sub-Contract Conditions for use between the main contractor and domestic sub-contractors, known as IN/SC. It is likely that the Tribunal will itself produce and publish a domestic form of sub-contract based in large part on the above form.

IFC 98 follows the general layout of the Minor Works Form, the clauses being grouped under very similar section headings, but there the similarity ends. The clauses are far more detailed and comprehensive although, by its very name, obviously less so than JCT 98.

IFC 98 is published in one edition only for both private and local authority use, and the text allows for a number of alternative documents to be included as contract documents. In addition to the drawings, the supporting documents can comprise bills of quantities, specification or schedule of works and one or more of these will be priced by the contractor. Where the drawings are supported by an unpriced specification only, then the contractor must supply either a schedule of rates or a contract sum analysis in support of his tender. It should be noted that in this contract, when there are no bills of quantities, the employer is responsible for the accuracy of any quantities given in the specification or schedule of works and in this respect it differs from JCT 98 (Without Quantities).

Practice Note 20 issued by the Tribunal (revised August 1993) gives guidance on the appropriate form of JCT main contract to use. In dealing with the use of IFC 84 (now IFC 98) it states in paragraph 13 as follows:

'...The Intermediate Form has been prepared so as to be suitable for contracts for which the more detailed provisions of the 1980 Standard Form are considered by the Employer or by his professional consultants to be unnecessary in the light of the foregoing criteria. The Form would normally be the most suitable form for use, subject to these criteria, where the contract period is not more than twelve months and the value of the works is not more than £280,000 (1992 prices), but this must be read together with paragraph 14 on the money limits within which the use of the Minor Works Form may be appropriate.

The Intermediate Form may however be suitable for somewhat larger or longer contracts, provided the three criteria referred to in the endorsement are met, but Employers and their professional consultants should bear in mind that the provisions of the Intermediate Form are less detailed than in the 1980 Standard Form and that circumstances may arise, if it is used for unsuitable works, which could prejudice the equitable treatment of the parties.'

From a consideration of the three criteria set out earlier regarding the nature of the building works for which IFC 98 is designed (see page 2) in conjunction with these two paragraphs of Practice Note 20, it is clear that the three criteria should govern the situation. It is not therefore the contract value or contract period which is the determining factor. It is the nature of the work which is undertaken. So for instance, there seems no reason at all why IFC 98 should not be used for the construction of a large estate of conventional dwelling houses whatever the total value of the works.

The second quoted paragraph above from Practice Note 20 refers to the fact that circumstances may arise, if IFC 98 is used for unsuitable works, which could prejudice the equitable treatment of the parties. When the form is scrutinised, it is almost certain that the reference to unsuitable works is a reference to the use of named sub-contractors to carry out specialist work of a complex nature. If such specialist works, e.g. complex service installations, form a significant proportion of the total cost of the contract and take up a significant part of the construction period, it might be argued by some that the increased risks which a main contractor takes under IFC 98 for named sub-contractors when compared with the responsibility of the main contractor under JCT 98 for nominated sub-contractors, is such that it is an unfair burden on the contractor in such circumstances. Others may well disagree. Under IFC 98 it is of course the main contractor who takes the risk of delay and disruption caused by named sub-contractors. This is in some respects different from the case of nominated sub-contractors under JCT 98.

Furthermore, in the event of a named sub-contractor's employment being determined, one of the options for the architect is to require the contractor to finish off the outstanding work of the named sub-contractor. This may be regarded as unfair in relation to highly specialist complex work. There is no equivalent provision in relation to nominated sub-contractors under JCT 98.

There are other risks of inequitable treatment of the parties where specialist or complex work is concerned. For example, the provision in JCT 98 clause 8.1.1 which provides relief for the contractor where materials or goods of the kinds and standards described in the contract documents are not procurable, is absent in IFC 98. Another example is to be found in JCT 98 clause 11 which provides for the architect to have access to off-site workshops or other places of the contractor and sub-contractors where work is being prepared. Again there is no such provision in IFC 98.

Probably the main difference between JCT 98 and IFC 98 is that in the latter there is no provision for nomination. IFC 98 allows the employer to 'name' a sub-contractor to execute work but does not permit the employer to nominate a sub-contractor in the way in which he can do so under JCT 98. The work may be specified in the main contract tender document, the tender submitted by the named sub-contractor being furnished to tendering main contractors in order that they can themselves price the work. This means that tenders from prospective 'named' sub-contractors must be obtained prior to inviting main contract tenders.

Alternatively, if it is intended to 'name' a sub-contractor after the main contract has been placed, this can be done by including a provisional sum for the work in the main contract tender document. This second method of naming is con-siderably closer to JCT 98 nomination, particularly in relation to the adjustment of the provisional sum in a similar way in which the prime cost sum is adjusted under JCT 98 in connection with nominated sub-contractors. There is no provision for the use of prime cost sums in IFC 98. As with nomination under JCT 98, the Tribunal has published related documents which have to be used in the naming procedure.

It appears to be the case that IFC 84 was well received by the industry. It is one of the hazards of drafting a form which attempts to achieve a reasonable balance of responsibility between the parties that, at the end, neither side feels entirely happy with the outcome. However, it has proved to be a very useful contract. Certainly it has been in very frequent use.

The comments on the clauses in IFC 98 on the following pages are dealt with in the same order as they appear in the contract. The contract places each clause within a section and each section is dealt with as a separate chapter in this book, except for section 3 which, due to the extensive treatment given to naming, has been divided into two parts. Each chapter commences with a summary of the general law appertaining to the matters covered by the particular section of the contract. This is followed by the full text of each clause included in the section, together with a general commentary on the clause and notes on specific extracts. For convenience the term 'the architect' has been used in place of 'the architect/ the contract administrator' and should therefore be read as including the contract administrator unless the context otherwise requires.

In spite of the simplicity of the building works for which it is intended, IFC 98 should be handled with care and the respective responsibilities of the parties should be clearly understood. In this respect it is hoped that the summaries of the relevant law, together with the commentaries on the various clauses, followed by the notes, will assist towards a better understanding of the contract.

Chapter 2
The agreement, recitals and articles

CONTENT

This chapter deals with the form of agreement, recitals and articles.

SUMMARY OF GENERAL LAW

The recitals, if included in an agreement, usually begin with the word 'Whereas'. The recitals are not within the operative part of a document and they cannot generally be used as an aid to the construction of that part. However, if there is a doubt about the construction of the operative part of the contract, the recitals may be looked at in order to see if they assist in determining the true construction. The recitals really introduce what it is that the parties are intending to achieve by their contract. They are in the nature of an introduction and background.

The articles of agreement will often commence with words such as 'Now it is hereby agreed as follows:' and these words introduce the operative part of the agreement.

This Agreement

is made the _____ day of _____ 19 _____

BETWEEN _____

_____

_____

of (or whose registered office is at) _____

_____

_____

(hereinafter called the 'Employer') of the one part

AND _____

_____

_____

_____

of (or whose registered office is at) _____

_____

_____

(hereinafter called the 'Contractor') of the other part.

Whereas

Recitals   **The Works**

First   The Employer wishes the following work _____

_____

_____

_____

to be carried out under the direction of the Architect/the Contract Administrator [a] named in article 3 hereunder and has caused the following documents showing and describing that work to be pre-pared:

the Contract Drawings numbered _____
[b]   the Specification,
[b]   the Schedules of Work,
[b]   Bills of Quantities.

[c]   and, in respect of any work described and set out therein for pricing by the Contractor and for the execution of which the Con-tractor is required to employ a named person as sub-contractor in accordance with clause 3.3.1 of the Conditions annexed hereto, has provided all the particulars of the tender of the named person for that work in a Form of Tender and Agreement NAM/T with Sec-tions I and II completed together with the Numbered Documents referred to therein.

Pricing by the Contractor [d]

Second   A.   the Contractor has priced the Specification/priced the Sche-dules of Work/priced the Bills of Quantities (as priced, called 'the Contract Bills') [b] and the total of such pricing is the Contract Sum as mentioned in article 2 hereof;

and such priced documents and the Contract Drawings, both signed by or on behalf of the parties (together with, where applicable, the particulars, referred to in the First recital, of the tender of any named person in a certified copy of a Form of Tender and Agreement NAM/T with Sections I and II completed, also signed by or on behalf of the parties hereto), the Agreement and the Conditions annexed hereto are here-inafter called the 'Contract Documents';

the Contractor has provided the Employer with a priced Activity Schedule; [e]

Second   B.   the Contractor has –

stated the sum he will require for carrying out the work shown on the Contract Drawings and described in the Specification (which sum is the Contract Sum as mentioned in article 2 hereof) and these documents, both signed by or on behalf of the parties (together with, where applicable, the particulars, referred to in the First recital, of the tender of any named person in a certified copy of a Form of Tender and Agreement NAM/T with Sections I and II completed, also signed by or on behalf of the parties hereto), the Agreement and the Conditions annexed hereto are hereinafter called the 'Contract Documents',

and
supplied to the Employer either a Contract Sum Analysis or a Schedule of Rates on which the Contract Sum is based;

the Contractor has provided the Employer with a priced Activity Schedule; [e]

CDM Regulations

Third   the extent of application of the Construction (Design and Management) Regulations 1994 (the 'CDM Regulations') to the work referred to in the First recital is stated in the Appendix; [f]

Information Release Schedule

Fourth   the Employer has provided the Contractor with a schedule ('Information Release Schedule') which states what information the Architect/the Contract Administrator will release and the time of that release; [e]

Bonds

Fifth   if the Employer requires any bond to be on terms other than those agreed between the JCT and the British Bankers' Association, the Contractor has been given copies of these terms;

---

[a] See clause 8.4 and article 3. Delete 'the Architect' or 'the Contract Administrator'.

[b] Delete as appropriate.

[c] Delete if no items specifying a named person are included in the documents.

[d] Delete alternative A or alternative B as appropriate.

[e] Delete if not provided.

[f] See the notes on the JCT 80 Fifth recital in Practice Note 27 'The application of the Construction (Design and Management) Regulations 1994 to Contracts on JCT Standard Forms of Contract' for the statutory obligations which must have been fulfilled before the Contractor can begin carrying out the Works.

Now it is hereby agreed as follows

Article 1

Contractor's Obligations

For the consideration mentioned in article 2 the Contractor will upon and subject to the Contract Documents carry out and complete the work briefly described in the First recital and shown upon, described by or referred to in the Contract Documents and including any changes made to that work in accordance with this Contract (hereinafter called 'the Works').

Article 2

Contract Sum

The Employer will pay to the Contractor the sum of _____

_____

_____ (£_____ .____) exclusive of VAT (hereinafter called 'the Contract Sum') or such other sum as shall become payable hereunder at the times and in the manner specified in the Conditions.

Article 3

The Architect/ The Contract Administrator

[a] The term 'the Architect'/'the Contract Administrator' in the Conditions shall mean the person referred to in the First recital namely:

_____

of _____

_____

or in the event of his death or ceasing to be so appointed for the purpose of this Contract such other person as the Employer shall within 14 days of the death or cessation nominate for that purpose *not being a person to whom the Contractor shall object for reasons thought to be sufficient by a person appointed pursuant to the procedures under this Contract relevant to the resolution of disputes or differences.* [g] Provided that no person subsequently so appointed under this Contract shall be entitled to disregard or overrule any certificate or instruction given by any person for the time previously appointed.

---

[g] Strike out the words in italics in article 3, when the Architect/the Contract Administrator, or in article 4, when the Quantity Surveyor, is an official of the local authority.

Article 4 [h]

The Quantity
Surveyor

The term 'the Quantity Surveyor' in the Conditions shall mean

_____

of _____

_____

or, in the event of his death or ceasing to be the Quantity Surveyor for the purpose of this Contract, such other person as the Employer shall nominate for that purpose, *not being a person to whom the Contractor shall object for reasons considered to be sufficient by a person appointed pursuant to the procedures under this Contract relevant to the resolution of disputes or differences.* [g]

Article 5 [i]

Planning
Supervisor

A. [j] The term 'the Planning Supervisor' in the Conditions shall mean the Architect/the Contract Administrator

B. [j] The term 'the Planning Supervisor' in the Conditions shall mean

_____

of _____

or in the event of the death of the Planning Supervisor or his ceasing to be the Planning Supervisor such other person as the Employer shall appoint as the Planning Supervisor pursuant to regulation 6(5) of the CDM Regulations.

Article 6 [i]

Principal
Contractor

The term 'the Principal Contractor' in the Conditions shall mean the Contractor, or, in the event of his ceasing to be the Principal Contractor, such other contractor as the Employer shall appoint as the Principal Contractor pursuant to regulation 6(5) of the CDM Regulations.

Article 7 [d]

Completion of
Works by
Sections

A. The modifications to the Conditions listed in the Sectional Completion Supplement to the Intermediate Form of Building Contract are incorporated in this Contract and the provisions of the Articles of Agreement and the annexed Conditions shall have effect as so modified.

B. Sectional completion does not apply.

Article 8

Dispute or
difference –
adjudication

If any dispute or difference arises under this Contract either Party may refer it to adjudication in accordance with clause 9A.

Article 9A

Dispute or
difference –
arbitration

Where the entry in the Appendix stating that "Clause 9B applies" has not
been deleted then, subject to article 8, if any dispute or difference as to
any matter or thing of whatsoever nature arising under this Contract or
in connection therewith, except in connection with the enforcement of
any decision of an Adjudicator appointed to determine a dispute or
difference arising thereunder, shall arise between the Parties either
during the progress or after the completion or abandonment of the
Works or after the determination of the employment of the Contractor,
except under Supplemental Condition A7 *(value added tax)* or Supplemental Condition B8 (*statutory tax deduction scheme*), it shall be referred to
arbitration in accordance with clause 9B and the JCT 1998 edition of the
Construction Industry Model Arbitration Rules (CIMAR). [k]

Article 9B

Dispute or
difference –
legal
proceedings

Where the entry in the Appendix stating that "Clause 9B applies" has
been deleted then, subject to article 8, if any dispute or difference as to
any matter or thing of whatsoever nature arising under this Contract or
in connection therewith shall arise between the Parties either during the
progress or after the completion or abandonment of the Works or after
the determination of the employment of the Contractor it shall be
determined by legal proceedings.

---

[h] If the Architect/the Contract Administrator is to exercise the
functions ascribed by the Conditions to the Quantity Surveyor, his
name should be inserted in article 4.

[i] Delete articles 5 and 6 when only regulations 7 and 13 of the CDM
Regulations apply (see Appendix under the reference to the Third
recital).

[j] Delete alternative A or alternative B as appropriate (retaining the
final paragraph).

[k] The JCT 1998 edition of the Construction Industry Model Arbitration Rules (CIMAR) contains procedures for beginning an arbitration and the appointment of an arbitrator, the consolidation or
joinder of disputes including related disputes between different
parties engaged under different contracts on the same project, and
for the conduct of arbitral proceedings. The objective of CIMAR is
the fair, impartial, speedy, cost-effective and binding resolution of
construction disputes. The JCT 1998 Edition of the Construction
Industry Model Arbitration Rules (CIMAR) includes additional
rules concerning the calling of preliminary meetings and supplemental and advisory procedures which may, with the agreement of
the parties, be used with Rule 7 (short hearing), 8 (documents only)
or 9 (full procedure).

Notes

[A1] For Agreement executed under hand and NOT as a deed

[A1]   **AS WITNESS THE HANDS OF THE PARTIES HERETO**

[A1]   Signed by or on behalf of the Employer _____ in the presence of:

[A1]   Signed by or on behalf of the Contractor _____ in the presence of:

. . . .

[A2] For Agreement executed as a deed under the law of England and Wales by a company or other body corporate: insert the name of the party mentioned and identified on page 1 and then use *either* [A3] and [A4] *or* [A5].
If the party is an *individual* see note [A6].

[A3] For use if the party is using its common seal, which should be affixed under the party's name.

[A4] For use of the party's officers authorised to affix its common seal.

[A5] For use if the party is a company registered under the Companies Acts which is not using a common seal: insert the names of the two officers by whom the company is acting *who MUST be either a director and the company secretary or two directors,* and insert their signatures with 'Director' or 'Secretary' as appropriate. *This method of execution is NOT valid for local authorities or certain other bodies incorporated by Act of Parliament or by charter if exempted under s.718(2) of the Companies Act 1985.*

[A2]   **EXECUTED AS A DEED BY THE EMPLOYER**
hereinbefore mentioned namely _____

*[A3]   by affixing hereto its common seal

[A4]   in the presence of:

OR . . . .

[A5]   acting by a director and its secretary*/two directors* whose signatures are here subscribed:

namely _____

[Signature] _____ DIRECTOR

and _____

[Signature] _____ SECRETARY*/DIRECTOR*

[A2]   **AND AS A DEED BY THE CONTRACTOR**

hereinbefore mentioned namely _____

[A3]   by affixing hereto its common seal

[A6] If executed as a deed by an *individual*: insert the name at [A2], delete the words at [A3], substitute 'whose signature is here **[A5]** subscribed' and insert the individual's signature. The individual MUST sign in the presence of a witness who attests the signature. Insert at [A4] the signature and name of the witness. Sealing by an individual is not required.
Other attestation clauses are required under the law of Scotland.

[A4] in the presence of:

* **OR** . . . .

acting by a director and its secretary*/two directors* whose signatures are here subscribed:

namely _____

[Signature] _____ DIRECTOR

and _____

[Signature] _____ SECRETARY*/DIRECTOR*
* Delete as appropriate

Recitals

COMMENTARY

The recitals include a brief description of the work to be constructed and indicate the documents which are to be included as contract documents. Unlike the JCT Standard Forms of Building Contract, where contracts based on differing documentation are published separately, for example with and without quantities, IFC 98 has been designed to allow the maximum degree of flexibility which can be accommodated in one document. Hence the various types of alternative documentation referred to.

The first recital allows for the insertion of a brief description of the work to be carried out. The first recital also makes reference to the person under whose direction the works are to be carried out (that person being named in article 3) and, depending on whether the person so named is an architect, i.e. a person registered under the Architects Act 1997, either the words 'the Architect' or the words 'the Contract Administrator' should be deleted. The deletion made will then be deemed to have been made throughout the conditions (clause 8.4).

The first recital also includes provision for identifying the documents prepared by or on behalf of the employer and '...showing and describing that work...'. These will include the contract drawings and one or more of the three documents listed. The contractor then uses the contract drawings and the other documents in order to compile his tender bid.

The contractor must either price one of the three documents mentioned in the first recital, i.e. the specification, the schedules of work or the bills of quantities (see A of the second recital), following which it will become one of the contract documents; or alternatively, where neither schedules of work nor bills of quantities are used and the specification is unpriced, the contractor must provide the necessary pricing information in a contract sum analysis or schedule of rates (see B of the second recital). Once the contract sum analysis or schedule of rates has been supplied, the unpriced specification will become one of the contract documents. In practice it might occasionally be convenient to have, on the one hand, part of the works covered by a priced specification or schedules of works or bills when alternative A of the second recital will apply, and on the other, part of the work covered by an unpriced specification or contract sum analysis or schedule of rates when alternative B of the second recital will apply. Unfortunately footnote [d] to the second recital (see page 7) only contemplates the use of either alternative A or B for the whole of the works.

The second recital also makes optional provision for the contractor to provide the employer with a priced activity schedule. If this is to be provided then it must be so indicated in the appendix to the conditions stating that the activity schedule is attached to the appendix (clause 4.2.1). The activity schedule is defined as a schedule of activities with each activity priced and with the sum of the prices equal to the contract sum excluding provisional sums and the value of work for which approximate quantities is included in the contract documents (clause 8.3). The activity schedule is then used to value the works for the purposes of making interim payments. The value of the work payable in interim certificates will be valued by taking the value for that work as stated in the activity schedule and applying to it the percentage of the work in that activity which has been properly executed (clause 4.2.1). The valuation of provisional sums and work for which approximate quantities is included is dealt with separately. Furthermore, any payments due as a result of variations will be treated separately. The use of a contract sum analysis does not dispense with the need for a priced document as referred to earlier in this paragraph. For those interested in seeing an example of a priced activity schedule, the Tribunal has provided an example in JCT 80 Amendment 18 Guidance Notes (April 1998).

The drawings must be identified by listing the drawing numbers. Where any work is to be carried out by a named person who has been named as a sub-contractor in the main contract documentation (dealt with in detail in Chapter 5), the employer must also provide all of the particulars of the tender of the named person in respect of that work in a form of tender and agreement known as NAM/T with sections I and II completed together with the numbered documents referred to therein. Alternative A and alternative B of the second recital provide for what are to be the contract documents, namely the agreement, the conditions, the contract drawings, a priced bill of quantities (which becomes known as 'the Contract Bills'), and/or a priced specification, and/or a priced schedule of work, or an unpriced specification together with, where a named person is being appointed as a sub-contractor, the form of tender and agreement NAM/T. Presumably the reference to the conditions is enough to bring in also the appendix to the conditions and the supplemental conditions. However 'Conditions' is nowhere defined and it is interesting to note that in clause 1.3 dealing with priority of contract documents, reference is made separately to the conditions, supplemental conditions and the appendix. The matter is therefore not free from doubt.

A valuation of a variation is based on the 'priced document' (see clause 3.7). Where alternative A in the second recital is applicable, the priced document will be the priced specification and/or schedule of works and/or the contract bills. Where alternative B in the second recital is applicable, the priced document will be the contract sum analysis or the schedule of rates (see clause 3.7.1).

Further, where alternative A is applicable, the total of the pricing equals the contract sum referred to in article 2. Where alternative B is applicable, then the contract sum in article 2 is based on the contract sum analysis or schedule of rates. The contract sum analysis means an analysis of the contract sum provided by the contractor in accordance with the stated requirements of the employer (see clause 8.3). In practice it may be little more than a breakdown of the contract sum into elements, in which case it would have only limited use in the valuation of variations but may be of greater use for the purpose of interim valuations. Alternatively it may take the form of an elemental breakdown with separately priced items showing in detail the breakdown of costs within each element. The greater the refinement of the document, the greater will be its value both in terms of evaluating the tender and subsequently as a means of valuing variations. JCT Practice Note 23 gives guidance on the use and structure of contract sum analyses. This Practice Note, at page 1, refers to the purposes for which a contract sum analysis is required as:

'1. the valuation of variation and provisional sum work insofar as reference to the Contract Sum Analysis for this purpose is required by:
   clause 13.5 in the Standard Form Without Quantities,
   and
   clause 3.7 in the Intermediate Form;
 2. in the Standard Form Without Quantities where clause 40 (use of the price adjustment formulae) is included in the Conditions, to enable the operation of this clause in accordance with the Formulae Rules.
   A further purpose, which is not specifically required by the Conditions, is to facilitate the determination of the amounts to be stated in Interim Certificates under clause 30 in the Standard Form Without Quantities and for Interim Payments under clauses 4.2 and 4.3 in the Intermediate Form.'

It can be seen that so far as JCT 98 without quantities is concerned, the contract sum analysis enables the use of price adjustment formulae in accordance with the Tribunal's Consolidated Main Contract Formula Rules where no bills of quantity have been used. This was specifically catered for by Amendment 3 to JCT 80 Without Quantities Edition (issued March 1987).

A similar amendment to IFC 84 and IFC 98 could have been made but was not. Accordingly as presently drafted IFC 98 does not provide for the use of the Consolidated Main Contract Formula Rules unless priced bills of quantities have been used.

However, if it is desired to use IFC 98 without priced bills but with the price adjustment formula, then it should be quite possible with careful drafting to incorporate similar amendments into IFC 98. This should make IFC 98 even more flexible.

An alternative to the use of a contract sum analysis, where alternative B is applicable, is the use of a schedule of rates which should not be confused with a schedule of work. The latter will generally include quantities, whereas the former will not and will usually take the form of a list of priced items described in a similar fashion to work included in a bill of quantities, and the unit of measure on which the prices are based will be stated. It is likely to have only limited use in the evaluation of the contractor's tender. Furthermore, unless it is possible to have access to the detailed figures building up to the tender sum, it will also be very difficult to say with any certainty whether the rates are those on which the tender was based. That may not necessarily be important providing one can be satisfied that the rates are, in any case, reasonable and that they would form a fair basis for the pricing of variations. It could of course be argued that there are advantages in including a schedule of rates priced much higher than those on which the contract sum is based. The argument would be that it would discourage variations which can have such a disastrous effect on the contractor's efforts to organise the contract works efficiently. That view will clearly not be shared by everyone and it will therefore be prudent to give careful thought to the level of rates included in a schedule of rates when this type of document is to be used in preference to a contract sum analysis. For the reason stated above however, a priced document with quantities with the rates extended and totalled to equal the contract sum will generally be preferable to a schedule of rates.

The contract drawings will always be provided but the extent of detailing included in those drawings will depend on the choice of the other documents. Where contract bills are used it is likely that only a limited number of drawings will be included since the contractor will look to the contract bills to indicate the quality and quantity of work (clause 1.2) and will have priced accordingly. As the work proceeds, the architect will then issue further information in the form of drawings or details. This will be provided either in accordance with the information release schedule, if any (clause 1.7.1), or in relation to information not contained in the schedule then at a time that will enable the contractor to carry out and complete the works in accordance with the conditions (clause 1.7.2).

It can be seen therefore that the contract allows considerable flexibility in the type of document on which the contract sum is to be based. Where the contract sum is based on drawings and a specification or schedule of works, a more comprehensive set of drawings will be required than would be the case if bills of quantities were incorporated as a contract document. The briefer the specification or schedule of works, the greater the reliance which must be placed on the drawings. Further drawings and details will still be issued during the course of the contract but, in the case of dispute, it may be more difficult to decide whether these further drawings and details simply amplify the contract drawings or whether they represent a departure which would qualify as a variation. The same problem would not normally present itself where bills of quantities were used since the quality and quantity of work will be defined in the bills and not on the drawings.

The third recital together with an item in the appendix to the conditions, indicates the extent to which the Construction (Design and Management) Regulations 1994 (the CDM Regulations – and see definition in clause 8.3) apply to the work to be carried out pursuant to the contract. The appendix item provides that either all of the CDM Regulations apply or alternatively that only Regulations 7 and 13 apply. A very brief summary of the CDM Regulations will be found in Chapter 8 dealing with statutory obligations (see page 256). See also clause 5.7 and the commentary thereon (page 270).

In terms of IFC 98, article 5 deals with the appointment of the planning supervisor (see below) and article 6 deals with the appointment of the principal contractor (see below). The IFC 98 conditions make provision for the impact of the Regulations in various ways, e.g. the notification to the contractor of the re-appointment of planning supervisor or principal contractor (clause 1.12); relevant events (clause 2.4); practical completion (clause 2.9); loss and expense matters (clause 4.12); statutory obligations (clause 5.7); determination (clause 7); and the appendix to the conditions. The matters raised in these clauses will be dealt with in the commentaries to the various clauses throughout this book.

The fourth recital which is to be deleted if not applicable states that the employer has provided the contractor with an information release schedule stating what information the architect will release together with the time of that release. Where therefore this recital is included, clause 1.7.1 places an obligation on the architect through the employer to ensure that the information referred to in the schedule is released at the time stated. This topic is discussed in more detail in the commentary to clause 1.7.1 (see page 51).

The fifth recital refers to two forms of bond which are now catered for under this contract. The first relates to the option which the employer has (unless a local authority) to make an advance payment to the contractor, often referred to as a mobilisation payment. If this option is exercised the employer can, through an item in the appendix to the conditions, require an advance payment bond to be provided by the contractor (see clause 4.2(b)). Secondly, a form of bond has been provided to cater for the situation where the employer pays for materials or goods or prefabricated items prior to delivery to or adjacent to the works (clause 4.2.1(c)). The provision of a bond is an option, chosen through the appendix to the conditions, where the payment is in respect of uniquely identified items, but it is expressed in obligatory terms where the payment is in respect of items which are not uniquely identified. If payment for off-site goods and materials etc. is to be provided for in this contract, it is achieved by the provision of a list supplied by the employer to the contractor and annexed to the contract bills, specification or schedules of work. In respect of both types of bond, the contract provides that it shall be from a surety approved by the employer and on terms agreed between the Tribunal and the British Banker's Association. It is then annexed to the conditions. However, both clause 4.2(b) and clause 4.2.1(c) provide for the situation where a bond in different terms is required by the employer. If this is the case then the fifth recital requires the contractor to have been given copies of the terms of the required bond.

Articles

COMMENTARY

CONTRACTOR'S OBLIGATIONS

Article 1 provides that for the consideration (the contract sum) mentioned in article 2, '...the Contractor will upon and subject to the Contract Documents carry out and complete the work briefly described in the first recital and shown upon, described by or referred to in the Contract Documents and including any changes made to that work in accordance with this Contract (hereinafter called 'the Works')'. The definition of 'the Works' therefore encompasses the original description of the works together with any changes (variations) made to that work pursuant to the provisions of the contract.

CONTRACT SUM

Article 2 requires the insertion by the contractor of a sum which becomes 'the Contract Sum' which is capable of adjustment and is payable at the times, and in the manner specified, in the conditions of contract. The contract is a lump sum contract in the sense that it is a contract to complete the whole work for a lump sum. The contractor must carry out and complete the works in accordance with the contract documents (see clause 1.1). Whilst the contractor is entitled to interim payments in accordance with the conditions, these payments are on account of the finally adjusted contract sum. An interim payment is not therefore a final payment in respect of the work to which its value relates.

THE ARCHITECT/THE CONTRACT ADMINISTRATOR

Article 3 provides for a named architect or contract administrator to be inserted, or in the event of his death or his ceasing to be so appointed, such other person as the employer nominates within 14 days of the death or cessation. In relation to subsequent appointments the contractor has a right to object for reasons thought to be sufficient by a person appointed pursuant to the procedures for adjudication, arbitration or litigation set out in articles 8 and 9. The new nominee is not entitled to disregard or overrule any certificate or instruction given by his predecessor.

The use of the title 'architect' is restricted by the Architects Act 1997 which consolidated the enactments relating to architects including the Architects (Registration) Act 1931, the Architects Registration Act 1938, the Architects Registration (Amendment) Act 1969 and relevant parts of the Housing Grants, Construction and Regeneration Act 1996. No one may practise or carry on business under any name, style or title containing the word 'Architect' unless he is a person registered by the Architects Registration Board. However, a body corporate, firm or partnership can carry on the business under the style or title of 'Architect' provided that certain conditions are fulfilled as to the business being under the control and management of a registered architect. It is permissible therefore for the name of an architectural practice to be inserted in article 3 as well as that of an individual ('person' is defined in clause 8.3 to include a partnership or body corporate). It is submitted that this is so, despite the evidential difficulties which may arise in any dispute where the architect's opinion is called into question and which he is required to defend or explain. In practice it does not appear to impose insuperable problems. Provision is also made in this legislation for enrolment of visiting EEC architects.

It is quite common for local authority employers to insert the name of a chief officer as either architect or contract administrator, although actual decisions and opinions are given and expressed by others acting in the name of the chief officer concerned. The chief officer in person may subsequently be called on to justify a decision or opinion which in truth he may not hold. This can be embarrassing.

If the person appointed to administer the contract is not entitled to describe himself as an 'Architect' the reference to 'the Architect' in article 3 should be deleted, and in accordance with clause 8.4, the term 'the Architect' shall be deemed to have been deleted throughout the contract, and of course the reverse applies where the words 'the Contract Administrator' are deleted in article 3.

THE QUANTITY SURVEYOR

Article 4 requires completion with the name and address of the quantity surveyor. Similar provisions apply as in article 3 in the event of death or a cessation of the appointment. The architect or contract administrator can act as the quantity surveyor and if this is the case then his name should be inserted in article 4 – see footnote (h) to Article 4 (reproduced on page 10 of this book).

PLANNING SUPERVISOR

Article 5 deals with the appointment of the planning supervisor. It states that the planning supervisor shall mean the architect or, if it is to be someone else, then space is provided for the name and address of the person to be appointed planning supervisor. There is also provision for a further appointment in the event of the death of the planning supervisor or his ceasing to be a planning supervisor.

PRINCIPAL CONTRACTOR

Article 6 provides that the principal contractor shall mean the contractor under the contract or in the event of the contractor ceasing to be the principal contractor such other contractor (not being the contractor who is a party to the IFC 98 contract) as the employer shall appoint.

The appointment of both a planning supervisor and principal contractor is made pursuant to Regulation 6(5) of the CDM Regulations.

Clause 8.3 (definitions) provides definitions for 'CDM Regulations' which includes any remaking or amendment; 'Health and Safety Plan', 'Planning Supervisor' and 'Principal Contractor'. See also Chapter 8 for a commentary on the CDM Regulations.

COMPLETION OF WORKS BY SECTIONS.

Article 7 deals with the completion of works in predetermined sections. If the works are to be divided into sections in this way, this article ensures that the modifications to the conditions required as a result of using the Sectional Completion Supplement to IFC 98 are incorporated into the contract. More is said about the use of the Sectional Completion Supplement at the end of Chapter 4 (see page 111).

ADJUDICATION

Article 8 provides that if any dispute or difference arises under the contract either party may refer it to adjudication in accordance with clause 9A.

The Housing Grants, Construction and Regeneration Act 1996 in section 108(1) provides that a party to a construction contract has the right to refer a dispute arising under the contract for adjudication. Section 108 goes on to state that the parties can provide for a procedure in the contract itself. To be valid such a procedure must meet certain requirements set out in sub-sections (2), (3) and (4) of section 108. By section 108(5) if the contract does not comply with these requirements, the adjudication provisions of the statutory Scheme for Construction Contracts applies. IFC 98 in clause 9A purports to provide an adjudication procedure which meets the requirements of section 108. This will be looked at when dealing with section 9 (page 382).

The 1996 Act deals not only with adjudication but also with a number of other important matters such as: the right to interim payments; the provision of an adequate payment mechanism; the regulation of the right to withhold payment; the right to suspend performance for non-payment; and a partial prohibition on conditional payment (pay when paid) provisions. Where the Tribunal thought it appropriate for IFC 84 to be amended in the light of the Act, these were included in Amendment 12 issued in April 1998 and are now included in IFC 98. These matters will be dealt with in the commentary to the relevant clauses.

However it is appropriate to provide a separate note in relation to the application and structure of the Act and this is at the end of this chapter.

ARBITRATION

Article 9A deals with arbitration. Article 9B provides for the resolution of disputes or differences by litigation. One or other article will apply to the resolution of disputes if adjudication has not been used or if it has not resulted in a satisfactory resolution of the dispute or difference. Bear in mind however that as with the litigation option in article 9B, the disputes or differences referred to relate to '...any matter or thing of whatsoever nature arising under this Contract or in connection therewith...' whereas article 8 dealing with adjudication provides that 'If any dispute or difference arises under this Contract...'. The difference in wording could be significant. This is discussed in Chapter 12 dealing with the settlement of disputes (see page 382).

The choice of arbitration or litigation is made through an entry in the Appendix to the Conditions. A similar option was incorporated into JCT 80 by Amendment 18 issued in April 1998 (now consolidated into JCT 98). The Tribunal in its Guidance Notes to Amendment 18 to JCT 80 (pages 53 and 54) sets out the basic advantages of each method of resolving disputes and differences and generally gives some advice on the choice between arbitration and litigation. To delve into this matter is outside the scope of this book and reference should be made to the guidance note and to standard texts on arbitration, e.g. Mustill and Boyd *Commercial Arbitration*, second edition, Chapter 2.

Article 9A provides that where the parties have chosen to resolve disputes or differences by arbitration then, subject to article 8 (dealing with adjudication) '...any dispute or difference as to any matter or thing of whatsoever nature arising under this Contract or connection therewith, except in connection with the enforcement of any decision of an Adjudicator ... shall be referred to arbitration in accordance with clause 9B and the JCT 1998 edition of the Construction Industry Model Arbitration Rules (CIMAR)'.

Apart from the question of enforcement of the adjudicator's decision, which must by this provision be processed through the courts and not through arbitration, two other types of dispute or difference are excluded from arbitration under IFC 98:

- Concerning any employer's challenge in regard to value added tax claimed by the contractor in which case the matter is dealt with in accordance with clause A7 of Supplemental Condition A (Value Added Tax) incorporated into IFC 98 by virtue of clause 5.5 of the Conditions;
- Concerning the statutory tax deduction scheme under the Income and Corporation Taxes Act 1988 and the Income Tax (Sub-contractors in the Construction Industry) Regulations 1993 SI No 743 to the extent that the Act or Regulations or any other Act of Parliament or statutory instrument rule or order provides for some other method of resolving such dispute or difference – see clause 5.6 and Supplemental Conditions B8.

Arbitration is dealt with in clause 9B of the Conditions and reference should be made to Chapter 12 which deals with settlement of disputes by arbitration (see page 392)

In addition, it should be borne in mind that the issue of the final certificate can effectively restrict the matters which can be referred to arbitration (see clause 4.7).

LITIGATION

Article 9B provides, by means of an entry in the Appendix to the Conditions, that disputes or differences may be referred to the courts for resolution. This will be after the adjudication process has been used, if either party wishes to adjudicate. Note however the difference of wording in relation to which disputes and differences can be referred (discussed above in connection with arbitration under article 9A). Article 9B is considered along with clause 9C (see Chapter 12 and the commentary on clause 9C on page 400).

The attestation clause

The two parties to the contract are to either sign with witnessed signatures or alternatively execute the articles as a deed. Since the coming into force on 31 July 1990 of section 1(1)(b) of the Law of Property (Miscellaneous Provisions) Act 1989, deeds may be executed by individuals without the need for a seal to be affixed. Companies incorporated under the Companies Act can affix a common seal if they wish. Alternatively they need not affix a common seal. Any document signed by a director and secretary of the company or by two directors and expressed to be executed by the company has the same effect as if executed under the common seal of the company (see section 36A of the Companies Act 1985). The articles of agreement provide useful margin notes to assist the parties in completing the attestation clauses.

The two main characteristics of a deed are that it does not require consideration in order to be enforceable and that under limitation legislation, a right of action on a contract under seal is, in general, barred 12 years after its accrual, whereas in respect of a contract under hand, the period is six years. This fact prompts many employers to require the contractor to create a deed.

THE HOUSING GRANTS, CONSTRUCTION AND REGENERATION ACT 1996

Introduction

Part II of the 1996 Act, dealing with construction matters, came into force on 1 May 1998. It implemented certain recommendations in the report of Sir Michael Latham published in July 1994 called *Constructing the Team*. This report contained 30 main recommendations and many minor proposals in respect of changes in construction industry practice. For present purposes we are concerned with contractual issues, but the report took a much wider perspective of the industry dealing with such matters as education and training, procurement methods, tendering practice and construction costs. The report recommended legislation in certain areas and the Conservative Government of the day confirmed that they would be prepared to legislate provided:

- There was a broad industry consensus in support of the contents of the legislation
- There were practicable legislative solutions
- Sufficient parliamentary time was available.

This led to discussions from which Part II of the Act resulted.

The report also identified some key principles which, it recommended, should be found in construction contracts. The Tribunal where appropriate amended IFC 84, now consolidated in IFC 98, to cater for the legislation and also went beyond this to implement some of the report's key principles for construction contracts. These are dealt with throughout this book in the commentaries to the relevant contract clauses.

Part II of the Act marks an unprecedented statutory excursion, some would say incursion, into the field of construction contracts. It covers adjudication; payment matters including a right to interim payments; the provision of an adequate payment mechanism; notice of proposed payments; notice of intention to withhold payment; and a partial prohibition on the inclusion of pay and paid clauses in construction contracts. In addition, the Part II of the Act deals with the service of notices etc. and the reckoning of periods of time. Finally, Part II of the Act provides for the making by statutory instrument of the Scheme for Construction Contracts. So far as England and Wales is concerned, this is to be found in 'The Scheme for Construction Contracts (England and Wales) Regulations 1998 – SI 1998 No 649' which came into force on 1 May 1998. Scotland has a separate scheme.

It is proposed to deal briefly at this point with the scope and application of Part II of the Act and then to look at the substantive provisions. Where the provisions of Part II of the Act have led specifically to the insertion of contractual provisions in IFC 98, these are cross-referenced at the appropriate point.

Scope

Part II of the Act applies to 'construction contracts', which are defined in section 104(1) as agreements for:

- The carrying out of 'construction operations'
- Arranging for the carrying out of construction operations by others, e.g. sub-contractors
- Providing labour for the carrying out of construction operations.

Construction operations are defined in section 105 of the Act to include all normal building and civil engineering activities and will thus encompass both new work and refurbishment. Preparatory operations such as site clearance, scaffolding and access works are also expressly included, as are painting and decorating.

The professions

Section 104(2) of the Act applies Part II to agreements to carry out architectural design or surveying work or to provide advice on building, engineering, interior or exterior decoration or on the laying out of landscapes, in relation to construction operations. Accordingly, most terms of engagement with architects, consulting engineers and surveyors will be included.

Exclusions

Section 105(2) of the Act expressly excludes certain activities from the definition of construction operations, in particular:

- The extraction of oil or natural gas
- Mineral extraction
- Certain process plant operations.

Apart from these excluded categories, if a contract is partly for construction operations and partly for other purposes, section 104(5) of the Act provides that it applies to that part of the contract which relates to the construction operations. This is likely to cause problems in the practical application of the provisions of Part II of the Act.

It can be seen from this brief summary that where IFC 98 is used, the Act is likely to apply. However, there is one particular exception which should be noted. Section 106 of the Act provides that it does not apply to a construction contract with a residential occupier. A construction contract with a residential occupier means a construction contract which principally relates to operations on a dwelling which one of the parties to the contract occupies, or intends to occupy, as his residence. However, as many of the provisions of Part II of the Act have in any event now been embodied in the IFC 98 Conditions, much of the substance of what is contained in the Act will still apply to such residential construction contracts.

Exclusion Order

After the definition of a construction contract had been drafted and considered by Parliament, it was decided that certain types of agreement which would otherwise be caught by the legislation, should be expressly excluded. An Exclusion Order was therefore made by Parliament. For England and Wales it is the Construction

Contracts (England and Wales) Exclusion Order 1998. It came into force on 1 May 1998. It provides that certain agreements will not be covered by the Act.

Firstly it excludes certain statutory agreements, e.g. under section 38 (power of highway authorities to adopt by agreement) and section 278 (agreements as to execution of works) of the Highways Act 1980; an agreement under section 106 (planning obligations), section 106(A) (modification or discharge of planning obligations) or section 299(A) (Crown planning obligations) of the Town and Country Planning Act 1990; an agreement under section 104 of the Water Industry Act 1991 (agreements to adopt sewer, drain or sewage disposal works); and an externally financed development agreement within the meaning of section 1 of the National Health Service (Private Finance) Act 1997 (power of NHS Trusts to enter into agreements).

Secondly it excludes construction contracts entered into under the private finance initiative as described under paragraph 4(2) of the Exclusion Order. Effectively the exclusion only applies to the first level or concession contract between the government department or other private finance initiative procurer and the special purpose vehicle. The special purpose vehicle will in turn let a construction contract for the construction of the capital asset which is required for the special purpose vehicle to fulfil its obligations. This contract will be covered by the Act in the normal way as will any other contracts connected with it such as sub-contracts or professional terms of engagement.

Finally, certain other contracts are specifically excluded such as finance agreements and development agreements (that is, an agreement which includes provision for the grant or disposal of freehold or leasehold for a period which is to expire no earlier than 12 months after the completion of the construction operations under the contract). If such provision is included within a construction contract then the whole of the contract is excluded from the provisions of Part II of the Act including the construction element.

Contracts to be in or evidenced in writing

Section 107 of the Act provides that Part II only applies where the construction contract is in writing and this includes agreements made in writing whether or not signed by the parties, agreements made by exchange of communications in writing or agreements which are evidenced in writing. Section 107(3) provides that where the parties agree otherwise than in writing, e.g. orally or by conduct, by reference to terms which are in writing, they make an agreement in writing. In other words, if the IFC 98 conditions are incorporated orally by reference, this will be sufficient for Part II of the Act to apply.

Structure of the Act

Part II of the Act adopts an interesting approach in the regulation of contract terms. It provides that certain terms, such as those relating to adjudication, must be included in all contracts which fall within the scope of the Act. If no such clause is included, or if one is included which does not comply with the provisions of the Act, then the fall-back provisions contained in the statutory Scheme for Construction Contracts will automatically be implied into the contract. In many of the areas covered by Part II therefore the parties are, in certain cases within stated constraints, entitled to make their own arrangements in relation to those matters, and it is only if they fail to do so, or fall outside the constraints, that statutorily implied terms will become part of the contract between the parties. The way in which this works can be seen below in the summary of the substantive provisions of Part II of the Act.

Substantive provisions of Part II of the Act

Adjudication

The coming into force of Part II of the Act marks the historic introduction into construction contracts of a right for either party to call for adjudication as a first tier method of dispute resolution in respect of disputes and differences. Section 108 provides as follows:

'(1) A party to a construction contract has the right to refer a dispute arising under the contract for adjudication under a procedure complying with this section.

  For this purpose "dispute" includes any difference.

(2) The contract shall –
  (a) enable a party to give notice at any time of his intention to refer a dispute to adjudication;
  (b) provide a timetable with the object of securing the appointment of the adjudicator and referral of the dispute to him within 7 days of such notice;
  (c) require the adjudicator to reach a decision within 28 days of referral or such longer period as is agreed by the parties after the dispute has been referred;
  (d) allow the adjudicator to extend the period of 28 days by up to 14 days with the consent of the party by whom the dispute was referred;
  (e) impose a duty on the adjudicator to act impartially; and
  (f) enable the adjudicator to take the initiative in ascertaining the facts and the law.

(3) The contract shall provide that the decision of the adjudicator is binding until the dispute is finally determined by legal proceedings, by arbitration (if the contract provides for arbitration or the parties otherwise agree to arbitration) or by agreement.

  The parties may agree to accept the decision of the adjudicator as finally determining the dispute.

(4) The contract shall also provide that the adjudicator is not liable for anything done or omitted in the discharge or purported discharge of his functions as adjudicator unless the act or omission is in bad faith, and that any employee or agent of the adjudicator is similarly protected from liability.

(5) If the contract does not provide for a procedure which complies with the requirements of subsections (1) to (4), the adjudication provisions of the Scheme for Construction Contracts apply.

(6) For England and Wales the Scheme may apply the provisions of the Arbitration Act 1996 with such adaptations and modifications as appear to the Minister making the scheme to be appropriate...'

Summary of main features

The main features of section 108 can be summarised as follows:

(1) The right to adjudicate is just that, a right rather than a duty. The parties can therefore agree not to adjudicate and to proceed directly to a further tier of dispute resolution such as arbitration or litigation. This ability to choose not to adjudicate could well cause difficulties in terms of when and how the parties contract out of adjudication.

(2) The contracting parties can have their own procedures for adjudication, as indeed does IFC 98 in section 9 (see clause 9A). However, if the contractual adjudication scheme is to be valid it must:

 • Enable a party to give notice at any time
 • Provide a time-table with the object of securing the appointment of an adjudicator and referral of dispute within 7 days of the notice
 • Provide for a decision within 28 days (or longer if the parties agree)
 • Provide for the adjudicator to have the right to extend time for up to 14 days with the consent of the referring party
 • Provide for a duty on the adjudicator to act impartially
 • Provide for the adjudicator to take the initiative in ascertaining facts and law.

(3) The contract must also provide for the decision to be binding until determined by the courts or in arbitration or by the parties agreeing to accept the decision as finally determining their dispute. As to the approach of the courts to the enforcement of adjudicators' decisions, see *Macob Civil Engineering Ltd v. Morrison Construction Ltd* (1999), discussed in some detail in Chapter 12 at page 388.

Section 108(4) provides that the adjudicator is not liable for anything done or omitted in the discharge or purported discharge of his functions as adjudicator unless the act or omission is in bad faith, and that any employee or agent of the adjudicator is similarly protected from liability. This immunity for the adjudicator will clearly bind the parties to the contract. However it may well not bind other parties who may be affected by a negligent decision of the adjudicator, provided of course that such a party can clear the very considerable hurdle of establishing that they were owed a duty of care by the adjudicator and that the adjudicator's breach of that duty caused that party loss. (See also page 391.)

Clearly, clause 9A of IFC 98 is intended to satisfy the requirements of section 108 and probably does so. However as will be seen when the clause is considered in detail later (see page 383), it is not entirely without its difficulties in this regard. Should the contractual adjudication provisions fail to meet the requirements of section 108, the statutory scheme will apply. As it is intended to proceed on the basis that the clause 9A adjudication scheme meets the requirements of section 108, Part I (adjudication) of the statutory scheme is not considered in this book.

Payment procedures

Introduction

Part II of the Act imposes a basic payment structure on all construction contracts. It achieves this by requiring contracts to include provisions dealing with certain payment matters, and stating that if they do not, provisions from the statutory scheme will be implied.

A right to interim payments

Section 109 of the Act provides that a party to a construction contract is entitled to interim payments. This can take the form of payment by instalments, stage payments or any other periodic payment arrangement. There is a wide variety of payment mechanisms which can fit within this basic structure. The conventional systems of monthly certification, milestone payments and payment schedules would all satisfy the statutory requirement.

Section 109(2) makes it clear that the parties retain the freedom to agree the amounts of the payments and the intervals at which or circumstances in which they become due. It could be argued that this provision will therefore make little practical difference to payment practices in the industry, as there is nothing to prevent the paying party imposing onerous requirements on the performing party which must be satisfied before payments become due. For instance, a stage payment arrangement which entitled the contractor to payment of only, say, 5% of the contract sum during the progress of the works with 95% payable on completion would still meet the requirements of section 109.

The right to interim payments does not apply where the contract specifies that the duration of the work is to be less than 45 days, or the parties agree that the duration of the work is estimated to be less than 45 days. This will mean that section 109 will not apply to some main contracts and many sub-contracts of short duration, even where they may be of high value.

If the parties fail to agree an interim payment mechanism, the relevant provisions from the statutory scheme will be implied into the contract. This provides for monthly payments based on the value of work executed.

IFC 98 already provides for interim payments and satisfies the requirements of section 109 in this regard.

An adequate mechanism for determining payments

Section 110(1) of the Act requires all construction contracts to provide for an 'adequate mechanism' for determining:

• When payment becomes due under the contract; *and*
• What payments become due.

What amounts to an 'adequate mechanism' is not defined but it is apparent from the context that the question of whether the amount or the timing of payments under the contract is reasonable is not relevant in determining what is adequate. The adequacy appears to relate to the certainty with which the date and amount can be ascertained rather than its fairness. It is almost certain that standard forms such as those produced by the Tribunal, including IFC 98, which provide quite elaborate payment mechanisms, will satisfy the statutory requirements.

The IFC 98 payment provisions (section 4 of the Conditions – see Chapter 7) have been amended to incorporate the language used in section 110 and in an effort to ensure that IFC 98 payment mechanisms are indeed 'adequate'.

Under section 110(1)(b), construction contracts must also provide a final date for payment in relation to any sums which become due. The use of the phrase 'final date for payment' should not be confused with final payments, final accounts or final certificates. It refers to a final date in respect of any payment, including interim payments, falling due under the contract.

The parties are free to agree the interval between the due date and the final date for payment, although if they do not do so, this will be as provided in the statutory scheme, where the relevant period is 17 days from the date when the payment becomes due.

Under IFC 98, the due date for payment is the date of the architect's certificate and the final date for payment in respect of interim certificates is 14 days thereafter. In respect of the final certificate, the final date for payment is 28 days thereafter. More is said about this when considering clauses 4.2 and 4.6 respectively (see pages 204 and 213).

Notice of sums due

Under section 110(2) of the Act, the paying party must give notice of the amount proposed to be paid not later than five days after the due date. He must also specify the basis on which the amount is calculated: it is unclear whether this will require the sum to be broken down, or whether a simple reference back to the relevant provisions in the contract as to how the sums due are to be calculated will be sufficient.

Section 110(2) is not easy to interpret. It requires notice to be given 'not later than five days after the date on which payment becomes due ... or would have become due ... if the other party had carried out his obligations under the contract, and ... no set-off or abatement was permitted by reference to any sum claimed to be due under one or more other contracts...'. On an extreme interpretation, this would appear to mean that if no payment at all is due because, for example, defective work is discovered which diminishes the cumulative value of the works, the party who would have received payment must be informed that no sums are in fact proposed to be paid. Whilst the intention behind the provision, namely to improve communications between the parties and to give the party who is due to receive payment adequate notice of the sums it is proposed paying, is laudable, the inept drafting runs the risk of producing an unwanted and unnecessary bureaucratic burden.

IFC 98 has therefore sought to reflect this provision in a simple statement in clause 4.2.3(a) (interim payments), clause 4.3(b) (interim payment on practical completion) and clause 4.6.1.2 (payment in relation to final certificate) to the effect that the basis on which the proposed payment is calculated is to be given.

Set-off

Section 111 provides that if a party intends to withhold payment after the final payment date of a sum due under the contract then an effective notice of intention to withhold payment must be given to the other party. This notice can be combined with the section 110(2) notice referred to above. To be an effective notice it must specify the amount proposed to be withheld and the ground or grounds for withholding payment. It must be given not later than the 'prescribed period' before the final date for payment. This prescribed period can be agreed, failing which the statutory scheme requires at least seven day's notice. IFC 98 provides for advance notice of at least five days (clause 4.2.3(b) (interim payments); clause 4.3(c) (interim payment on practical completion); and clause 4.6.1.3 (payment in relation to final certificate)).

If an effective notice is given, the merits of any such withholding can be the subject of adjudication. If the adjudicator decides that all or some of the money is to be paid then it becomes payable not later than seven days from the date of the adjudicator's decision or the relevant final date for payment under the contract, whichever is the later.

It should be noted that section 111 refers to a withholding of a 'sum due'. This means that if the reason for not paying is because the sum was never 'due' under the contract, these notice requirements will not apply. The obvious example of this is where the paying party abates on the basis that all or part of the amount included, e.g. in the payment certificate, is for work which was not properly carried out and which therefore is not properly due under the contract at all. There may be a risk that paying parties will seek to categorise any withholding of sums stated as due under interim certificates on this basis even where it is a pretext. Loss or damage alleged to have been suffered as a result of a breach of contract causing disruption or delay, while likely to be capable of being set off subject to the appropriate notice, is not capable of being treated as an abatement, unless, unusually, the disruption or delay affects the actual value of the work – see *Mellowes Archital Ltd* v. *Bell Projects Ltd* (1997) in which Lord Justice Hobhouse said:

> 'It is therefore clear that, for a party to be able to rely upon the common law right to abate the price which he pays for goods supplied or work done, he must be able to assert that the breach of contract has directly affected and reduced the actual value of the goods or work – "the thing itself".'

Under IFC 98, a notice of intention to withhold is already a requirement in respect of deducting liquidated and ascertained damages. The relationship between the statutory requirement and this contractual requirement is considered later when dealing with the deduction of liquidated damages pursuant to clause 2.7 (see page 96).

Suspension of performance for non-payment

Introduction

There is no right under English common law to suspend performance of contractual obligations when the other party is in breach of contract in not paying sums due. If the breach is serious enough the innocent party can bring the contract completely to an end but does not have the option of simply suspending performance while keeping the contract alive (see for example *Lubenham Fidelities and Investments Ltd* v. *South Pembrokeshire District Council* (1986)).

This issue was also referred to by Lord Justice Staughton in *Euro Tunnel* v. *TML* (1992) when he remarked that:

> 'There is a further potential dispute of considerable importance. The Contractors maintain that they are entitled to suspend work on the cooling system (although they have not yet done so), by reason of Eurotunnel's breaches of contract described above. If it were solely a question of English law, this argument would face some difficulty. It is well established that if one party is in serious breach, the other can treat the contract as altogether at an end; but there is not yet any established doctrine of English law that the other party may suspend performance, keeping the contract alive.'

This case went on appeal to the House of Lords in 1993 but the speeches of the Law Lords did not add to or detract from the above statement of the law.

Some construction contracts provide an express right to suspend, e.g. the DOM/1 Form of Domestic Sub-contract for use with JCT 98 and the NSC/C Form for use with nominated sub-contractors under JCT 98. In addition the Form NAM/SC for use with named sub-contractors and IN/SC for use with domestic sub-contractors under IFC 98 contain such provisions. However, IFC 84 prior to Amendment 12 did not provide for such a right. There is now a statutory right to suspend pursuant to section 112 of the Act. This provides that if a sum is due and is not paid in full by the final date for payment in circumstances where no effective notice to withhold payment has been given, the person to whom the sum is due has a right to suspend performance of his obligations under the contract. The right to suspend can only be exercised upon giving at least seven days notice of intention to suspend stating the grounds on which it is intended to do so. The right to suspend then ceases when the party in default makes the payment in full of the amount due.

Any period during which performance is suspended is to be reflected in a statutory extension of time to the completion date.

It should be noted that the right to suspend is in respect of 'performance of his contractual obligations' and is not limited to just construction of the works. It will extend to such matters as site security and even payment of insurance premiums which would otherwise fall due for renewal during a period of suspension.

The statutory extension of time provision in section 112(4) is not well worded. On one construction, it would appear to equate the period of the extension to the contract completion date with the period of suspension for non-payment. It may well be the case that a period of suspension due to non-payment, e.g. 10 days, could affect the contractual completion date by either more or less than that period. IFC 84 in Amendment 12 provided expressly in clause 2.4.18 for an extension of time for completion of the works based on the delay to the date for completion caused by the suspension. This is now consolidated in IFC 98. A comparison of the statutory and contractual extensions of time is considered further when dealing with clause 2.4.18 (see page 94).

Section 112 does not deal with the question of costs associated with the exercise of the right to suspend performance. Presumably this could be recovered as a claim for damages for breach of contract. IFC 98 expressly provides that if the exercise of this right causes the regular progress of the works to be materially affected, causing the contractor to incur direct loss and/or expense, this is recoverable, provided the suspension is not frivolous or vexatious (see clause 4.12.10).

Services of notices etc.

Section 115 of the Act provides that the parties are free to agree the manner of service of any notice or other document required or authorised to be served pursuant to the contract or any other purposes of Part II of the Act. In the event that the parties do not make such an agreement then fall-back provisions are included by which service can be by any effective means. 'Effective means' is not defined, though an example is given in section 115(4) which states that if a notice or document is addressed, prepaid and delivered by post to the addressee's last known principal residence or if in business to his last known principal business address or, if a body corporate, to its registered or principal office, it is to be treated as effectively served.

IFC 98 deals with the service of notices in clause 1.13 and endeavours to follow the provisions of section 115. It should be noted that provision is made in clause 1.16 for the parties to agree to the use of electronic document interchange (EDI) as a means of communicating and this could extend to the service of notices, subject however to the limits set out in the supplemental provisions for EDI set out in annex 2 to the conditions. This is discussed in more detail in the commentary to clause 1.16 (see page 65).

Reckoning periods of time

Section 116 of the Act provides that for the purposes of Part II of the Act, where an act is required to be done within a specified period after or from a specified date, the period begins immediately after that date. It then goes on to provide that where such a period would include Christmas Day, Good Friday or a day which under the Banking and Financial Dealings Act 1971 is a bank holiday in England and Wales or, as the case may be, in Scotland, that day shall be excluded.

IFC 98 in clause 1.14 deals with the matter of reckoning periods of time in a rather more limited way than it is dealt with in section 116 of the Act in that clause 1.14 relates only to the reckoning of periods of time where they are expressed in days rather than more generally. This matter is considered again in the commentary to clause 1.14 (see page 63).

Finally, by section 117, Part II of the Act applies to contracts entered into by or on behalf of the Crown and the Duchy of Cornwall.

Chapter 3
Intentions of the parties

CONTENT

This chapter looks at section 1 which is an assemblage of topics under the general heading 'Intentions of the parties'. They comprise the following:

Contractor's obligations
Quality and quantity of work
Priority of contract documents
Instructions as to inconsistencies, errors or omissions
Bills of quantities and standard method of measurement
Custody and copies of contract documents
Further drawings and details: information release schedule
Limits to use of documents
Issue of certificates by the architect
Unfixed materials or goods: passing of property, etc.
Off-site materials and goods: passing of property, etc.
Reappointment of Planning Supervisor or Principal Contractor – notification to contractor
Giving or service of notices or other documents
Reckoning periods of days
Applicable law
Electronic data interchange.

With such a mixture of topics, it is difficult to give them a suitable composite heading. The heading given, 'Intentions of the parties', could apply equally to each and every clause throughout IFC 98. However, the precedent for this particular grouping was set with the JCT Agreement for Minor Building Works and it has been repeated in IFC 84 and IFC 98.

SUMMARY OF GENERAL LAW

(A) Contractor's obligations

The obligation to complete

In any building contract, the contractor's prime obligation is to carry out and complete the contract works in accordance with the contract documents. This usually takes the form of an express term in the contract to this effect. However, even without such an express term, this obligation would be implied.

The contractor's obligation to complete can have quite extreme consequences unless the contract concerned in some way limits or qualifies the obligation. In its unqualified form, the obligation will mean that if the contract works are damaged or destroyed, even on the last day before completion, the contractor will generally be responsible for reinstating or rebuilding them at his own cost. For the contractual limitations placed on this general obligation in IFC 98 – see note [1] on page 41. However, even apart from qualifications in the contract itself, there are certain legal excuses for a failure by the contractor to perform this prime obligation. Briefly they are as follows.

(i) FRUSTRATION

It is not possible in a book of this kind to consider in detail the law relating to the frustration of contracts. Suffice it to say that there must be some intervening event of a fundamental nature which renders continued performance of the contract impossible or of a totally different nature to that which was envisaged when the contract was entered into. Only very rarely will a contractor be excused performance of his obligation to complete the contract works by reason of the contract becoming frustrated.

The mere fact that performance of the contract turns out to be more difficult or expensive than envisaged by the contractor at the outset will not be sufficient to frustrate the contract. Furthermore, if the contract itself expressly caters for the eventuality concerned, then it cannot be a frustrating event. A relatively modern example of a frustrating event in relation to a building contract occurred in the case of *Wong Lai Ying and Others* v. *Chinachem Investment Co. Ltd* (1979), heard by the Judicial Committee of the Privy Council on appeal from the Court of Appeal of Hong Kong. It involved a landslip which took with it a block of flats of 13 storeys together with hundreds of tons of earth which landed on the site of partly completed buildings completely obliterating them. It was accepted for the purposes of argument that the landslip was an unforeseeable natural disaster. Following the landslip it was uncertain whether the partly completed contract could ever be completed and even if it could, it was uncertain when it could be completed. This was held to be a frustrating event.

In the case of *Davis Contractors Ltd* v. *Fareham UDC* (1956), a fixed price contract to build 78 houses in eight months was held not to be frustrated when, owing partly to severe and unforeseeable shortages of labour and materials, completion took 22 months. Lord Radcliffe in this case said:

'...it is not hardship or inconvenience or material loss itself which calls the principle of frustration into play. There must be as well a change in the significance of the obligation that the thing undertaken would, if performed, be a different thing from that contracted for.'

In the case of *McAlpine Humberoak Ltd* v. *McDermott International Inc. (No. 1)* (1992) the judge at first instance had held that a contract for providing seven steel pallets forming part of a weather deck structure for a tension leg platform had become frustrated. Judge John Davies QC had held that a contract, which was originally based on 22 drawings but which eventually became based on 161 drawings, had been transformed to such an extent that they were not 'changes' within the contract provision for extras and variations. The Court of Appeal overturned this decision, Lord Justice Lloyd saying:

> 'The revised drawings did not "transform" the contract into a different contract, or "distort its substance and identity". It remained a contract for the construction of ... pallets....'

The contract had, properly construed, provided for such an eventuality even if exceptional in extent. Having said this, there must come a moment when, taking a typical contract provision for variations, the number or extent of variations reaches a point where it so changes the nature of the contract as to be outside the scope of the clause. In such a case the contractor would be contractually entitled to refuse to proceed, or to agree to do so only on renegotiating the contract.

It has been said that:

> 'Frustration is a doctrine only too often invoked by a party to a contract who finds performance difficult or unprofitable, but it is very rarely relied on with success. It is in fact a kind of last ditch.'
>
> (Lord Justice Harman in *Tsakiroglou & Co. Ltd* v. *Noblee Thorl GMB* (1961))

A contract may become frustrated if the government prohibits or restricts the work contracted for: see *Metropolitan Water Board* v. *Dick, Kerr & Co. Ltd* (1918).

In the event that the contract is frustrated, the financial position between the parties will generally be governed by sections 1 and 2 of the Law Reform (Frustrated Contracts) Act 1943, which gives the court discretion to award reasonable sums for work done or benefits conferred by the parties to the contract.

(ii) EXCLUSION CLAUSES AND TERMS INTERFERING WITH COMMON LAW RIGHTS

The contract between the parties may seek to relieve the contractor from liability to perform under the contract or from liability for a tort connected with the contract. It may well be that in many instances, particularly where the parties are of equal bargaining power, such exclusions or restrictions do no more than apportion the risk between the parties which will, in turn, be reflected in the make-up of the tender sum. The more the contractor is able to exclude, limit or define a risk, the more likely it is that a lower tender sum will result. It is not possible in this book to deal with exclusion clauses in any detail and reference should be made to text books which deal with the topic, e.g. *Chitty on Contracts*, 27th edition, Chapter 14.

What can be said at this point is that, apart from statutory controls (referred to in the next paragraph), it is possible, provided the exclusion clause is appropriately worded in all the circumstances, to exclude liability even for fundamental breaches of contract. There is no rule of law that a fundamental breach committed by one party to a contract inevitably deprives that party of the right to rely on an exclusion clause: see *Photo Production Ltd* v. *Securicor Transport Ltd* (1980), a House of Lords case disapproving the Court of Appeal's decision in *Harbutt's Plasticine Ltd* v. *Wayne Tank and Pump Co Ltd* (1970). The House of Lords unanimously rejected the view that a breach of contract by one party, accepted by the other as discharging him from further performance of his obligations under the contract, brought the contract to an end and, together with it, any exclusion clause. It all depends on the construction of the particular contract as to whether or not an exclusion clause is adequate to limit what would otherwise be the liability of the party at fault.

There is some statutory control of exclusion clauses in business and consumer contracts by virtue of the Unfair Contract Terms Act 1977 and in consumer contracts by the Unfair Terms in Consumer Contracts Regulations 1994. The 1977 Act and the 1994 Regulations provide some control over contract terms which exclude or restrict liability for breach of certain terms implied by statute or at common law into building contracts. They also control some contract terms which purport to entitle one of the parties to render a contractual performance substantially different from that reasonably expected of him. Generally speaking, where a consumer transaction takes place, i.e. one of the parties is acting as a 'consumer' and the other contracting party is selling or providing the work in the course of a business, then the exclusion or restriction of liability in respect of such implied terms is rendered absolutely ineffective. On the other hand, if it is a business transaction, then the exclusion or restriction clause is likely to be effective but only in so far as it satisfies the requirement of reasonableness contained in the 1977 Act; the 1994 Regulations have no application to business transactions.

It is worth noting that the Regulations and the Act differ somewhat in how they define what amounts to a consumer transaction. For instance, under the Act, an individual entering into a contract incidental to the carrying on of his business, say an architect buying a car as a business asset, might well be dealing as a consumer, whereas this would not be so under the Regulations.

The 1994 Regulations apply to any term in a contract between a seller or supplier and a consumer where the term has not been individually negotiated.

Where a consumer uses an independently produced standard form such as IFC 98 which is not therefore offered by the contractor, there may just be an argument that the 1994 Regulations still apply for the consumer's benefit as the Regulations are not, at any rate by express words, limited to standard forms of contract proffered by the supplying party. If this is so the provisions of IFC 98 would need to satisfy the requirements of good faith under Regulation 4 (see also Schedules 2 and 3). Candidates for challenge in consumer transactions might be thought to include a number of those introduced as a result of Part II of the Housing Grants, Construction and Regeneration Act 1996, e.g. requirement of advance notice before being entitled to set-off (clauses 4.2.3(b), 4.3(c) and 4.6.1.3; contractor's right to suspend for non payment (clause 4.4A). The 1996 Act is not of course directed at consumers and in some respects might be thought to be objectionable when considered from a consumer viewpoint. Section 106 provides that the Act does not apply to construction contracts with residential occupiers and it will be rare for a consumer when acting as such to be a party to any construction contract other than as a residential occupier, though it is not impossible, e.g. a parent having a house built for his or her child or vice versa.

IFC 98 does not distinguish between consumer and business transactions. If the contract is with a residential occupier the 1996 Act will not apply but its provisions will still be reflected in the contract. In the event of any such terms being held to be unfair, they would be unenforceable.

In the rare event of the 1996 Act applying even though it is a consumer transaction so far as the 1994 Regulations are concerned, e.g. the parent/child example above, a difficult situation may arise. On the one hand, as it is a construction contract there will be terms implied as a statutory requirement; on the other hand, it will be a consumer contract falling within the Regulations and certain of these implied terms may be arguably 'unfair'. Could a court decide that a term implied as a matter of law is unfair under the Regulations and therefore unenforceable? There might appear to be a clear conflict between two pieces of legislation. As the Regulations implement EEC Council Directive 93/13 which provides in Article 10 that member states are required to bring into force laws or regulations to comply with the Directive, it might be thought likely that the 1994 Regulations would prevail as a matter of construction: see section 2[4] European Communities Act 1972 and, by way of illustration, *Shields* v. *E. Coomes (Holdings) Ltd* [1979]. However, Article 1.2 of the Directive provides that:

> '...contractual terms which reflect mandatory statutory or regulatory provisions ... shall not be subject to the provisions of this Directive.'

Accordingly it is submitted, the Regulations would not apply to such statutory implied terms where the IFC contract was a 'construction contract' within the meaning given to that term under the 1996 Act; whereas the Regulations would apply to the same contract terms in the IFC contract where it is not classified as a 'construction contract' because one of the parties is a 'residential occupier'.

The 1977 Act and the 1994 Regulations are considered again later in this chapter (see page 43).

(iii) BREACH OF CONTRACT BY EMPLOYER

Where the employer is guilty of a sufficiently serious breach of contract, the contractor is entitled to treat the employer's breach as a repudiation which releases the contractor from any further obligation to perform the contract. It is not every breach that will have this effect. It must be of a very serious nature. If it is not, it will entitle the contractor to claim damages for breach of contract but will not entitle him to decline further performance of his contractual obligations.

(iv) MISREPRESENTATION

A misrepresentation is an untrue statement of fact made by one contracting party to the other at, or before, the time of entering into the contract and which acts as an inducement to that other party to enter into the contract.

Where such a statement made by one party is relied on by the other party to its detriment, then that other party may, in appropriate circumstances, be able to obtain a rescission of the contract and also, depending on the circumstances, to claim damages under the Misrepresentation Act 1967. If the statement concerned also becomes a term of the contract, then the contractor will be entitled to sue for breach of contract. For a detailed discussion of misrepresentation the reader is referred to the appropriate text books covering this topic, e.g. *Chitty on Contracts,* 27th edition, Chapter 6.

(v) ACCORD AND SATISFACTION

There is nothing to stop the employer and contractor agreeing to excuse one another from any further performance of their contractual obligations upon agreed terms, e.g. the employer may be running short of funds and the contractor may have entered into a very unprofitable contract and they may well both be happy to terminate their obligations. The parties are of course always free to agree whatever terms they wish to end the contract before completion of the contract work. This is known as an accord and satisfaction.

Implied terms

It has been stated above that, in the absence of an express term requiring the contractor to carry out and complete the contract works in accordance with the contract documents, such a term would be implied. Building contracts, as with all other contracts, are likely to have incorporated within them certain implied terms apart from their express terms.

Certain terms are implied as matters of law, e.g. that in a contract for work and materials the goods will be of satisfactory quality; that to the extent that the builder's skill and judgment is being relied on, the work will be reasonably fit for its purpose; and that reasonable skill and care will be used in relation to the workmanship employed. This last implied term was held to be incorporated into a sub-contract relationship where the main contract used was IFC 84 in the case of *Barclays Bank Plc* v. *Fairclough Building Ltd and Others* (1995) and was held to extend to the care and skill necessary to perform the work required safely.

The fitness for purpose implied term will not operate unless it was reasonable for the employer to rely on the builder's skill and judgment. So, where the employer seeks to specify precisely the materials which are believed to be suitable for his requirements, the implied term will have no room in which to operate. See for example the employer's specification of hardcore requirements in the case of *Rotherham Metropolitan Borough Council* v. *Frank Haslam Milan and Co. Ltd and M.J. Gleeson (Northern) Ltd* (1996) where the main contractor was held not liable for providing hardcore which, while it met the specification was nevertheless (unknown to the parties) unsuitable for use in confined spaces due to its propensity to expand on hydration.

These implied terms were the product of the common law. They are now however embodied in statute law – see sections 4 (as amended by the Sale and Supply of Goods Act 1994) and 13 of the Supply of Goods and Services Act 1982. This Act also incorporates other implied terms where the contract is silent, e.g. the time for performance – see section 14 of the Act. The ability of a contracting party to exclude or restrict these implied terms is limited by the provisions of the Unfair Contract Terms Act 1977 (see section 7) and, where a 'consumer' is involved, the Unfair Terms in Consumer Contracts Regulations 1994. This can have particular significance where, despite carrying out the work in accordance with the contract documents, the contractor is in breach of building regulations, e.g. as to the adequacy of the foundations. It has been held that in this situation a contractor may, despite having observed the strict terms of the contract, nevertheless be liable to the building owner – see the case of *Street and Another* v. *Sibbabridge Ltd and Another* (1980) in which it was held that there was an implied term in a building contract that the contractor would comply with building regulations even if his failure to do so was because he did no more than comply with the architect's design. A further, somewhat revolutionary, example of statutory implied terms is to be found in the Housing Grants Construction and Regeneration Act 1996 which provides in section 114(4) that where any provision of the Scheme for Construction Contracts applies by virtue of Part II of the Act, in default of a contractual provision agreed by the parties, such provisions have effect as implied terms of the contract concerned. For a brief summary of the Act's provisions see Chapter 2 (page 22).

There can be many other non-statutory implied terms in relation to building and other contracts. Examples are implied terms as to co-operation if this is required to enable the other party to carry out the works in a regular and orderly manner and similarly in relation to not hindering the other contracting party: see for example *Allridge (Builders) Ltd* v. *Grandactual Ltd* (1996). Such implied terms can, where the situation warrants it, extend to, for example, a contractor providing information to a sub-contractor in such manner and at such times as is reasonably necessary to enable the sub-contractor to fulfil its obligations under the sub-contract: see *J. & J. Fee Ltd* v. *The Express Lift Co.* (1993) considering the DOM/2 Form of Domestic Sub-contract for use with the JCT 81 (now 98) with Contractor's Design form of contract.

There are other implied terms which are not implied as a matter of law but are implied as a matter of fact. Where the express terms of a contract do not cover a particular situation, the court may imply a term based on the imputed intentions of the parties gleaned from the actual circumstances of the case, where this is required to give the contract what is called 'business efficacy'. The extent and nature of such implied terms will clearly vary from case to case.

Statutory obligations

The contractor's obligation to carry out and complete the works is bound to be subject to a greater or lesser degree to statutory control, for example, in relation to the safety aspects of equipment and methods of working, the Health and Safety at Work etc. Act 1974. Furthermore, under the Construction (Design and Management) Regulations 1994 made pursuant to section 15 of the 1974 Act, the contractor will have significant obligations in relation to health and safety. A summary of these provisions is in Chapter 8, page 256.

Furthermore, the quality or standard of work to be achieved may be affected by statute, e.g. the building regulations under the Building Act 1984 or the Defective Premises Act 1972. Obligations under the building regulations and under the Defective Premises Act 1972 are dealt with in more detail under the heading 'Summary of general law' in Chapter 8 when dealing with clause 5 of IFC 98 (statutory obligations etc – see page 253).

(B) The passing of ownership in goods and materials

For the sake of the contractor's cash flow it is important that he receives as much as possible as soon as possible. It is therefore of considerable benefit to him to be paid for materials and goods which are intended to be incorporated into the works before actual incorporation, e.g. as soon as they are delivered to the site or even off-site if they have been allocated to the particular contract concerned. Many building contracts provide for this.

On the other hand, the employer wishes to be sure that if he pays for materials or goods before they are incorporated into the works, the ownership will vest in him. These are often conflicting interests and difficulties can arise.

Once materials or goods are physically incorporated into the building there will be no problem as the property will pass to the landowner. This legal rule carries the Latin tag *quicquid plantatur solo, solo cedit* (the property in all materials and fittings, once incorporated in or affixed to a building, will pass to the freeholder). Unless perhaps they are readily removable without causing damage, the employer can safely pay for them.

Before such incorporation there is a risk. Simply to state in a contract that property is to pass from contractor to employer is not a full protection for the employer – for instance, where at the time that the contract says the property is to pass, e.g. at the time of payment for the goods or materials, the contractor does not own them. There is a general legal rule that no one can transfer a better title than he himself possesses. The Latin tag is *nemo dat quod non habet*. There are exceptions to this rule, most of which relate to contracts for the sale of goods rather than a building contract. A building contract is a contract for work and materials which is not governed by the law relating to the sale of goods. However, these exceptions will nevertheless be relevant to the position between the contractor and his suppliers and this in turn can determine the ability of the contractor to pass title to the employer even though he himself may not own the materials or goods. It is not possible in a book of this kind to deal with this rule and its exceptions in detail. Reference should be made to the appropriate text books, e.g. *Chitty on Contracts*, (27th edition, Vol. 2 Specific Contracts, section 41 paragraphs 150 to 181).

If a contractor accepts payment from the employer under a contract which expressly provides that the property in materials or goods is to pass to the employer upon payment, when the contractor does not own such materials or goods at that time, clearly he is in breach of contract. If there is no express term in the contract relating to title, there will nevertheless be an implied term as to title by virtue of section 2 of the Supply of Goods and Services Act 1982 which applies, *inter alia*, to building contracts. Section 2 provides that:

'In a contract for the transfer of goods ... there is an implied condition on the part of the transferor that in the case of a transfer of the property in the goods he has the right to transfer the property and in the case of an agreement to transfer the property in the goods he will have such right at the time when the property is to be transferred.'

There is also an implied warranty that:

(1)   The goods are free and will remain free from any charge or incumbrance not disclosed or known to the transferee before the contract is made; *and*

(2)   The transferee will enjoy quiet possession of the goods except so far as it may
      be disturbed by the owner or other person entitled to the benefit of any
      charge or incumbrance disclosed or known.

There are statutory restrictions on the ability of a contracting party to exclude or
limit liability for breach of this implied condition or these implied warranties (see
Unfair Contract Terms Act 1977 section 7(3A)) and in relation to contracts with
consumers see the Unfair Terms in Consumer Contracts Regulations 1994 paragraph 4.

While it is clear therefore that an employer will have a right of redress against a
contractor who is in breach of an express or implied term concerning the transfer
of title, such a remedy is of little value if the contractor becomes bankrupt or goes
into liquidation.

The most common reason why a contractor does not own the materials or goods
at the time of their delivery to site is that the contract between the contractor and
his supplier provides that title is not to pass to the contractor until payment in full
has been received by the supplier. Such clauses are of various types and degrees of
elaboration, e.g. purporting to deal with the situation where the materials or
goods are mixed with others or become part of other materials or goods; making
ownership dependent on the discharge of all debts from that contractor to the
supplier, and so on. These retention of title clauses are sometimes called Romalpa
clauses after the leading case of that name, *Aluminium Industrie Vaassen BV* v.
Romalpa Aluminium Ltd (1976).

A successful retention of title clause will mean that if the contractor becomes
bankrupt or goes into liquidation or a receiver is appointed, without the supplier
having been paid, the supplier's title will hold good against the trustee in bankruptcy, liquidator or receiver and also as against the employer even though he
may have paid the contractor for them. However, such clauses, to be effective,
require very careful drafting.

Where the sub-contract is not for the mere supply of materials or goods under a
sale of goods contract but is a sub-contract for works and materials, the position
can be even more precarious from the employer's point of view as the sub-contract
may well not expressly state when title in the materials or goods is to pass to the
contractor, if at all. It could well be that on a true construction of the sub-contract
the title may never be intended to pass at all to the contractor. For instance, if the
sub-contract assumes that the materials or goods will be incorporated into a
building before payment is made to the sub-contractor by the contractor, then the
title will transfer directly from the sub-contractor to the landowner. The sub-
contract may however provide for payment on delivery to the site, and for the title
to pass when delivery or payment is made. If there is no express contractual term
dealing with the passing of title, then it will be a question of determining from the
circumstances when the parties intended the title in the property to pass: see also
the case of *Dawber Williamson Roofing Co. Ltd* v. *Humberside County Council* (1979)
page 121.

CONSIDERATION OF THE RELEVANT CLAUSES OF IFC 98

Clause 1.1

Contractor's obligations
1.1 The Contractor shall carry out and complete[1] the Works in a proper and workmanlike manner[2] and in accordance with the Contract Documents[3] identified in the Second recital and with the Health and Safety Plan: provided that where and to the extent that approval of the quality of materials or of the standards of workmanship is a matter for the opinion of the Architect/the Contract Administrator such quality and standards shall be to the reasonable satisfaction of the Architect/the Contract Administrator.[4]

COMMENTARY ON CLAUSE 1.1

This clause covers the basic obligation of the contractor to carry out and complete the contract works generally in a proper and workmanlike manner and specifically in accordance with the contract documents and using materials of the quality and workmanship to the standards specified in those documents. As between the employer and the contractor, the employer is generally responsible for design. There is therefore no reference to any design obligation falling upon the contractor so that he is not responsible for the suitability of the materials and goods specified. The contract simply requires him to carry out the works as designed. While there is probably no implied contractual or tortious duty on the contractor to inform the employer of any design defect, the contractor would be well advised to notify the architect in writing if he considers that either the design or the specification is in any way deficient.

Furthermore, if the contractor discovers that there is a divergence between any statutory requirements in relation to the works and the contract documents or any instruction of the architect, he is required immediately to give the architect written notice specifying the divergence (clause 5.2).

The architect's function in inspecting the works is to ensure that the contractor carries them out in accordance with the contract documents and he cannot insist on a higher standard of workmanship or a higher quality of materials than is expressly stated in those documents, unless he issues an instruction under clause 3.6 requiring a variation, in which case there will be an adjustment to the contract sum.

Where the contract documents provide that the approval of the quality of materials or the standards of workmanship is to be a matter for the opinion of the architect, then there is more discretion, as the quality or standards must then be to the reasonable satisfaction of the architect. This particular topic is dealt with in more detail later under note [4], page 44.

NOTES TO CLAUSE 1.1

[1] '...carry out and complete...'
For a general discussion of this obligation see earlier, page 32.

For a general discussion on the time for completion see Chapter 4 dealing with possession and completion.

This obligation to complete generally requires the contractor to carry out and complete the works for the contract sum so that damage or loss, including total destruction of the works, is the contractor's risk. However, this will not be the case in relation to those risks to be covered by 'All Risks Insurance' (see later page 308) where clause 6.3B or 6.3C.2 applies or those risks known as 'Specified Perils' where clause 6.3C.1 applies, as in these cases the restoration, replacement or repair of loss or damage is treated as if it were a variation with the contractor being paid accordingly. This aspect is dealt with in more detail in Chapter 9 dealing with injury to persons and property, insurance and indemnity, and insurance of the works. It is also possible for either party to seek determination of the contractor's employment under the contract following such loss or damage where it is just and equitable to do so where clause 6.3C applies (see clause 6.3C.4.3) or, where the loss or damage is due to a specified peril, if it causes a prolonged suspension of the works (see clause 7.13.1(b)).

[2] '... in a proper and workmanlike manner...'
These words impose an express obligation introduced into the IFC 84 conditions in Amendment 6 issued in July 1991. It was prompted by a review of the contractors' responsibilities under the contract and took account of the case of *Greater Nottingham Co-Operative Society Ltd* v. *Cementation and Others* (1985). This case, when at first instance (4 CLR page 57), held, perhaps surprisingly, that words such as these contained in the contract bills, or alternatively implied, added nothing to the liability provisions set out in the relevant indemnity clause in that contract (clause 18(2) of JCT 63). That being so, it appeared to condition those words by reference to a requirement of negligence omission or default on the part of the contractor. When the same case came before the Court of Appeal (41 BLR page 43) this finding or assumption was left undisturbed. These words are now introduced as an express requirement and are an additional obligation to that contained in any of the other contract documents or to any statutory implied term that workmanship is to be carried out with reasonable care and skill. The reference to 'proper' can clearly relate to how the work is to be carried out, e.g. without trespassing with an overhanging tower crane or without adversely affecting the soil conditions as in the *Greater Nottingham* case itself.

A breach of this obligation by the contractor can have potentially serious consequences. Not only will he be liable for common law damages for breach of contract, the architect by clause 3.14.2 is given the power to issue such instructions, at the contractor's expense, as are reasonably necessary as a consequence of the contractor's failure. This point is considered further in the commentary on clause 3.14.2 (see page 184).

[3] '... in accordance with the Contract Documents...'
The contract documents will be the contract drawings together with either a priced specification, and/or priced schedules of work and/or the contract bills, or alternatively the contract drawings together with an unpriced specification as well as the agreement (recital and articles) and contract conditions (see second recital IFC 98).

The contract documents will usually specify the materials or goods. They may also specify the standard of workmanship. Where they do not so specify, there will be an implied term that the materials or goods will be of satisfactory quality and that the workmanship will be carried out with reasonable care and skill (see sections 4 (as amended by the Sale and Supply of Goods Act 1994) and 13 of the Supply of Goods and Services Act 1982). Even where the contract documents do specify there will still be room for the statutory implied term to apply unless it is inconsistent with the specified requirements and in any event, the express obligation relating to 'a workmanlike manner' will apply (see previous note [2]). Furthermore, even if it were to be argued that a specified standard of workmanship set out in the contract documents required something less than '...a proper and workmanlike manner...' this would not override these express words in clause 1.1 – see clause 1.3 to the effect that the other contract documents shall not override or modify the application or interpretation of the conditions.

While clause 1.2 effectively provides that the specification/schedules of work/ contract bills will determine quality of work, it is submitted that the reference in clause 1.1 to a proper and workmanlike manner is not limited to the quality of the work but extends to the manner in which it is carried out.

The Unfair Contract Terms Act 1977 places restrictions on the ability of a contracting party to exclude or limit the operation of the implied terms as to quality. Where the contract falls within the definition of 'business transaction' any such exclusion or restriction must satisfy 'the requirement of reasonableness'; see sections 7 and 11 and Schedule 2 to the Act.

So far as a contract involving one party 'dealing as consumer' is concerned, the implied terms as to quality cannot be excluded or restricted at all: see section 12 of the Act. Additionally, where the contract is with a consumer, then in relation to terms which have not been individually negotiated, the Unfair Terms in Consumer Contracts Regulations 1994 apply. By virtue of paragraph 4 of the Regulations, any term which is contrary to the requirement of good faith and causes a significant imbalance in the parties' rights and obligations to the detriment of the consumer, will not be binding. In deciding whether a term meets the requirements of good faith relevant considerations include the parties' respective bargaining positions; whether the consumer had an inducement to accept the term; whether the work was provided to meet the consumer's special order; and the extent to which the supplier has dealt fairly and equitably with the consumer.

The majority of contracts let under IFC 98 will form part of a business transaction. In most contracts where the contract documents specifically require a lesser standard than that which would be implied under the Supply of Goods and Services Act 1982, this will be the requirement of the employer, and there is no exclusion or restriction upon which the Unfair Contract Terms Act 1977 can operate. It is also possible that a standard form such as IFC 98 could be covered by the Unfair Contract Terms Act 1977 as being the employer's written standard terms of business. See, for example, a case concerning the British International Freight Association Standard Trading Conditions: *Overland Shoes Ltd* v. *Schenkers Ltd* (*The Times*, 26 February 1998) in which it was held that the section 3 requirement of reasonableness could apply to any contractual term seeking to exclude or restrict the employer's liability in respect of any breach by the employer of a contract term. Such terms are likely to be few and far between in IFC 98 but a possible example could be the conclusive effect of the final certificate on contractors' claims for breach of contract (see clause 4.7 – last inset).

Where a 'consumer' under the 1994 Regulations uses an independently produced standard form such as IFC 98, even though not being put forward for use by the contractor, there may still be an argument that the Regulations still apply for the consumer's benefit as the Regulations are not, at any rate by express words, limited to standard forms of contract proffered by the supplying party. In such a case the conclusiveness of the final certificate against the employer (consumer) could be open to challenge: see especially Schedule 3 item (q) which includes as an indicative unfair term, any term

> 'excluding or hindering the consumer's right to take legal action or exercise any other legal remedy ...'

[4] '... where and to the extent that approval of the quality of materials or of the standards of workmanship is a matter for the opinion of the Architect/the Contract Administrator such quality and standards shall be to the reasonable satisfaction of the Architect/the Contract Administrator'

Where this proviso applies, it is clearly sensible for the contractor to seek the architect's written expression of his reasonable satisfaction. This is particularly so as in some limited circumstances the quality of materials and standards of workmanship are subject to the conclusiveness provisions of the final certificate in clause 4.7 (dealt with in detail in the commentary to that clause, page 213).

It is thought that when this type of provision was originally drafted (in JCT 63) the intention was to limit the application of the proviso to matters of quality and standards for which there was no objective criterion set out in the contract documents so that the expression of satisfaction by the architect was a means by which the contractor knew that he had met his contractual obligations. However, in the case of *Colbart Ltd* v. *H. Kumar* (1992) Judge Thayne Forbes QC sitting on Official Referee's Business decided that the proviso to clause 1.1 was not restricted to such materials and workmanship as are expressly reserved by the contract to the opinion of the architect for approval, but included all materials and workmanship where approval of such matters was inherently something for the opinion of the architect.

Such opinion could of course be expressed by the architect in the ordinary exercise of his certification functions. If an architect issues an interim certificate for payment in respect of work carried out, he must have formed a view and have in a sense expressed his satisfaction before allowing such value to be included in an interim certificate. Even more so there is an approval of a kind when the architect issues the certificate of practical completion and also the certificate of making good defects. Under the particular edition of IFC 84 (prior to Amendment 9 – July 1995) it was held, under the version of clause 4.7 being considered in that case, that the effect was to exclude the employer's remedy in respect of such matters.

This approach, suggesting that this type of proviso applies to all work and materials whether or not expressly stated in the contract documents to be reserved for the expression of satisfaction on the part of the architect, and whether or not the contract documents also provided a relevant objective criterion by which to judge standards or quality, was reinforced in the Court of Appeal case of *Crown Estate Commissioners* v. *John Mowlem & Co. Ltd* (1994). This interpretation coupled with a repetition of the proviso in the final certificate clause led to the conclusiveness of it being applied in situations never intended by the Tribunal. This in turn led to Amendment 9 changing the wording of the final certificate provisions in clause 4.7. This is dealt with again when considering that clause (page 213).

Presumably the reference to materials in this proviso includes goods.

As the quality or standards are to be to the reasonable satisfaction of the architect, the expression of satisfaction by the architect can no doubt be challenged by the employer as well as by the contractor, provided the appropriate steps stated in clause 4.7 are taken.

If the architect's expression of satisfaction or his lack of it is challenged, it is a matter for argument as to whether the price contained in the contract for the materials or work is a relevant factor. In reliance on the case of *Cotton* v. *Wallis* (1955) it is possible to argue that the architect can take the price into account as being a relevant factor in determining whether or not he should express his satisfaction. However, the architect's first point of reference is the description in the contract documents and if this assists in determining the quality or standards then it would be wrong for him to take the price or rate for the work into account. If however the contract documents leave the architect genuinely in doubt, he may bring into his consideration the price or rate. It was confirmed in the case of *Brown and Brown* v. *Gilbert-Scott and Another* (1992) that the architect can have regard to the price in determining matters of standards and quality. He cannot be expected to ignore the fact that the works may have been 'built down' to a price. Even so, the line has to be drawn somewhere between work which exhibits a low quality or standard and that which can be regarded as defective when compared with the contract specification.

The Supply of Goods and Services Act 1982 (as amended by the Sale and Supply of Goods Act 1994), which applies (*inter alia*) to building contracts, states in relation to the implied condition that goods supplied under such a contract are to be of satisfactory quality, and that they are of satisfactory quality if they meet the standard that a reasonable person would regard as satisfactory, taking account of any description of the goods, the price (if relevant) and all the other relevant circumstances (see section 4(2A)). This provision to some extent, therefore, begs the question. If the price is a relevant factor, i.e. where it is clearly intended to have a direct bearing on quality or standards, then the architect is, it is submitted, obliged to take the price into account.

There is no specified limit as to the time within which the architect must express any dissatisfaction. Clearly to express it belatedly could result in much good work becoming abortive. It might be argued that there is an implied term that the architect will express any dissatisfaction within a reasonable time of the work being carried out. There is no provision in IFC 98 equivalent to clause 8.2.2 of JCT 98 which provides for the architect to 'express any dissatisfaction within a reasonable time from the execution of the unsatisfactory work'.

Clauses 1.2 and 1.3

Quality and quantity of work
1.2 Where or to the extent that quantities are not contained in the Specification/Schedules of Work and there are no Contract Bills, the quality and quantity of the work included in the Contract Sum (stated in article 2) shall be deemed to be that in the Contract Documents

taken together; provided that if work stated or shown on the Contract Drawings is inconsistent with the description, if any, of that work in the Specification/Schedules of Work then that which is stated or shown on the Contract Drawings shall prevail for the purpose of this clause.

Where and to the extent that quantities are contained in the Specification/Schedules of Work, and there are no Contract Bills, the quality and quantity of the work included in the Contract Sum for the relevant items shall be deemed to be that which is set out in the Specification/Schedules of Work.

Where there are Contract Bills, the quality and quantity of the work included in the Contract Sum shall be deemed to be that which is set out in the Contract Bills.

Priority of Contract Documents

1.3   Nothing contained in the Specification/Schedules of Work/Contract Bills shall override or modify[1] the application or interpretation of that which is contained in the Articles, Conditions, Supplemental Conditions or Appendix.

COMMENTARY ON CLAUSES 1.2 AND 1.3

These two clauses deal with priority within the contract documents in relation to the quality and quantity of work. The position can be summarised as follows:

(1)   If contract bills are used then the quality and quantity of work is deemed to be that set out in the bills; but if there are no contract bills, then

(2)   If there is a specification/schedules of work which contain quantities then the quality and quantity of work shall be deemed to be that set out therein for the relevant item; but where and to the extent that there are no quantities contained therein, then

(3)   The quality and quantity of work is deemed to be that in the contract documents taken together, but if work shown on the contract drawings is inconsistent with any description in the specification/schedules of work then the contract drawings prevail.

NOTES TO CLAUSES 1.2 AND 1.3

[1] *'...override or modify...'*

It appears on the face of the contract that, except where the quality and quantity of work is concerned, no specification/schedule of works or contract bills can override or modify the conditions of contract. Great care must therefore be taken if it is intended in any way to alter the conditions of contract. This should not be done other than by inserting the amendment into the contract itself or possibly in a separate note of amendments which must itself be made a contract document in its own right. It is sensible practice to add an additional article to the articles of agreement setting out any changes to the conditions or referring to a schedule of amendments or to the fact that the amendments are made on the face of the contract conditions themselves.

The courts have considered the effect of these or similar words in a number of cases:

Gold v. *Patman and Fotheringham Ltd* (1958)
Gleeson v. *Hillingdon LBC* (1970)
North-West Metropolitan Regional Hospital Board v. *T. A. Bickerton & Sons Ltd* (1970)
English Industrial Estates Corporation v. *George Wimpey & Co. Ltd* (1972)

Henry Boot Construction Ltd v. *Central Lancashire New Town Development Corporation* (1980)
E. Turner & Sons Ltd v. *Mathind Ltd* (1986).

In the Scottish case of *Barry D Trentham Ltd* v. *Robert McNeil* (1995) the court had occasion to look at this issue. In this case some amendments were made in the bills of quantities which sought to override or modify the contract conditions (being JCT 80). The changes under consideration related to requirements for sectional completion without the use of the standard form of sectional completion supplement. The court held that while generally such amendments were to be of no effect because of the operation of a similar provision to that under consideration here, nevertheless the appendix to the conditions which indicated sections with their own possession dates, completion dates and liquidated damages provisions, was capable of overriding or modifying the conditions in the contract which were inconsistent with this, e.g. provision for a single possession or completion date, or one sum in respect of liquidated damages for failing to complete the works as a whole. This was so even though the only place in which the appendix was to be found was in the contract bills. The appendix prevailed as it was still a contract document even if located in the contract bills. This case was based on JCT 80 under which it is clear that the appendix is a contract document. Mention has been made earlier in connection with IFC 98 of the slight possibility that the appendix is not a contract document (see page 14).

It may be said that, whilst lip service is paid to the dominant effect of words such as these in preventing any special provisions in the contract bills (or specification/schedule of works) affecting the contract conditions, nevertheless in practice the courts seem, even if by dint of semantic gymnastics, to take heed of what these other documents say even though they may be inconsistent with the contract conditions themselves.

Whether it is a sound policy to give standard printed conditions priority over specially formulated documents is doubted by some. However, it does help prevent those specially formulated documents accidentally affecting the conditions of contract. If amendments are necessary, then provided those using the contract are familiar with its terms, the conditions themselves may be amended in the proper way.

Clauses 1.4 and 1.5

Instructions as to inconsistencies, errors or omissions

1.4 The Architect/The Contract Administrator shall issue instructions in regard to the correction:

- of any inconsistency in or between the Contract Documents or drawings and documents issued under clause 1.7.1 (*Information Release Schedule*) or 1.7.2 (*Provision of further drawings or details*); or
- of any error in description or in quantity or any omission of items in the Contract Documents or in any one of such Documents; or
- of any error or omission in the particulars provided by the Employer of the tender of a person named in accordance with clause 3.3.1 (*Named sub-contractors*); or
- of any departure from the method of preparation of bills of quantities[1] included in the Contract Documents referred to in clause 1.5 (including any error in or omission of information in any item which is the subject of a provisional sum for defined work*);

and no such inconsistency, error or departure shall vitiate this Contract. Where the Contract Documents include bills of quantities and the description therein of a provisional sum for defined work does not provide the information required by General Rule 10.3 in the Standard Method of Measurement the correction shall be made by correcting the description so that it does provide such information.

If any such instruction changes the quality or quantity of work deemed to be included in the Contract Sum as referred to in clause 1.2 (*Quality and quantity of work*) or changes any obligations or restrictions imposed by the Employer, the correction shall be valued under clause 3.7 (*Valuation of Variations*).

---

*See footnote **[bb]** to clause 8.3 (*Definitions*).

Bills of quantities and SMM

1.5   Where the Contract Documents include bills of quantities, those bills, unless otherwise expressly stated therein in respect of any specified item or items, are to have been pre-pared in accordance with the Standard Method of Measurement of Building Works, 7th Edition, published by the Royal Institution of Chartered Surveyors and the Building Employers Confederation (now Construction Confederation).

COMMENTARY ON CLAUSES 1.4 AND 1.5

These clauses can be summarised as follows:

(1) Where the contract documents include bills of quantities then, unless they expressly state otherwise, it is assumed that the Standard Method of Measurement of Building Works, 7th edition (SMM 7), published by the Royal Institution of Chartered Surveyors and the Building Employers Confederation (now the Construction Confederation) has governed their preparation.

(2) If the contract bills depart in any particular from SMM 7, such departure is to be regarded as an error and must be corrected by an instruction from the architect unless the bills expressly deal with the departure in relation to any specified item or items. Such instruction, if it changes the quality or quantity of the work or any obligations or restrictions imposed by the employer shall be valued as a variation under clause 3.7. Even where, therefore, there has been an intentional departure from SMM 7 in respect of an item, if this is achieved by necessary implication rather than expressly, it nevertheless appears to be the case that the contractor may be able to disregard it so far as his rates are concerned and then require a correction which will be valued, if appropriate. This seems unfair if no one was misled. Perhaps in an appropriate case, e.g. where it is established that the contractor had included in his rates for an omitted item, or where the departure has no practical effect on the work to be carried out, the valuation of the instruction would be nil.

(3) The following shall be corrected by an instruction and will, if it changes the quality or quantity of work or changes obligations or restrictions imposed by the employer, be valued under clause 3.7 as a variation:

(a)   Inconsistencies in or between the contract documents or further drawings or details issued under clause 1.7 or drawings or other information under clause 3.9 in relation to setting out; *or*

(b)   Errors in description or in the quantities or omissions of items in the contract drawings; *or*

(c) Errors or omissions in the particulars provided by the employer in relation to the tenders of named sub-contractors (see clause 3.3.1);

(d) Departures from SMM 7 in the preparation of bills of quantities included in the contract documents including any error in or omission of information in any item which is the subject of a provisional sum for defined work.

Where there is an error in or omission of information in any item which is the subject of a provisional sum for defined work, the description of the item is to be corrected so that the mistake or omission is made good.

The phrase 'provisional sum' is defined in clause 8.3 (see page 370) to include defined and undefined work. SMM 7 in General Rules 10.1 to 10.6 includes a description of what amounts to a provisional sum for defined or undefined work (see footnote [bb] to clause 8.3 page 370).

If work is to be classified as defined work certain minimum information is required in accordance with General Rule 10.3:

(a) The nature and construction of the work
(b) A statement of how and where the work is fixed to the building and what other work is to be fixed thereto
(c) A quantity or quantities which indicate the scope and extent of the work
(d) Any specific limitations and the like identified in Section A35 of SMM 7.

If such information cannot be given, the work will be treated as a provisional sum for undefined work. The significance of this is in relation to the contractor's programming, planning and pricing of preliminaries. If the work is classified as a provisional sum for defined work the contractor will be deemed to have made due allowance for all of these matters; whereas if the work is classified as undefined the contractor will be deemed not to have made any such allowance.

The importance of classification is clearly relevant in relation to the valuation of preliminary items (see clause 3.7.7 and page 170), extensions of time (see clause 2.4.5) and loss and expense (see clause 4.12.7).

Clause 1.4 attempts to make it clear that if the description in the bills amounts to an attempt to provide the information required by General Rule 10.3 of SMM 7 but fails to do so as a result of an error in or omission of information, it is nevertheless still a provisional sum for defined work and will be corrected accordingly. The item is not to be treated, as a result of the inaccurate or omitted information, as a provisional sum for undefined work giving rise to a valuation of preliminaries, an extension of time or reimbursement of loss or expense on that basis.

Even so there may be situations where not all of the information required under General Rule 10.3 can be given, rather than where it can but has been omitted by accident. In such cases, e.g. where there is an item for a reception desk and all the information has been given other than how and where it is to be fixed, the contractor will nevertheless to a large extent be able to make an accurate allowance in his programming, planning and pricing preliminaries and is very likely to reflect this in his actual tender. It seems absurd if he were then able to demand that the item be treated as a provisional sum for undefined work resulting in his tender being 'deemed' to exclude any such allowance.

(4) No such inconsistencies, errors or departures will vitiate the contract. However, if the inconsistency etc. is truly fundamental, the correction of which would

render performance of the contract radically different from what the contractor could have reasonably envisaged, then it may vitiate the contract and entitle the contractor to rescind the contract and to be paid for work done on a *quantum meruit* basis, that is, as much as it is worth, rather than in accordance with the contract documents. Such situations will be very rare indeed.

NOTES TO CLAUSES 1.4 AND 1.5

[1] '...bills of quantities...'
Amendment No 4 to IFC 84 issued in July 1988 substituted these words for the former words 'Contract Bills'. This is to ensure that clauses 1.4 and 1.5 apply also to bills of quantities forming part of any named sub-contract documents which include bills of quantities. Such bills of quantities are not within the meaning of 'Contract Bills' as described in the second recital; alternative A, to the agreement recitals and articles for IFC 98.

Clause 1.6

Custody and copies of Contract Documents
1.6   The Contract Documents shall remain in the custody of the Employer so as to be available at all reasonable times for the inspection of the Contractor.
        The Architect/The Contract Administrator without charge to the Contractor shall provide the Contractor with one copy of the Contract Documents certified on behalf of the Employer and two further copies of the Contract Drawings and the Specification/ Schedules of Work/Contract Bills.

COMMENTARY ON CLAUSE 1.6

This clause calls for no comment.

Clause 1.7

Information Release Schedule
1.7.1   Except to the extent that the Architect/the Contract Administrator is prevented by the act or default of the Contractor or of any person for whom the Contractor is responsible, the Architect/the Contract Administrator shall ensure that 2 copies of the information referred to in the Information Release Schedule are released at the time stated in the Schedule provided that the Employer and Contractor may agree, which agreement shall not be unreasonably delayed or withheld, to vary any such time.

Provision of further drawings or details
1.7.2   Except to the extent included in the Information Release Schedule the Architect/the Contract Administrator as and when from time to time may be necessary without charge to the Contractor shall provide him with 2 copies of such further drawings or details which are reasonably necessary to explain and amplify the Contract Drawings and shall issue such instructions (including those for or in regard to the expenditure of provisional sums) to enable the Contractor to carry out and complete the Works in accordance with the Conditions. Such provision shall be made or instructions given at a time when, having regard to the progress of the Works, or where, in the opinion of the Architect/the Contract Administrator, Practical Completion of the Works is likely to be achieved before the Date for Completion or within any extension of time made under clause 2.3, having regard to such Date for Completion or extension of time, it was reasonably necessary for

the Contractor to receive such further drawings or details or instructions. Where the Contractor is aware and has reasonable grounds for believing that the Architect/the Contract Administrator is not so aware of the time when it is necessary for the Contractor to receive such further drawings or details or instructions the Contractor shall, if and to the extent that it is reasonably practicable to do so, advise the Architect/the Contract Administrator of the time sufficiently in advance of when the Contractor needs such further drawings or details or instructions to enable the Architect/the Contract Administrator to fulfil his obligations under clause 1.7.2.

COMMENTARY ON CLAUSE 1.7

Clause 1.7.1 makes provision for an information release schedule to be provided by the architect. This is an option which, if exercised, is anticipated in the wording of the fourth recital to the articles of agreement. If used, it must therefore have been provided to the contractor during the tender period.

The schedule will indicate the information to be provided and the time at which it is to be released. The architect is to provide two copies of this information. If the architect fails to provide the information in accordance with the schedule this will often amount to a breach of contract by the employer. Additionally, provision is made for an extension of time for completion (clause 2.4.7.1) and for reimbursement of direct loss and expense (clause 4.12.1.1).

However, there are two situations in which the architect's non-compliance with the schedule will not expose the employer to the risk of lost liquidated damages and/or a claim for reimbursement of loss and expense. Firstly, the architect's obligation to comply with the schedule does not apply to the extent that he is prevented by the act or default of the contractor or of any person for whom the contractor is responsible, from releasing the information. For example, the architect may on occasions require and be kept waiting for information or details from the contractor before he can comply with the requirements set out in the schedule; for instance, where the architect needs to know the contractor's sequences of working in order to provide the information in an appropriate form.

Secondly, the clause provides that the parties may agree to vary any such time. This agreement is not to be unreasonably delayed or withheld. It should be noted that the agreement is made between employer and contractor rather than architect and contractor. However, it will often be the case that the employer's agreement will depend on the architect's particular situation. The most likely sphere for the operation of this proviso is where progress of the works moves ahead of or alternatively falls behind that which would result in completion on the contractual completion date. Where progress is ahead of that required to complete by the completion date, it is suggested that the employer's consent would often not be being unreasonably withheld where the anticipated time for completion would be before the original completion date as, in such circumstances, the architect may well not be contractually obliged to his employer client to provide information on such a basis. Clearly the matter depends on all the circumstances.

Clause 1.7.2 covers the situation where further information is to be provided which is not included within the information release schedule. It is clear that the amount of information to be provided by way of the schedule is a matter for decision by the employer together with his architect. It is not assumed in IFC 98 that all information necessary to construct the works is to be included in the schedule.

These clauses taken together provide for what is almost invariably the case, that is, that the contract documents as defined must be supplemented with additional drawings or details to enable the contractor to carry out and complete the works.

The drawings included in the contract documents are often small-scale drawings sufficient for the contractor to assess the scope of the works for tendering purposes. In order to construct the works it is necessary for the architect to supply the contractor with construction details, reinforcement bending schedules and finishing schedules etc. If these depart in any way from the scope of the works envisaged by the contract documents taken as a whole, then they should be accompanied by an instruction varying the works under clause 3.6, specifying the extent to which the further drawings or details depart from those provided in the contract documents.

Where further drawings or details reasonably necessary to explain and amplify the contract drawings are not catered for in the schedule, then this is to be provided by the issue of such instructions (including those for expenditure of provisional sums) as will enable the contractor to carry out and complete the works in accordance with the conditions. It is the duty of the architect on behalf of the employer to provide the necessary further drawings or details at an appropriate time, and failure to do so is likely to amount to a breach of contract by the employer, whether or not the contractor has requested the architect to issue them.

The provision of such further drawings or details is generally linked to the progress of the works rather than to any contractual completion date. This means that if the contractor is overrunning, the provision of information by the architect can be geared to actual progress rather than artificially to a contractual completion date which may already have passed. However, if the progress of the works is such that the architect forms the view that the contractor is likely to complete ahead of any contractual completion date, the architect is not bound to provide information geared to actual progress. The basis for this exception is that to require the employer, through the architect, to provide information against a construction programme which was in advance of the contractual completion date, would change the employer's obligations from that envisaged when the contract was entered into. In addition, it would lack reciprocity in that the architect may on the one hand strive to provide such information, only for the contractor to re-programme his works so that he completes later but still by the contractual completion date. In this situation therefore clause 1.7.2 provides for the information to be based on the contractual completion date rather than actual progress. No doubt, generally, it will be in the interests of the employer as well as the contractor to complete early and the architect will be willing or encouraged to provide information based on actual progress.

There will be situations where the architect is unaware of the time when the contractor needs the further information but where the contractor is himself aware of when it is required. In these circumstances and provided the contractor has reasonable grounds for believing that the architect is not aware of when the information is required, clause 1.7.2 requires the contractor, to the extent that it is reasonably practicable to do, to advise the architect of the time sufficiently in advance of when the contractor needs the further information to enable the architect to fulfil his obligations under the clause. While therefore on the one hand the architect has an obligation to provide further information in due time and a failure to do so will often be a breach of contract by the employer; nevertheless on occasions the contractor himself will be in breach of contract in not giving advance notice. It appears that these two requirements are written independently so that even if the contractor is in breach of his obligations it will not necessarily mean that the employer is not in breach by reason of the architect's failure. If this is a correct analysis of these obligations, it means that while the contractor may be able to obtain an extension of time and reimbursement of direct loss or expense the employer can claim damages for breach of contract including as part of the claim the recovery of both the lost liquidated damages due to the extension of time and any loss and expense paid out.

Where the further information required by the contractor takes the form of instructions for or in regard to the expenditure of undefined provisional sums it will often be the case that the contractor will not be in a position to give any advance warning to the architect of when the information will be required pursuant to the last sentence of clause 1.7.2.

Where the main contractor funnels information from a sub-contractor to the architect for him to approve it and issue it back to the contractor, the late supply of this information from the sub-contractor through the main contractor to the architect may not be a breach of contract by the employer. It was held in the case of *H. Fairweather & Co. Ltd* v. *London Borough of Wandsworth* (1987), a case on the JCT 63 form of contract, that where a nominated sub-contractor was late in supplying installation drawings (held in this case not to be design drawings) to the main contractor for onward transmission to engineers on behalf of the architect, who in turn approved them and formally issued them back to the main contractor, this was not a breach by the employer of clause 3(4) of JCT 63 (which was broadly similar for these purposes to clause 1.7.2 of IFC 98). This was so despite the fact that the main contractor had dealt with them promptly. The only delay was on the part of the nominated sub-contractor. Accordingly the contractor could not claim for an extension of time or reimbursement of any direct loss or expense based on late receipt of information under clauses 23(f) and 24(1)(a) of JCT 63 (similar to clauses 2.4.7 and 4.12.1 of IFC 98).

The reasoning is to be found in the following extract from the judgment of Judge Fox-Andrews QC:

'Whatever means of conveying and approving the drawings may have been agreed upon Conduits (the named sub-contractor) had a contractual obligation to Fairweather to supply installation drawings. It necessarily follows that Fairweather had a similar obligation to Wandsworth.

   If without faults on the part of the architect or engineer Fairweather were delayed in their work by reason of Conduits' failure to supply installation drawings in good time, it would be surprising if Wandsworth were liable to Fairweather for such delay. It may be that if and to the extent that delay was due to the architect or engineer Wandsworth would be liable for breach of an implied term to consider and approve such installation drawings with reasonable expedition (see for example *London Borough of Merton* v. *Stanley Hugh Leach Ltd* (1985) 32 BLR 51) or for breach of a collateral contract.

   I am satisfied, however, that the arbitrator was correct in rejecting a claim under clause 3(4) of the JCT contract.'

There are many, however, who argue that this was an exceptional case and that if the information from the sub-contractor to the architect for approval and issue to the main contractor is typically information which it is the employer's responsibility (through the architect) to provide, then delay on the part of the sub-contractor will be the responsibility of the employer and not the contractor. Almost invariably, the information under consideration will be in the form of drawings or details reasonably necessary to explain and amplify the contract drawings and as such will fall within the architect's obligation under clause 1.7.2.

Bearing in mind that under the main contract in the *Fairweather* case, it was for the architect to furnish the contractor with such drawings or details as were necessary to enable the contractor to carry out and complete the works in accordance with the contract conditions rather than requiring the contractor to obtain such information directly from a nominated sub-contractor, the decision is in some respects surprising, particularly as, when using JCT 63 as the main contract, there was available a 'Grey Form' of warranty between the employer and the nominated sub-contractor under which the sub-contractor promised to provide information to the architect in such manner as not to give the contractor an entitlement to an extension of time or reimbursement of loss and expense on the basis of late receipt of information from the architect.

Turning to IFC 98, if the contract is set up on the basis that installation drawings are to be issued by the architect to the main contractor, the fact that for administrative convenience the flow of information from a named sub-contractor to the architect is channelled through the main contractor should not, without more, relieve the employer of responsibility if the information is delivered late due to the named sub-contractor's default. The situation might be different if there is a clear contractual obligation in the sub-contract requiring the sub-contractor to provide the information to the main contractor. The lesson for the contractor is clear: any information which has to find its way from the named sub-contractor to the architect should at best by-pass the contractor and at worst flow through the main contractor on the explicit understanding that it is for administrative convenience only and is not the subject of any express or implied sub-contractual obligation. It should be borne in mind that clause 3.2 of Agreement ESA/1 (for use between the employer and a named sub-contractor under IFC 98) places an obligation on the sub-contractor to provide the architect with such further information as is reasonably necessary to enable the architect to provide the main contractor with sufficient information to, in turn, enable the main contractor to complete the main contract works in accordance with the IFC 98 Conditions, which of course includes the sub-contract works themselves.

Clause 1.8

Limits to use of documents

1.8   None of the documents mentioned in clauses 1.3 (*Priority of Contract Documents*) or 1.7.1 (*Information Release Schedule*) or 1.7.2. (*Provision of further drawings or details*) shall be used by the Contractor for any purpose other than this Contract, and neither the Employer, the Architect/the Contract Administrator nor the Quantity Surveyor shall divulge or use except for the purposes of this Contract any of the rates or prices in the Contract Documents or, where the Second recital, alternative B, applies, in the Contract Sum Analysis or in the Schedule of Rates.

COMMENTARY ON CLAUSE 1.8

This clause calls for no comment.

Clause 1.9

Issue of certificates by Architect/Contract Administrator

1.9 Except where provided otherwise any payment or other certificate to be issued by the Architect/the Contract Administrator shall be issued to the Employer and a duplicate shall at the same time be sent to the Contractor.

COMMENTARY ON CLAUSE 1.9

This clause calls for no comment.

Clause 1.10

Unfixed materials or goods: passing of property, etc.

1.10 Unfixed materials and goods delivered to, placed on or adjacent to the Works and intended therefor shall not be removed[1] except for use upon the Works unless the Architect/the Contract Administrator has consented in writing to such removal which consent shall not be unreasonably delayed or withheld.

Where the value of any such materials or goods has in accordance with clause 4.2.1(b)[2] (*Materials and goods delivered to Works*) been included in any payment certificate under which the amount properly due to the Contractor has been discharged[3] by the Employer, such materials and goods shall become the property[4] of the Employer, but subject to clause 6.3B and 6.3C.2 to .4[5] (*Insurance by Employer*) (if applicable), the Contractor shall remain responsible for loss or damage to the same[6].

COMMENTARY ON CLAUSE 1.10

For a summary of the general law in relation to the passing of property see page 39.

This clause seeks, *inter alia*, to pass the ownership in materials and goods delivered to the site, but not yet incorporated into the works, from the contractor to the employer once their value has been included in any payment certificate under which the amount properly due to the contractor has been paid or otherwise discharged by the employer. Problems can arise where the employer pays but the contractor does not own the materials or goods. This is a breach of contract by the contractor but if the contractor is or becomes insolvent, the employer's remedy for breach of contract, i.e. generally to sue for damages, may be worthless.

The contractor naturally enough wishes to be paid as early as possible. The employer needs to be as certain as he reasonably can be that, having paid for the unfixed materials or goods, they will become his property; otherwise, between the period of payment for the materials or goods and either ownership passing to the contractor and thence to the employer, or incorporation into the works (whichever is the sooner), there is a risk that the true owner might recover the materials or goods and the employer will then receive no value for the money paid.

If the architect is aware that the contractor is not in a position to transfer the title in the materials or goods, their value ought not be included in the payment certificate. This could of course result in the contractor having to pay his supplier or his sub-contractor before being paid by the employer and this in turn could involve the contractor in cash flow problems. Should this occur to any considerable extent in the industry, it could have an effect on tender prices.

In very many instances, it will not be practicable for the architect or, where appointed, the quantity surveyor, to check whether or not the contractor is in a position to transfer the title in materials or goods to the employer. In many cases even detailed investigation would not necessarily clarify the position and even if it did, the administrative load would be out of proportion to the level of risk being taken. However, where there are deliveries to the site of materials or goods having a particularly high value then the lengths to which the architect, or the quantity surveyor as the case may be, should go to check the contractor's ability to transfer the title will depend on a number of factors, e.g.:

(1)   The value of such materials or goods
(2)   His knowledge of the contractor and the contractor's methods of conducting business, e.g. does he usually pay for materials or goods before being paid
(3)   The terms of the supply contract or the sub-contract
(4)   The financial state and status of the contractor.

For high-value items it is suggested that the least that the architect should do is to check, so far as he reasonably can, that the contractor's own contract for the supply of the materials or goods does not purport to retain title.

Even this limited step is of course of little use if the title still vests in a sub-supplier or sub-sub-contractor. Where a named sub-contractor is engaged under NAM/SC, by virtue of clause 19.5.2 the named sub-contractor will be stopped from denying the title of the main contractor to the materials or goods delivered to site by the named sub-contractor, provided the main contractor has been paid under the payment certificate including the value of the materials or goods; see also on this the commentary to clause 3.2 page 119. This is of some limited protection to the employer but does not of course protect him where the named sub-contractor does not own the materials or goods and is not in a position to transfer title to them.

The question of title remains a difficult area and it is necessary to strike a reasonable balance between, on the one hand, going to inordinate lengths in trying to fully protect the employer and, on the other, to accepting that the practical difficulties of establishing title to all materials and goods before their value is included in a payment certificate is such that some element of risk must be taken. It is in the end a matter of using professional judgment in all the circumstances of the situation: see also commentary to clause 3.2 under the heading 'Materials and goods' page 121.

NOTES TO CLAUSE 1.10

[1] '. . .shall not be removed . . .'
This restriction against removal is a necessary safeguard to the employer's interests. Once materials or goods have been paid for, it would rarely be proper for the architect to unconditionally allow their removal. Although this clause, by

its terms, prevents the contractor from removing materials and goods delivered to the site, even if delivered prematurely, as they will not have been paid for, the architect would have less reason to withhold his consent to their removal, though there may still be circumstances in which he could reasonably object; e.g. consider the case where the type of materials or goods concerned are in short supply and the contractor wishes to remove them for use on another contract thus causing potential delays to completion.

Furthermore, if too large a quantity of materials or goods is brought on to the site so that some deliveries are premature but others are not, and a proportion is included in the valuation for a payment certificate, it may not be possible to differentiate between those in which the property has passed to the employer and those in which the property has not e.g. a stock pile of bricks. It may be totally impracticable to separately store or identify the two categories and in such circumstances the architect will be acting reasonably in refusing consent to any of them being removed from the site.

[2] '... in accordance with clause 4.2.1(b) ...'
By clause 4.2.1(b), in order to qualify for payment in a payment certificate, the materials or goods must have been reasonably and properly and not prematurely delivered and must be adequately protected against weather and other casualties.

[3] '... has been discharged ...'
It is clear that such materials and goods are, so far as the contract can ensure it, to become the employer's property once the certificate has been duly discharged – not necessarily that its full amount has been paid. It is the amount properly due which must have been paid or otherwise discharged. This need not equal the amount shown as due in the certificate, e.g. where the employer deducts liquidated damages.

[4] '... shall become the property ...'
The transfer of title is dealt with earlier on page 39.

[5] '... subject to clause 6.3B and 6.3C.2 to .4 ...'
By virtue of clause 6.3B and 6.3C.2 to .4 the employer is required to take out 'All Risks Insurance' (see clause 6.3.2 for definition) for the benefit of the employer and contractor in a 'Joint Names Policy' (see clause 6.3.2 for definition). This will cover the risk of physical loss or damage to 'Site Materials' (see clause 8.3 for definition). The payment structure within clauses 6.3B and 6.3C in relation to replacement or repair of site materials which have been lost or damaged as a result of one or more of the insured risks is such that the cost falls on the employer rather than the contractor (see clauses 6.3B.3.2, 6.3B.3.5, 6.3C.4.1 and 6.3C.4.4). However, the employer may of course suffer other losses not covered by such insurance, e.g. business interruption and other purely financial losses. If the cause of the loss or damage was attributable to a breach of contract or negligence on the part of the contractor, no doubt the employer could claim against the contractor.

[6] '... the Contractor shall remain responsible for loss or damage to the same.'
Subject to what was said under Note (5) above, the contractor is liable to replace lost or damaged materials or goods even though they belong to the employer. If

clause 6.3A applies, the contractor must insure against the risk of loss or damage by taking out all risks insurance in a joint names policy for the benefit of both contractor and employer. Any shortfall of insurance monies will not relieve the contractor of his obligation to replace or repair the lost or damaged site materials.

Clause 1.11

Off-site materials and goods: passing of property, etc.

1.11   Where as provided in clause 4.2.1(c)[1] (*Off-site materials and goods – 'the listed items'*) the value of any listed items has been included in any payment certificate under which the amount properly due to the Contractor has been paid[2] by the Employer, such listed items shall become the property[3] of the Employer and thereafter the Contractor shall not, except for use on the Works, remove or cause or permit the same to be moved or removed[4] from the premises where they are, but the Contractor shall nevertheless be responsible for any loss thereof or damage thereto and for the cost of storage, handling and insurance[5] of the same until such time as they are delivered to and placed on or adjacent to the Works whereupon the provisions of clause 1.10 (except the words 'Where the value' to the words 'the property of the Employer, but') shall apply thereto.

COMMENTARY ON CLAUSE 1.11

For a general summary of the law relating to the transfer of title, see earlier page 39.

This clause deals with payment for materials or goods intended for the works which have been listed by the employer in a list supplied to the contractor and annexed to the contract bills, specification or schedules of work and for which it is intended that payment should be made before delivery to or adjacent to the works. The architect does not therefore, as with the version of this clause in IFC 84 prior to the issue of Amendment 12 in April 1998, have a discretion as to whether to include the value of off-site materials and goods in payment certificates. The decision is taken by the employer at the time of entering into the contract.

The employer is actively encouraged to provide such payments by the facility in clause 4.2.1(c).1 and .2 for the contractor to provide a bond in a standard form attached to the contract conditions. More will be said about this in the commentary to clause 4.2.1(c) (see page 206). However, it can be pointed out here that the bond is an automatic contractual requirement in respect of those listed items which are not 'uniquely identified'; whereas the requirement for a bond will only apply if so indicated in the appendix to the conditions in respect of uniquely identified listed items. The reason for this differing treatment is obvious. The purpose of the bond is to provide protection for the employer in the event that listed items are not delivered to or adjacent to the works. This will most likely happen where the contractor becomes insolvent coupled with an inability on the part of the employer to prove title. Where the listed items are uniquely specified it is likely to be that much easier for the employer to prove title and therefore it might be that a bond will not be required in such circumstances; whereas title will almost certainly be more difficult to establish where the listed items are not uniquely specified.

Clause 4.2.1(c) also makes it a requirement of payment that certain conditions have been fulfilled in respect of the listed items, such as reasonable proof that the contractor owns them; that they are in accordance with the contract; that they have been set apart and identified etc.; and that they are properly insured.

Compared with the pre-Amendment 12 version of IFC 84, clause 4.2.1(c) provides much better protection for the employer, not only in relation to the facility for requiring what amounts to an 'on demand' bond but also in relation to the specific requirement for the contractor to provide reasonable proof of ownership.

Provided that the property in the listed items vests in the contractor, then once the employer has paid any sum due in respect of them following their value being included in a payment certificate, the listed items will become the property of the employer.

NOTES TO CLAUSE 1.11

[1] '... clause 4.2.1(c) ...'
This clause is much expanded and is considered in detail under the commentary to clause 4.2 (page 206).

[2] '... has been paid ...'
This should be compared with Note [3] to clause 1.10 earlier on page 57 which refers to '... has been discharged ...'. Under clause 1.10, property passes when the amount properly due has been discharged; whereas under clause 1.11 property passes when the amount properly due has been paid. There is no sound reason for this distinction. In IFC 84 prior to Amendment 12 (issued April 1998) the word 'discharge' was also used in clause 1.11. This was changed to 'paid' by Amendment 12 in this clause and in some others, particularly clauses 4.2, 4.3 and 4.6 dealing with payment of sums included in certificates. It is likely that the replacement of an obligation to discharge by an obligation to pay was prompted in relation to the payment clauses by virtue of clauses 110 to 112 of the Housing Grants, Construction and Regeneration Act 1996 which expressly refers to 'payment' and 'paid'. The change in terminology probably helps to confirm that the IFC 98 payment provisions properly provide for periodic payments and for an adequate mechanism for determining what payments become due in accordance with the 1996 Act. However, this begs the question of what 'payment' and 'paid' mean in the Act. For example, in section 112 (and see clause 4.4A of IFC 98), the contractor's right to suspend performance of his obligations under the contract depends on the employer not having '... paid in full ...'. Does this mean literally not paid rather than the employer having failed to meet his obligation to pay? In other words, a sum may be due under a payment certificate in favour of the contractor, and the employer may decide, quite properly, to deduct liquidated damages from the certified sum. While this may be a proper discharge of the employer's obligation to pay, it is not physical payment. It is probably the apprehension that the Act is talking about physical payment rather than a proper discharge of the obligation to pay which has led to this change in terminology.

Having made this change in the payment provisions it appears that an attempt has been made to repeat the changes in other clauses, including clause 1.11. On this basis the retained use of the word 'discharge' in clause 1.10 is probably a drafting oversight. This view is reinforced by considering clause 3.2.2(b) and (c) in both of which the word 'discharge' has been replaced by 'paid' and in both of which it has been assumed that this has also been done in clause 1.10. It seems most unfortunate that it was felt appropriate to make these changes which could now cause some confusion in interpretation. For instance, if the obligation to pay is discharged, e.g. by the deduction of liquidated damages so that there is no physical payment, does it now mean that property does not pass under clause 1.11? This clearly ought not to be the case but the point might well be argued, particularly while there is still a reference to 'discharge' in clause 1.10.

It is submitted that in relation to the payment clauses of IFC 98, the references to 'payment' and 'paid' do not on a true interpretation of the contract require physical payment as opposed to a proper discharge of the obligation to pay. However the matter could have been put beyond doubt by retaining references to 'discharge'.

[3] '... shall become the property...'
The transfer of title is dealt with earlier on page 39.

[4] '... the same to be moved or removed...'
While under both clause 1.10 and clause 1.11 the materials or goods are not to be removed, under clause 1.11 the listed items may not even be moved within the premises in which they are stored. This is sensible from the employer's point of view as he needs to ensure that such listed items are stored separately and are thereafter kept separate from other materials or goods.

There is no provision, as in clause 1.10, for listed items to be removed with the consent of the architect. Such consent could still be requested and given either directly by the employer or by the architect on his behalf if so authorised. This could be necessary, e.g. in the event of damage to the premises in which the listed items are stored.

[5] '... insurance...'
This should be carefully checked by the architect to ensure that the cover is adequate and that any conditions required by the insurance policy in connection with storage and protection have been complied with.

Clause 1.12

Reappointment of Planning Supervisor or Principal Contractor – notification to Contractor
1.12   If the Employer pursuant to article 5 or article 6 by a further appointment replaces the Planning Supervisor referred to in, or appointed pursuant to, article 5 or replaces the Contractor or any other contractor appointed as the Principal Contractor, the Employer shall immediately upon such further appointment notify the Contractor in writing of the name and address of the new appointee.

COMMENTARY ON CLAUSE 1.12

This clause deals with any further appointment of a planning supervisor or principal contractor following their having already been named in article 5 and article 6. In the event of such replacement, the employer must immediately notify the contractor in writing of the name and address of the new appointee. It should be appreciated that the contractor will most likely have been named as the principal contractor under article 6 but that it is open to the employer to replace him as principal contractor. His position as contractor under IFC 98 will remain unaffected by such a replacement appointment. In the unlikely, but possible, event that the contractor has also been appointed as the planning supervisor under article 5, the position is the same. Accordingly, as the IFC 98 contract does not deal with the terms of such appointments, the contractor should have a separate agreement with the employer regarding any such appointment. Clearly, if the contractor under IFC 98 is not also the principal contractor for the purposes of the CDM Regulations, there are likely to be interface problems with possible delay and disruption.

Clause 1.13

Giving or service of notices or other documents
1.13 Where the Contract[1] does not specifically state the manner[2] of giving or service[3] of any notice or other document required or authorised in pursuance of this Contract such notice or other document shall be given or served by any effective means[4] to any agreed address. If no address has been agreed then if given or served by being addressed pre-paid and delivered by post to the addressee's last known principal business address or, where the addressee is a body corporate, to the body's registered or principal office it shall be treated as having been effectively given or served.

COMMENTARY ON CLAUSE 1.13

This clause is based on the provisions of section 115 of the Housing Grants, Construction and Regeneration Act 1996 dealing with the service of notices or other documents. It applies to all notices or other documents which the construction contract requires or authorises to be served. The Act, by section 115(6) states that a notice or document *includes* any form of communication in writing. Under the Act therefore, documents are not limited to communications in writing and could take the form of, for example, drawn information. However, clause 1.13 does not include a provision such as that in section 115(6). Its exclusion is ambivalent in terms of interpreting the rest of the clause. There is no doubt however that the clause covers the giving or service of drawn information.

It is open to the parties to agree different means of service for different purposes. For example, notices under clause 7 dealing with determination are to be in writing and given by actual delivery, special delivery or recorded delivery but not by ordinary post unless actually delivered (clause 7.1); and notice of referral to an adjudicator under clause 9A is to be given by actual delivery or by fax or by special delivery or recorded delivery. On the other hand, it might be provided in the contract bills or specification that interim payment certificates will be sent to the contractor's regional headquarters; or that drawings will be issued to the contractor at the site address. Another example of an agreed manner of service is to be found in clause 9. Clause 9B dealing with arbitration incorporates the JCT 1998 edition of the Construction Industry Model Arbitration Rules (CIMAR) and these rules provide specific means of service of documents (Rule 14.2 and section 76(3) to (6) of the Arbitration Act 1996).

If no manner of giving or serving documents is indicated, this can be done by any effective means to any agreed address and if no address has been agreed then sending by ordinary post if properly addressed and pre-paid to the addressee's last known principal business address or, where a body corporate, its registered or principal office will be treated as having been effectively given or served.

Though the drafting is not absolutely clear, it is suggested that another means of service, provided it is effective, will suffice in addition to the stated method where no address has been agreed.

Finally, clause 1.16 enables the parties to communicate by electronic document interchange. This is discussed under the commentary to that clause on page 65.

NOTES TO CLAUSE 1.13

[1] '...Contract...'
This will include all the contract documents so that any of these could state the manner of giving or service of documents. Presumably, the agreement as to the manner of giving or service of documents could, however unwisely, be oral provided the oral statement was part of the contract, e.g. an oral acceptance by the employer of the contractor's tender, part of which oral acceptance states that applications for payment should be sent to the quantity surveyor's office address.

[2] '...the manner...'
This appears to include both the means of giving or service and also the place where it is to be effected.

[3] '...of giving or service...'
Section 115 of the Act refers only to the service of notices or other documents. Clause 1.13 refers both to the giving and service of notices or other documents. This is to cater for the fact that throughout the contract there are references to giving notices or other documents rather than serving them.

[4] '...by any effective means...'
This phrase is used both in section 115 of the Act and in clause 1.13. In neither is it defined. It clearly includes the use of ordinary post. It would also include the use of couriers provided the delivery was successful. It would also extend to giving or serving by facsimile transmission where transmission was successful. It is submitted that service through e-mail or other electronic means, even if received, may not be adequate as it does not guarantee receipt of a hard copy.

Clause 1.14

Reckoning periods of days
1.14   Where under this Contract an act is required[1] to be done[2] within a specified period of days after or from a specified date, the period shall begin immediately after that date. Where the period would include a day which is a Public Holiday that day shall be excluded.

COMMENTARY ON CLAUSE 1.14

Section 116 of the Housing Grants, Construction and Regeneration Act 1996 provides for the reckoning of periods of time without being specific in how that is expressed, whether in terms of days, weeks or months. In addition, section 116 applies only to periods of time for the purposes of Part II of the Act which concerns specific but limited matters. On the other hand, clause 1.14 applies to the whole of IFC 98 but restricts its application to situations where the contract refers to periods of days. In respect of both the Act and the clause, it only applies where an act is required to be done within a specified period. The period of days begins immediately after the relevant date from which the period of days is calculated. For example, where under clause 4.2(a) the employer must pay the contractor within 14 days from the date of issue of the payment certificate, the date of issue itself is excluded in making that calculation.

Where the period of days includes a public holiday, that public holiday is to be excluded from the calculation. Public holiday is defined in clause 8.3 to mean Christmas Day, Good Friday or a day which under the Banking and Financial Dealings Act 1971 is a bank holiday. This definition is broadly taken from section 116(3) of the Act. This definition can be amended if different public holidays are applicable.

NOTES TO CLAUSE 1.14

[1] '... required ...'
Although the Act and clause both talk about an act being required to be done, it is submitted that in this context it extends to situations where an act is merely authorised but where, if that authority is exercised, its exercise is required to be done within a specified period.

[2] '...an act is required to be done ...'
This attribute which must exist before applying the calculation, whether under the Act in respect of Part II thereof or, by clause 1.14, throughout the contract, is bound to give rise to difficulties. Firstly, so far as the contract is concerned, this provision extends to all of the contract documents and so will cover anything contained in contract documentation as well as in the conditions themselves. Secondly, the contract in many instances refers to periods of days where it is difficult to be sure whether an act is required to be done within the specified time. The following are given as examples:

- Clause 2.10 (notification of defects after expiry of defects liability period)
- Clause 3.13.1 (architect not having received contractor's statement following failure of work or materials – is it an act required to be done within seven days?)
- Clause 4.4(A) (is the employer being required to pay within the seven day period referred to?)
- Clause 7.2.2 (is the contractor being required to remedy the specified default within the 14 day period?)
- Clause 7.9.3 (is the employer being required to remedy the specified default within the 14 day period?).

If the act is required to be done before rather than after or from a specified date the calculation does not apply. For example, in clauses 4.2.3(b), 4.3(c) and 4.6.1.3 the employer may not later than five days before the final date for payment give written notice of withholding or deduction. Accordingly public holidays will not be excluded from the calculation.

Clause 1.15

Applicable law

1.15   Whatever the nationality, residence or domicile of the Employer, the Contractor or any sub-contractor or supplier and wherever the Works are situated the law of England shall be the law applicable to this Contract. [l]

---

[l] Where the parties do not wish the law applicable to the Contract to be the law of England appropriate amendments to clause 1.15 should be made.

COMMENTARY ON CLAUSE 1.15

This provides that the law applicable to the contract shall be English law whatever the nationality, residence or domicile of the employer, the contractor or any sub-contractor or supplier and wherever the works may be situated. It is therefore a very broad provision. It should be noted that IFC 98 is not intended for use in Scotland and there is no Scottish supplement applicable to it.

There is a footnote [l] reminding the parties that if they do not wish the law applicable to the contract to be the law of England then appropriate amendments to this clause should be made.

In passing it should be noted that in the rare event that even if, in relation to the carrying out of construction operations in England or Wales, the parties agree that they do not wish the applicable law to be the law of England, Part II of the Housing Grants, Construction and Regeneration Act 1996 will nevertheless still apply (see section 104(6) and (7)). See also clause 9B.5 in relation to arbitration proceedings under IFC 98 (and page 393).

Clause 1.16

Electronic data interchange

1.16   Where the Appendix so states, the 'Supplemental Provisions for EDI' annexed to the Conditions shall apply.

Annex 2 to the Conditions:
Supplemental Provisions for EDI
(Clause 1.6) (*sic* 1.16)
The following are the Supplemental Provisions for EDI referred to in clause 1.16 of the Conditions.

1   The Parties no later than when there is a binding contract between the Employer and the Contractor shall have entered into the Electronic Data Interchange Agreement identified in the Appendix ('the EDI Agreement'), which shall apply to the exchange of communications under this Contract subject to the following:

.1   except where expressly provided for in these provisions, nothing contained in the EDI Agreement shall override or modify the application or interpretation of this Contract;

.2   the types and classes of communication to which the EDI Agreement shall apply ('the Data') and the persons between whom the Data shall be exchanged are as stated in the Contract Documents or as subsequently agreed in writing between the Parties;

.3   the Adopted Protocol/EDI Message Standards and the User Manual/Technical Annex* are as stated in the Contract Documents or as subsequently agreed in writing between the Parties;

.4   where the Contract Documents require a type or class of communication to which the EDI Agreement applies to be in writing it shall be validly made if exchanged in accordance with the EDI Agreement except that the following shall not be valid unless in writing in accordance with the relevant provisions of this Contract:

    .4.1   any determination of the employment of the Contractor;

    .4.2   any suspension by the Contractor of the performance of his obligations under this Contract to the Employer;

    .4.3   the final certificate;

    .4.4   any invoking by either Party of the procedures applicable under this Contract to the resolution of disputes or differences;

    .4.5   any agreement between the Parties amending the Conditions or these provisions.

2   The procedures applicable under this Contract to the resolution of disputes or differences shall apply to any dispute or difference concerning these provisions or the exchange of any Data under the EDI Agreement and any dispute resolution provisions in the EDI Agreement shall not apply to such disputes or differences.

---

*The EDI Association Standard EDI Agreement refers to an Adopted Protocol and User Manual; the European Model EDI Agreement refers to EDI Message Standards and a Technical Annex. Delete whichever is not applicable.

COMMENTARY ON CLAUSE 1.16

Provision has been made for the parties to agree, by entering into an Electronic Data Interchange (EDI) Agreement to the electronic exchange of communications under the contract. An item in the Appendix to the Conditions will firstly indicate whether or not the supplementary provisions for EDI are to apply and if they are to apply whether the EDI Agreement is to be either the EDI Association Standard EDI Agreement or alternatively the European Model EDI Agreement.

The use of such means of communication is likely to become increasingly popular, and if it is properly regulated it can reduce paperwork and save time. The IFC 98 Conditions provide at Annex 2 for supplemental provisions for the use of EDI. The following points are particularly noteworthy:

- Nothing within the EDI agreement selected is to override or modify the application or interpretation of the contract unless it is expressly provided for within the supplemental provisions for EDI.
- The contract documents, e.g. the bills or specification, are to state the types and classes of communications and the persons between whom the data is to be exchanged. If nothing is contained in the contract documents then it can subsequently be agreed in writing between the parties.
- If the contract documents require a type or class of communication to which the chosen standard form of EDI agreement applies to be in writing it shall be

validly made if exchanged in accordance with the EDI agreement but there are exceptions to this where the communication must be in writing specifically in accordance with the relevant provisions of the contract, namely:
- any determination of the employment of the contractor;
- any suspension by the contractor of performance of his obligation under this contract;
- the final certificate;
- any invoking by either party of dispute resolution procedures;
- any agreement between the parties amending the contract conditions or the EDI supplemental provisions.
• The contract procedures applicable to the resolution of disputes or differences will also apply to any dispute or difference regarding the supplemental provisions for EDI or the exchange of any data under the selected EDI agreement. Accordingly any dispute resolution provisions which may be in the selected standard form of EDI agreement shall not apply to such disputes or differences.

It will be appreciated that while careful note must be taken of the exceptions to communicating by electronic data interchange, nevertheless it covers a wide range of communications including: interim certificates; the architect's notice specifying a default prior to the determination of employment notice itself; communications in connection with procedures applicable to the resolution of disputes or differences, with the exception of the notice of intention to refer to adjudication (clauses 9A.4.1 and 9A.4.2); and the notice commencing arbitration proceedings (see Rule 2.1 of the JCT 1998 Edition of the Construction Industry Model Arbitration Rules). The supplemental provisions for EDI, if incorporated, need therefore to be considered alongside clause 9A.4.2 in respect of adjudication and Rule 14 of the Construction Industry Model Arbitration Rules in connection with arbitration, both of which deal with the method of serving documents.

Where the parties choose to litigate disputes or differences, the provisions of the EDI agreement cannot override the provisions as to service contained in the Civil Procedure Rules 1998 which may apply, unless those rules themselves permit the parties to agree on the means of service.

The Tribunal has produced a comprehensive book explaining the system and how to implement it. It also provides a code of practice, glossary of terms, and appendices containing the JCT Supplemental Provisions for EDI and specimen forms of the EDI Association: Model Form of EDI Agreement and the European Model EDI Agreement.

Chapter 4
Possession and completion

CONTENT

Section 2 deals in 11 clauses with matters relating to possession and completion. It covers such important topics as:

The giving of possession
Liquidated damages and extensions of time
Practical completion and partial possession by the employer
Defects liability.

In dealing with section 2, clauses 2.3 to 2.8 inclusive have been grouped together. They all relate to the question of liquidated damages and extensions of time and must be considered as a whole. The commentary on this group of clauses therefore follows on after the print of clause 2.8 (see page 82). Similarly, clauses 2.1 and 2.2 dealing with possession are grouped together, as are clauses 2.9 dealing with practical completion and 2.10 dealing with defects liability. Finally, clause 2.11 provides the facility for partial possession prior to completion of the whole of the works. This is dealt with in a separate commentary at the end of this chapter (see page 111).

SUMMARY OF GENERAL LAW

(A) Possession

The obligation of the employer to give to the contractor possession of the site is of course fundamental. The degree of possession to be given by the employer will vary; for example, compare the case of a new building on a green field site with additions to an existing building which remains occupied.

Most forms of building contract restrict the degree of possession to be given to the contractor both in terms of physical extent and duration. The possession given must, however, be such as to reasonably enable the contractor to complete by the contractual completion date.

If a building contract failed expressly to provide for possession of the site to be given to the contractor, such a provision would be implied to the extent that possession was required to enable the contractor to complete by any agreed completion date. It would be necessary to imply such a term to give the contract business efficacy.

If the employer fails to give the appropriate degree of possession to enable the contractor to complete on time or probably to carry out the work in accordance with any agreed programme, this will be a breach of contract by the employer. This is so even though the employer's failure was due to circumstances completely outside his control.

In practice, although the employer may have entered into a contract with the very best intentions, some difficulty in giving possession may remain, e.g. a tenant who refuses to vacate without a court order. It is advisable therefore for the building contract to deal specifically with such problems.

The failure by the employer to hand over possession by an agreed date may, depending on its degree and duration, amount to a serious breach of contract entitling the contractor to treat the contract as at an end and to sue the employer for damages. Apart from this, unless there is an express provision for an extension of time for delay in giving possession, such delay, even if not amounting to a fundamental breach, will invalidate any liquidated damages clause: see *Rapid Building Group Ltd* v. *Ealing Family Housing Association Ltd* (1984). In such circumstances the employer may still claim general damages for breach of contract for delay if these can be proved. There has been some debate as to whether any such general damages claim is subject to a ceiling equal to the level of liquidated damages in the invalidated provision. The logic of such a contention is that otherwise, if the general damages were in excess of the liquidated damages figure, the employer would be profiting from his own act of delay or prevention. There is good persuasive authority that this ceiling argument is sound: see the case of *Elsley* v. *J. G. Collins Insurance Agencies Ltd* (1978) in the Supreme Court of Canada.

Once in possession, it is a question of some debate whether or not the contractor's licence to remain in possession is revocable or irrevocable while the contract period is still running. In other words, if the employer purports to determine the contractor's employment and this is disputed by the contractor, can he remain on site or is he compelled to leave?

In the case of *London Borough of Hounslow* v. *Twickenham Garden Developments Ltd* (1970) it was held that the contractor's licence to remain on site during the contract period was irrevocable. This decision was however criticised and not followed in the New Zealand case of *Mayfield Holdings Ltd* v. *Moana Reef Ltd* (1973). It is suggested that while the English decision may be the stronger authority, which courts, at any rate of first instance, may feel compelled to follow, the New Zealand decision is the more logical. Furthermore, the *Hounslow* case has expressly not been followed in Australia: see *Chermar Productions Proprietary Ltd* v. *Prestest Proprietary Ltd* (1989) in which it was confirmed that the contractor's licence to occupy the site may be determined and the contractor transformed into a trespasser even if the determination of the licence involved the employer in a breach of the contract. Even so, it appears that the *Hounslow* decision which concerned a JCT 63 contract has been followed in a case under the JCT 80 contract, *Vonlynn Holdings* v. *T. Flaherty* (1988). In a case under significantly different wording in the ICE 5th Edition Form of Contract, *Tara Civil Engineering Ltd* v. *Moorfield Developments* Ltd (1989), it was held that the contractor was bound to leave the site despite contesting the employer's right to determine. It should be noted that under clause 63 of the ICE Conditions, once the engineer has certified in writing to the employer that in his opinion the contractor has defaulted in one of the stipulated ways, the employer, after giving seven days notice to the contractor, is expressly entitled to enter the site of the works and expel the contractor therefrom. This is significantly different to the wording under JCT 80 and JCT 98.

However, IFC 98 clause 7.6(a) expressly requires the contractor to give up possession of the site in the event of the employer determining the contractor's employment under the contract. It is submitted that this requires the contractor to give up possession of the site even if the validity of the determination is challenged. This interpretation is also supported by the provisions of clause 6.3 in relation to the insurance of the works, which provides that the obligation to insure the works ceases upon a determination of the contractor's employment even if it is validly challenged (see clauses 6.3A.1, 6.3B.1 and 6.3C.2). (For further comments see pages 76 and 341.)

To allow the contractor to remain on site when there has been a complete breakdown of relationships seems a nonsense. Even if the employer is in the wrong, he cannot easily be compelled to make interim payments and the architect cannot be compelled to issue necessary drawings and other information etc., so there will be little or no progress. Also, the employer will be prevented, by the continued presence of the contractor, from employing others to finish off the work. It is submitted that the contractor should be required to leave the site and to pursue his claim for damages for breach of contract if he is wrongly removed from the site. Whilst it may be contended that compelling the contractor to leave, even though he may be the innocent party, could result in damage to his reputation, nevertheless permitting him to stay will cause a hopeless impasse in many situations. The *Hounslow* case is discussed again when dealing with determination of the contractor's employment – see page 319.

The site

Under the general common law, an employer does not impliedly warrant the condition of the site. This is of course subject to the express terms of the contract made between the parties. Much may depend on whether use is made of a specification or a bill of quantities. If a specification is used and nothing at all is stated so that the contractor is to satisfy himself as to the site conditions, then the risk of unforeseen difficulties will generally have to be borne by him, e.g. unforeseen ground conditions. On the other hand, if a bill of quantities is used then the express terms of the contract will have to be examined to decide the status of the bill of quantities. For example, the contract may say that the bills have been prepared in accordance with a particular standard method of measurement, in which case, to a greater or lesser extent, a warranty as to site conditions is given either by reason of what is stated in the bills by way of site investigation results, or assumptions to be made in the absence of such information. The precise extent of the warranties given in this way is difficult to assess, e.g. to what extent does information obtained from a trial pit or a bore hole amount to a warranty by the employer that the contractor can freely assume in pricing that such information is typical of the whole of the site?

If the site conditions are such that it is necessary for the employer to vary the design, then this will almost invariably amount to a variation to the contract works for which the contractor will generally be compensated by express terms of the contract.

On particular occasions, e.g. where the employer states the assumptions as to ground conditions on which the contractor is asked to design the works himself, there may be an implied warranty on the part of the employer as to site conditions: see *Bacal Construction (Midlands) Ltd* v. *Northampton Development Corporation* (1975).

(B) Completion

A building contract may or may not have a fixed date for completion. If it does not, then there will be a term implied at common law to the effect that the contractor will be obliged to complete within a reasonable time. Furthermore, section 14 of the Supply of Goods and Services Act 1982 provides that if the time for completion is not fixed or capable of being determined by the contract, there is an implied term that the work will be carried out within a reasonable time.

If the contract contains a fixed completion date, it is a matter of interpretation, in the absence of express words, as to whether the time stated for completion is of the essence of the contract so that failure to achieve it puts the contractor into breach of a fundamental term entitling the employer to treat the failure as a repudiation of the contract and as releasing him from his obligations under the contract. Historically, at common law, if the contract gave a completion date, this was then regarded as of the essence of the contract. However, courts of equity gave relief and did not generally regard time as of the essence and prevented the innocent party from treating the contract as at an end, though the right to claim damages for breach of contract remained.

Nowadays, unless it is otherwise clear from the terms of the contract or the surrounding circumstances, a fixed completion date will not make time of the essence of a building contract. The standard forms of construction contract in common use in the UK at the time of writing, whilst providing for a definite completion date, do not, by their terms, make time of the essence. This is demonstrated by the fact that such contracts envisage practical completion later than the contractual or extended contractual completion date and that they contain no express ground for determination of the contractor's employment for failure to complete by the contractual completion date: see *Gibbs* v. *Tomlinson* (1992) in relation to the JCT Minor Works Form 1980. However, it is possible for the parties to agree to the completion date being of the essence despite the contract also having a liquidated damages clause: see *Peak Construction (Liverpool) Ltd* v. *McKinney Foundations Ltd* (1970).

(C) Liquidated damages

Most standard forms of building contract provide for the payment of a sum by the contractor to the employer by way of liquidated damages for delay in completion of the contract works.

By means of such a provision an agreed sum is payable as damages for delay in completion. In building contracts the sum may be expressed simply as so much per week of delay, or alternatively, in a suitable case, e.g. a housing development, and where the contractual framework permits it, as so much per week per incomplete dwelling.

The main advantage to the employer of such a provision is that he is not required to prove his actual loss as a result of the delay.

The existence of a valid liquidated damages clause can be of particular benefit to an employer where the cost of delay cannot easily be measured, e.g. a fire station or a church. In such a case the employer would have difficulty in actually proving a loss beyond perhaps the loss of the use of capital money invested in the project with a delayed return together with something in respect of administration costs.

Provided that the stipulated sum represents, at the time of entering into the contract, a genuine attempt at a pre-estimate of the likely loss, it is not a penalty and will, other things being equal, be upheld by the courts: see *Dunlop Pneumatic Tyre Company Ltd* v. *New Garage and Motor Co. Ltd* (1915) and, more recently, *Philips (Hong Kong) Ltd* v. *A. G. of Hong Kong* (1993). This is so even if, in the result, no actual loss is suffered at all: see *BFI Group of Companies Ltd* v. *DCB Integration Systems Ltd* (1987). The courts have in recent years construed liquidated damages and extension of time clauses in printed forms of contract strictly *contra proferentem*, which means in effect against the employer as they are generally regarded as inserted primarily for the benefit of the employer: see *Peak Construction (Liverpool) Ltd* v. *McKinney Foundations Ltd* (1970). But more recently there has been evidence of a renewed robustness in the courts in support of upholding liquidated damages provisions wherever reasonably possible: see *Philips (Hong Kong) Ltd* v. *A. G. of Hong Kong* (1993). Further, when dealing with liquidated damages and extension of time clauses in respect of JCT 80, Mr Justice Coleman in *Balfour Beatty Building Ltd* v. *Chestermount Properties Ltd* (1993) said:

> 'In this respect the contract is not so ambiguous or so unclear as to call for application of the *contra proferentum* rule ...'

An employer may lose the right to deduct liquidated damages in a number of ways:

(1) Where the sum stipulated for is in truth a penalty. A penalty clause is unenforceable in English law. For instance, if the amount stipulated for goes beyond what could conceivably be a genuine pre-estimate of the likely loss due to delay; or where the same sum is stipulated for in differing circumstances where it is clear that the actual loss must vary so that the pre-estimate is not genuine, then it will amount to a penalty: see *Ford Motor Co.* v. *Armstrong* (1915).

(2) By the employer so delaying the contractor as to prevent the contractor completing by the agreed completion date. If, in such circumstances, the contract contains no provision for an extension of time covering such delay, then the employer cannot insist on completion by the agreed date and will lose his right to liquidated damages: see *Dodd* v. *Churton* (1897); *Peak Construction (Liverpool) Ltd* v. *McKinney Foundations Ltd* (1970); *Rapid Building Group Ltd* v. *Ealing Family Housing Association Ltd* (1984). In such a case, the fixed completion date will be lost and the contractor will have a reasonable time within which to complete the contract works. The employer may still

have a right to claim unliquidated damages, subject to such a claim being capped at the level of the invalidated liquidated damages clause (see earlier page 68).

(3)   By the breakdown of the contractual machinery for calculating liquidated damages. A liquidated damages clause needs to be operated from one specific date until another, namely from the expiry of the contractual completion date until actual completion is achieved. If therefore the starting date has not been specified or the finishing date is lost, it can lead to difficulties in calculation and thereby also in the operation of the liquidated damages provision.

(4)   Failure of the employer to comply with a condition precedent. For example, the employer may first require a certificate from the architect that the contractual completion date has passed and the works remain uncompleted: see *Token Construction Co. Ltd* v. *Charlton Estates* (1973).

(5)   Wilful failure by the architect or other third party certifier under the contract to properly administer the machinery of an extension of time clause: see *Peak Construction (Liverpool) Ltd* v. *McKinney Foundations Ltd* (1970). A number of cases decided in the middle to late 1980s suggest that the mere failure, even if due to incompetence and even possibly negligence, on the part of the architect in administering the extension of time machinery under the contract may not be enough to invalidate the liquidated damages provision. The contractor can generally seek immediate arbitration to put matters right: see for example *Temloc Ltd* v. *Errill Properties Ltd* (1987); *Lubenham Fidelities and Investment Co. Ltd* v. *South Pembrokeshire District Council and Another* (1986); *Pacific Associates Inc. and Another* v. *Baxter and Others* (1988). More recently there have been cases suggesting that perhaps there may be an implied term to the effect that the architect or other third party certifier will administer the extension of time clause in a reasonable, fair and impartial manner. If this is correct then any failure by the architect or other third party certifier to do so will amount to a breach of contract by the employer enabling the contractor to claim damages rather than simply seeking to have the architect's decision reviewed and altered: see *John Barker Construction Ltd* v. *London Portman Hotel Ltd* (1996); *Balfour Beatty Civil Engineering Ltd* v. *Docklands Light Railway Ltd* (1996); *Beaufort Developments Ltd* v. *Gilbert-Ash NI Ltd and Others* (1997).

Should the liquidated damages clause become invalidated the employer will still be able, subject to proof, to claim damages at common law if any delay in completion by the contractor amounts to a breach of contract: *Rapid Building Group Ltd* v. *Ealing Family Housing Association Ltd* (1984). But note the possible limitation on this referred to earlier (page 68).

(D) Extensions of time

Many building contracts make provision for extensions of time to the contractual completion date in certain events. Such clauses are directly related to any liquidated damages provision in the contract.

Extension of time provisions have the effect of extending the original contractual completion date where the delaying event falls within its terms. Where the delay is what might be called a neutral delay, i.e. not a direct fault of either the employer or the contractor, such as exceptionally adverse weather conditions, civil commotion, local combination of workmen, strike or lockout etc. (being, it is submitted, events which might otherwise be at the contractual risk of the contractor: see for example *Percy Bilton Ltd* v. *Greater London Council* (1982) 20 BLR 8 per Lord Fraser at pages 13 and 14), the extension of time clause benefits the contractor by relieving him of any liability to pay liquidated damages. On the other hand, if the event which causes the delay is the fault of the employer or someone acting on his behalf, e.g. the architect issuing late instructions, then the effect of extending the contractual completion date to a new later date will keep intact the liquidated damages clause for the benefit of the employer, which would not be the case if the employer's delay was not covered by the extension of time provision.

(E) Practical completion

Most standard forms of construction contract by their express terms treat completion as something less than entire completion. They do not require the contract works to be completed in every detail before contractual completion is achieved. They permit small details to be completed during a maintenance or defects liability period. Completion is qualified by reference to its being practical.

Practical completion probably means that the contract works have been completed to the extent that nothing important or which significantly affects their use for their intended purpose remains outstanding. In the case of *Emson Eastern Ltd (in receivership)* v. *E. M. E. Developments Ltd* (1991) Judge John Newey QC sitting on Official Referee's Business considered the meaning of practical completion in a JCT 80 contract. He reviewed a number of cases including *The Lord Mayor Aldermen and Citizens of the City of Westminster* v. *J. Jarvis & Sons Ltd and Another* (1970) and *H.W. Nevill (Sunblest) Ltd* v. *William Press & Son Ltd* (1981). He preferred his own view expressed in the *Nevill* case in which he said that the (similar) JCT 63 contract gave the architect a discretion to certify practical completion:

> '... where very minor *de minimis* work had not been carried out, but that if there were any patent defects ... the Architect should not have given a certificate of Practical Completion ...'.

The judge stressed that construction work was not like manufacturing goods in a factory. It was virtually impossible to achieve the same degree of perfection as could a manufacturer.

Once practical completion has been achieved and an appropriate certificate under the contract issued to that effect, it cannot later be cancelled by reason of the discovery of a defect which renders the works unusable for their intended purpose: see *The Lord Mayor Aldermen and Citizens of the City of Westminster* v. *J. Jarvis & Sons Ltd and Another* (1970).

It is not unusual for architects, whatever may be the proper meaning of the words 'practical completion', to issue a certificate of practical completion which is qualified by a reference, often by schedule, to defects which are required to be remedied. One or more of these, or all when taken together, may mean that the works have not legally achieved practical completion in accordance with the generally accepted meaning of that phrase in building contracts. In such a case, is it open to argument that in such circumstances no certificate of practical completion has been issued at all? If that argument were to succeed it would have significant implications in connection with contract insurances in respect of the works, liquidated damages, release of retention and the commencement of the defects liability period. Of all of these, the requirement for insurance of the works is probably the most important in that in relation to the others any dispute once resolved, by adjudication or otherwise, is likely retrospectively to be sorted out tolerably well. However in connection with the insurance, e.g. if the contractor ceases cover in respect of the works on the basis that the purported certificate of practical completion is valid, nothing which is done later by an adjudicator, arbitrator or judge following a finding that there was no certificate of practical completion issued, is likely to be able to adequately retrieve the situation if the works have been damaged by an insurable event.

It is submitted that such a qualified certificate of practical completion, even though not strictly envisaged by the wording of the contract, will nevertheless still be a certificate of practical completion under it. In the case of *George Fischer Holding Limited* v. *Multi Design Consultants Limited and Davis Langdon & Everest & Others* (1998) Judge Hicks QC considered this point among a number of points raised in the case. It was argued before him that the purported certificate of practical completion was not in fact such a certificate at all because it included the words 'subject to the enclosed Schedule of Defects and Reserved Matters'. The judge said as follows:

> 'It may first be observed that any submission that this was not a contractual certificate would lie ill in the mouth of a professional adviser who had chosen to use the words "certify" and "practical completion" and to describe it as issued "under the terms of the ... contract", even if the basis for such a submission were in other respects stronger than it is in this case.'

He rejected the substance of the argument and held that what had been issued was indeed a certificate of practical completion, although he emphasised that that was independent of the question whether practical completion had in fact been achieved.

(F) Defects liability period

A defects liability clause will usually refer to a stated period for which it is to run commencing with the date of practical completion. Often it is six or twelve months.

Often under building contracts such provisions require the contractor to remedy defects appearing within the period but do not expressly refer to the completion of any known outstanding or defective items which existed at the time when practical completion was achieved. This can lead to a strict interpretation of what amounts to practical completion, namely that nothing at all must remain to be done however minor in nature. Whilst this may be a literal interpretation it is not how the concept of practical completion has been interpreted by the courts in practice: see *Emson Eastern Ltd (in receivership)* v. *E. M. E. Developments Ltd* (1991).

Unless a contract by its terms expressly and unequivocally states to the contrary, the contractor's obligation and right to attend to defects etc. discovered within the defects liability period will not exclude or limit the employer's legal right to recover damages for losses suffered, if any, as a result of defective work. It simply means that the employer can call for the physical presence on site of the contractor to carry out the remedial work and that the contractor has a contractual right to remedy the defects. This can be of considerable importance to a contractor. Firstly, in the absence of such a right the employer could get the defects remedied by another contractor and the reasonable cost of so doing would be payable as damages by the contractor for breach of contract. The opportunity therefore for the contractor himself to attend to the defects can save him money. Secondly, it helps minimise the risk to the contractor of his reputation being tarnished if he can himself attend to the defects rather than having a third party examine his work.

If the employer unreasonably refuses to allow the contractor back to remedy defects, and has the remedial works carried out by others, this will be a failure to mitigate loss on the employer's part, which may prevent the employer recovering all of the costs of such remedial works from the contractor: see for example *City Axis Ltd* v. *Daniel P. Jackson* (1998), and also *Pearce and High Ltd* v. *John P. Baxter* (1999).

CONSIDERATION OF THE RELEVANT CLAUSES OF IFC 98

Clauses 2.1 and 2.2

Possession and completion dates

2.1   Possession of the site[1] shall be given to the Contractor on the Date of Possession stated in the Appendix[2]. The Contractor shall thereupon begin and regularly and diligently proceed with the Works[3] and shall complete the same on or before the Date for Completion stated in the Appendix[4], subject nevertheless to the provisions for extension of time in clause 2.3.

Possession by Contractor – use or occupation by Employer

For the purposes of the Works insurances the Contractor shall retain possession of the site and the Works up to and including the date of issue of the certificate of Practical Completion, and the Employer shall not be entitled to take possession of any part or parts of the Works until that date.

Notwithstanding the provisions of the immediately preceding paragraph the Employer may, with the consent in writing of the Contractor, use or occupy the site or the Works or part thereof whether for the purposes of storage of his goods or otherwise[5] before the date of issue of the certificate of Practical Completion by the Architect/the Contract Administrator. Before the Contractor shall give his consent to such use or occupation the Contractor or the Employer shall notify the insurers under clause 6.3A or clause 6.3B or clause 6.3C.2 to .4 whichever may be applicable and obtain confirmation that such use or occupation will not prejudice the insurance. Subject to such confirmation the consent of the Contractor shall not be unreasonably delayed or withheld.

Where clause 6.3A.2 or clause 6.3A.3 applies and the insurers in giving the confirmation referred to in the immediately preceding paragraph have made it a condition of such confirmation that an additional premium is required the Contractor shall notify the Employer of the amount of the additional premium. If the Employer continues to require use or occupation under clause 2.1 the additional premium required shall be added to the

Contract Sum and the Contractor shall provide the Employer, if so requested, with the receipt of the insurers for that additional premium.

Deferment of possession

2.2   Where this clause is stated in the Appendix to apply the Employer may defer the giving of possession for a time not exceeding the period stated in the Appendix calculated from the Date of Possession, which should not exceed 6 weeks[6].

COMMENTARY ON CLAUSES 2.1 AND 2.2

IFC 98 requires the employer to give to the contractor possession of the site on the date of possession stated in appendix to the contract. Clause 2.1 expressly limits the extent of possession to be given by the employer to the contractor by allowing the employer to use or occupy the site 'whether for the purposes of storage of his goods or otherwise...'. Quite apart from this express restriction, the IFC 98 Conditions elsewhere also make clear that the contractor's right to possession is not absolute, see for example clauses 2.2, 2.4.8, 2.4.14 and 3.11. The issue of late possession is considered in more detail in the commentary below on clause 2.2 and in the discussion of clause 2.4.14 on page 93.

The second paragraph of clause 2.1 was inserted into IFC 84 as part of Amendment 2 of 1984. It provides that, for the purposes of the works insurances, the contractor shall retain possession of the site and of the works up to and including the date of issue of the certificate of practical completion and that the employer shall not be entitled to take possession of any part of the works until that date. This express right is, it is submitted, to be read subject to clause 7.6(a) and to other clauses such as clause 2.2, 2.4.8, 2.4.14 and 3.11 as well as being qualified expressly by the paragraph of clause 2.1 which follows it. In addition the possession is expressly stated as being 'For the purposes of the Works insurances...'.

The third paragraph gives the employer certain limited rights to occupy the site or the works themselves or a part thereof for the storage of goods or otherwise prior to practical completion, but requires the consent in writing of the contractor. However, this consent must not be unreasonably withheld. The employer's use or occupation for storage or otherwise must be cleared by the works insurer and provided it is confirmed that the cover is not prejudiced the contractor must not unreasonably delay or withhold his consent.

If the works' insurers are prepared to maintain existing cover only on the payment of an additional premium, the employer, if he still wishes to take advantage of this provision, must meet the cost of this additional premium. If the relevant works insurance clause is 6.3A (contractor to insure in joint names) any additional premium will be added to the contract sum. This facility is clearly very useful for employers, e.g. where the contract is overrunning and the employer has to take delivery of furniture or equipment for the building.

The contract conditions do not provide for insurance against loss or damage to any stored goods etc. unless perhaps they are stored in a building which already existed when the contract was formed – see clause 6.3C.1. Otherwise the risk of loss or damage will be governed, if at all, by clause 6.1.2.

If the employer makes use of this provision he should be very careful to ensure that its use or occupation will not impede the contractor. There is no event to cover such a situation within clause 2.4 dealing with extensions of time, so that the employer could find that the liquidated damages provision is invalidated should the contractor be unable to complete by the completion date. Furthermore, there is no relevant matter in clause 4.12 entitling the contractor to reimbursement of direct loss or expense if the progress of the works is materially affected by the employer's use or occupation. As the contractor will have consented to the employer's use or occupation, it can hardly be a breach of contract by the employer, at any rate to the extent to which any disruption can be seen to be inevitable or even perhaps if it can be seen to be a likely consequence. In such circumstances the contractor could consider requiring suitable safeguards for recompense as a condition of giving his consent.

The degree of use or occupation required by the employer may prompt the contractor to suggest that the employer ought to operate the provisions of clause 2.11 (partial possession) so that there will be a deemed practical completion of the relevant part of the works. The situation in which clause 2.1 rather than clause 2.11 should be used will depend on the circumstances. It is submitted that the contractor's withholding of consent under clause 2.1 may on occasions be reasonable if at the same time he indicates that he would consent to a partial possession under clause 2.11.

Clause 2.2, where stated in the appendix to apply, gives the employer the power to defer the giving of possession of the site. It should be noted that it is not the architect that has this power but the employer. The maximum period recommended for which the deferment can be given is six weeks. Bearing in mind that IFC 98 is intended for contract periods of no more than 12 months, a period of six weeks is perhaps more than fair to the employer. In the event of such a deferment being made, the contractor may be entitled to an extension of time for completion and to reimbursement of loss and expense. Should the deferment exceed the period stated in the appendix, the employer will be in breach of contract entitling the contractor to damages and possibly, depending on the seriousness of the breach, entitling the contractor also to treat the employer's breach as a repudiation of the contract thereby excusing the contractor from further performance of his own contractual obligations. In any event it will, if it causes delay to completion, invalidate the liquidated damages provision as there is no event to cover this situation within clause 2.4 dealing with extensions of time.

NOTE TO CLAUSES 2.1 AND 2.2

[1] '...the site...'
Warranties in connection with site conditions will usually depend on whether on the one hand a specification is used or on the other hand a bill of quantities. If IFC 98 is used in conjunction with specification, without a bill of quantities, then the position depends on what is stated in the specification, if anything, as to conditions of the site. If nothing at all is stated, so that the contractor is to satisfy himself as to the site conditions, then the risk of unforeseen difficulties will generally have to be borne by him, e.g. unforeseen ground conditions.

On the other hand, if a bill of quantities is used, by virtue of clause 1.5 the contract bills will be treated as having been prepared in accordance with the Standard Method of Measurement of Building Works, 7th edition (SMM 7) published by the Royal Institution of Chartered Surveyors and the Building Employers Confederation (now the Construction Confederation). Accordingly, particulars will have been given of soil and ground conditions such as water level, trial pits or boreholes and over or underground services. If that information is not available then a description of the ground and strata which is to be assumed shall be stated (see section D20 of SMM 7). To this extent therefore a warranty as to site conditions is given. There are other examples within SMM. If the actual information given turns out to be inaccurate or misleading the contractor will be able to make a claim for extra costs incurred in overcoming the unforeseen problems. See clause 1.4 and page 48.

[2] '...Date of Possession stated in the Appendix...'
The appendix should contain a specified date. However, the insertion of a specified date may well be impracticable from the employer's point of view, bearing in mind such matters as the tendering procedures involved or last minute difficulties in obtaining possession of land by the employer. It may be appreciated that for one reason or another there may be difficulty in accepting a tender in sufficient time to give possession of the site to the contractor by a stipulated date. Often, therefore, the date will be stated in terms of some days or weeks from the date of acceptance of the contractor's tender. It is submitted that the insertion of such words as 'on a date to be agreed' at tender stage is fraught with danger for the employer. If such a phrase is used, it is vital that before the contractor's tender is accepted, the date for possession is agreed and inserted in the appendix. If this is not done at the time of acceptance of the contractor's tender and no date can subsequently be agreed, then presumably possession must be given within a reasonable time and this could arguably invalidate the liquidated damages provision in the contract. Further, the absence of agreement as to the date for possession of the site would deprive the employer of any power he might have had to defer possession under clause 2.2 even if stated in the appendix to apply.

[3] '...regularly and diligently proceed with the Works...'
Even without these express words it has been argued that such a term would be implied in a building contract: see *Hudson's Building and Engineering Contracts*, 11th edition, Chapter 4, paragraph 128 *et seq*. Failure by the contractor to so proceed can lead to a determination of the contractor's employment under this contract (see clause 7.2.1(b)). This requirement to regularly and diligently proceed with the works should be of much more practical use to the employer than the obligation on the contractor to complete by a certain date. In the latter case, generally no action can be taken by an employer until the completion date is past even though the contractor is clearly falling behind; whereas, in the former case, it is submitted that the appropriate notices etc. can be served in accordance with the contract and ultimately the contractor's employment can be determined if necessary.

    In the case of *Greater London Council* v. *Cleveland Bridge and Engineering Co. Ltd and Another* (1984), at first instance Mr Justice Staughton held that a failure by the contractor to execute the works with due diligence and expedition entitling the employer to discharge the contractor for such failure did not itself render the contractor in breach of contract and liable for damages. It was only a failure to use such diligence and expedition as would reasonably be required to meet the contractual deadlines, which would amount to a breach of contract by the contractor. It would seem possible to construct an argument (which it is submitted, would be erroneous) that while the employer's express right to determine the contractor's employment may be relied on, always assuming that the express contractual remedy of determination of the contractor's employment was drafted clearly enough to enable its use even where it might still be physically possible to complete on time, the employer would be taking a considerable risk in treating such a failure by the contractor before the completion date had expired as a repudiation and as a ground for determining the contract itself. On this basis, he would have to wait and see if the contractual completion date was met, or at least until it became absolutely clear that it would not be. Such an interpretation would virtually emasculate the value of such a provision. Further, the case of *West Faulkner Associates* v. *London Borough of Newham* (1994) makes it clear that the requirement to proceed regularly and diligently is not linked only to the completion date. In this case, Lord Justice Simon Brown looking at the words 'regularly' and 'diligently' said:

> 'Taken together the obligation upon the contractor is essentially to proceed continuously, industriously and efficiently with appropriate physical resources so as to progress the works steadily towards completion substantially in accordance with the contractual requirements as to time, sequence and quality of work'.

It can be readily appreciated therefore that the obligation involves not only the requirement to work steadily towards completion by the completion date but also having regard to any contractual requirements as to the sequence of working and also as to the quality of work. The obligation also extends to proceeding continuously, industriously and efficiently. Although there is no contractual requirement for a programme, if one is produced by the contractor, his failure to keep to it may be some evidence of a failure to proceed 'regularly and diligently'.

The words were considered by Mr Justice Megarry in the *Hounslow* case in relation to the determination provisions of JCT 63 (see page 377).

The words 'regularly and diligently' must be read together but the contractor is required to proceed in respect of both requirements. If he fails to proceed regularly even if diligently or alternatively if he fails to proceed diligently although regularly, he will be in breach of this provision – see the *West Faulkner* case.

[4] '...on or before the Date for Completion stated in the Appendix...'
The contractor is fully entitled to complete before the date for completion stated in the appendix and the architect will be obliged to issue a certificate of practical completion pursuant to clause 2.9. If the employer does not want practical completion until the date for completion is reached, then the proper method is to amend the contract conditions themselves rather than to insert such a provision elsewhere e.g. in the bills of quantities or specification (see clause 1.3).

A date for completion is to be stated in the appendix to the contract. The contract is therefore designed for a fixed date for completion. The fact that a fixed date for completion is provided for does not, under this contract, make the time for completion of the essence of the contract so that mere failure by the contractor to complete on time is not a sufficiently serious breach to entitle the employer to treat the breach as a repudiation of the contract by the contractor (see earlier page 70).

Subject to the extension of time provisions, it does however entitle the employer to liquidated damages under the contract.

[5] '...for the purposes of storage of his goods or otherwise...'
The reference to 'or otherwise' seems unnecessarily wide and vague. Coupled with the earlier reference to the use or occupation of the site or the works themselves it offers the possibility of extensive occupation by the employer while work is continuing and with works insurance remaining in place. Presumably, if the employer pushes too far in this direction, he will be faced with works insurers refusing to confirm that cover will be maintained in place, and with the contractor claiming that in truth there has been a partial possession under clause 2.11.

[6] '...should not exceed 6 weeks.'
These words make it clear that, while six weeks may be the recommended maximum, it is quite possible to insert a longer period in the appendix and this is likely to be valid even without any amendment to clause 2.2 or to the words '(period not to exceed 6 weeks)' contained in the appendix item.

Clauses 2.3 to 2.8

Extension of time

2.3   Upon it becoming reasonably apparent that the progress of the Works is being or is likely to be[1] delayed, the Contractor shall forthwith[2] give written notice of the cause of the delay to the Architect/the Contract Administrator, and if in the opinion of the Architect/the Contract Administrator the completion of the Works is likely[3] to be or has been delayed beyond the Date for Completion stated in the Appendix or beyond any extended time previously fixed under this clause, by any of the events[4] in clause 2.4, then the Architect/the Contract Administrator shall[5] so soon as he is able to estimate the length of delay beyond that date or time make in writing[6] a fair and reasonable[7] extension of time for completion of the Works.
If an event referred to in clause 2.4.5, 2.4.6, 2.4.7, 2.4.8, 2.4.9, 2.4.12, 2.4.15, 2.4.17 or 2.4.18 occurs after the Date for Completion (or after the expiry of any extended time previously fixed under this clause) but before Practical Completion is achieved the Architect/the Contract Administrator shall so soon as he is able to estimate the length of the delay, if any, to the Works resulting from that event make in writing a fair and reasonable extension of the time for completion of the Works.
At any time up to 12 weeks[8] after the date of Practical Completion, the Architect/the Contract Administrator may[9] make an extension of time in accordance with the provisions of this clause 2.3, whether upon reviewing a previous decision or otherwise[10] and whether or not the Contractor has given notice as referred to in the first paragraph hereof. Such an extension of time shall not reduce any previously made.
Provided always that the Contractor shall use constantly his best endeavours[11] to prevent delay and shall do all that may be reasonably required to the satisfaction of the Architect/the Contract Administrator to proceed with the Works[12].
The Contractor shall provide such information required by the Architect/the Contract Administrator as is reasonably necessary for the purposes of clause 2.3[13].

Events referred to in clause 2.3

2.4   The following are the events referred to in clause 2.3:

2.4.1 force majeure;

2.4.2 exceptionally adverse[14] weather conditions;

2.4.3 loss or damage caused by any one or more of the Specified Perils;

2.4.4   civil commotion, local combination of workmen, strike or lock-out affecting any of the trades employed upon the Works or any trade engaged in the preparation, manufacture or transportation of any of the goods or materials required for the Works;

2.4.5   compliance with the Architect's/the Contract Administrator's instructions under clauses
 1.4   (*Inconsistencies*), or
 3.6   (*Variations*), or
 3.8   (*Provisional sums*)
     except, where the Contract Documents include bills of quantities, for the expenditure of a provisional sum for defined work*[15] included in such bills, or
 3.15   (*Postponement*),
     or, to the extent provided therein, under clause
 3.3   (*Named sub-contractors*);

---

[*] See footnote [bb] to clause 8.3 (*Definitions*)

2.4.6   compliance with the Architect's/the Contract Administrator's instructions requiring the opening up or the testing of any of the work, materials or goods in accordance with clauses 3.12 or 3.13.1 (including making good in consequence of such opening up or testing), unless the inspection or test showed that the work, materials or goods were not in accordance with this Contract;

2.4.7   .1 where an Information Release Schedule has been provided, failure of the Architect/ the Contract Administrator to comply with clause 1.7.1;

2.4.7   .2 failure of the Architect/the Contract Administrator to comply with clause 1.7.2;

2.4.8   the execution of work not forming part of this Contract by the Employer himself or by persons employed or otherwise engaged by the Employer as referred to in clause 3.11 or the failure to execute such work;

2.4.9   the supply by the Employer of materials and goods which the Employer has agreed to supply for the Works or the failure so to supply;

2.4.10 where this clause is stated in the Appendix to apply, the Contractor's inability for reasons beyond his control and which he could not reasonably have foreseen at the Base Date to secure such labour as is essential to the proper carrying out of the Works;

2.4.11 where this clause is stated in the Appendix to apply, the Contractor's inability for reasons beyond his control and which he could not have foreseen at the Base Date to secure such goods or materials as are essential to the proper carrying out of the Works;

2.4.12 failure of the Employer to give in due time[16] ingress to or egress from the site of the Works or any part thereof through or over any land, buildings, way or passage adjoining or connected with the site and in the possession and control[17] of the Employer, in accordance with the Contract Documents, after receipt by the Architect/the Contract Administrator of such notice, if any, as the Contractor is required to give, or failure of the Employer to give such ingress or egress as otherwise agreed[18] between the Architect/ the Contract Administrator and the Contractor;

2.4.13 the carrying out by a local authority or statutory undertaker of work in pursuance of its statutory obligations in relation to the Works[19], or the failure to carry out such work;

2.4.14 where clause 2.2 is stated in the Appendix to apply, the deferment[20] of the Employer giving possession of the site[21] under that clause;

2.4.15 by reason of the execution of work for which an Approximate Quantity is included in the Contract Documents which is not a reasonably accurate forecast of the quantity of work required;

2.4.16 the use or threat of terrorism[22] and/or the activity of the relevant authorities in dealing with such use or threat;

2.4.17 compliance[23] or non-compliance by the Employer with clause 5.7.1;

2.4.18 delay arising from a suspension by the Contractor of the performance of his obligations[24] under the Contract to the Employer pursuant to clause 4.4A.

Further delay or extension of time
2.5   In clauses 2.3, 2.6 and 2.8 any references to delay, notice, extension of time or certificate include further delay, further notice, further extension of time, or further certificate as appropriate.

Certificate of non-completion
2.6   If the Contractor fails to complete the Works by the Date for Completion or within any extended time fixed under clause 2.3 then the Architect/the Contract Administrator shall issue a certificate to that effect.
       In the event of an extension of time being made after the issue of such a certificate such making shall cancel that certificate and the Architect/the Contract Administrator shall issue such further certificate under this clause as may be necessary[25].

Liquidated damages for non-completion
2.7   Provided:
       – the Architect/the Contract Administrator has issued a certificate under clause 2.6;[26] and
       – the Employer has informed the Contractor in writing[27] before the date of the final certificate that he may require payment of, or may withhold or deduct, liquidated and ascertained damages,
       the Employer may not later than 5 days before the final date for payment of the debt due under the final certificate

       either

2.7.1 require in writing the Contractor to pay to the Employer liquidated and ascertained damages at the rate stated in the Appendix for the period during which the Works shall remain or have remained incomplete and may recover the same as a debt

       or

2.7.2 give a notice pursuant to clause 4.2.3(b)[28] or clause 4.6.1.3 that he will deduct liquidated damages at the rate stated in the Appendix[29] for the period during which the Works shall remain or have remained incomplete;
       Notwithstanding the issue of any further certificate of the Architect/the Contract Administrator under clause 2.6 any written requirement or notice given to the Contractor in accordance with this clause shall remain effective[30] unless withdrawn by the Employer.

Repayment of liquidated damages
2.8   If after the operation of clause 2.7 the relevant certificate under clause 2.6 is cancelled the Employer shall pay or repay to the Contractor any amounts deducted or recovered under clause 2.7 but taking into account the effect of a further certificate, if any, issued under clause 2.6.

COMMENTARY ON CLAUSES 2.3 TO 2.8

Clauses 2.3, 2.4, 2.5, 2.6, 2.7 and 2.8 deal with the making of extensions of time by the architect and the deduction or recovery of liquidated damages by the employer from the contractor.

IFC 98 provides for the date for completion to be extended where completion of the works is likely to be or has been delayed by one or more of the events referred to in clause 2.4.

Extensions of time clauses probably share with variation clauses the doubtful honour of having caused more difficulty in building contracts over the years than any other factor. They cause difficulty not only between contractor and architect but also between the architect and employer since few employers can understand why so many events are built into the contract allowing the contractor to claim an extension of time. They also find it hard to understand the advantage of granting the contractor an extension of time for so called 'neutral' events which may be included in an extension of time clause, regarding such events as being matters which should be accepted as the contractor's responsibility and his 'commercial risk'. However, such a concept may not always work to the advantage of the employer as the contractor may cover such a risk in his tender prices only to find that they may not always occur to the degree which he had assumed.

Much difficulty experienced in relation to extension of time clauses has arisen firstly because contractors have failed to notify the architect when it has become reasonably apparent that the progress of the works has been or is likely to be delayed, thus depriving the architect of the opportunity to take, or to instruct the contractor to take, measures to alleviate the consequences; and, secondly, because the architect has often failed to make an extension of time sufficiently early to allow the contractor to plan completion of the works by the revised completion date.

The drafting of JCT 80 attempted to improve this position by laying down a timetable for response by the architect and at the same time making the contractor responsible for providing information on the anticipated effects and extent of the delay in order to assist the architect in making his assessment.

In the drafting of IFC 84 and therefore IFC 98 the Tribunal has for the most part abandoned the procedural matters contained in JCT 80 and JCT 98, reverting to the general concept to be found in the old JCT 63, but with some significant departures which are referred to later in this chapter. The abandonment of the procedural matters, including a strict timetable for response by the architect, was clearly an attempt to simplify the conditions rather than any recognition that these matters would be less important in a form of contract designed for less complex projects. Experience indicates that it is in fact just as important whatever the size of project and architects would be well advised to respond as rapidly as possible to a contractor's written notice of the cause of delay, notwithstanding the fact that the only reference to the timing of the architect's response in IFC 98 is that he should act 'so soon as he is able to estimate the length of delay'.

As already stated, the clause is more akin to the old JCT 63 than JCT 80 or JCT 98. Clause 25 of JCT 98 requires the contractor to provide specific information concerning the expected effects and an estimate of the delay or likely delay attributable to each and every relevant event together with updates. IFC 98 is less specific. It requires the contractor to provide such information required by the architect as is reasonably necessary for the purposes of operating clause 2.3. Whilst this may be an improvement over JCT 63, it may nevertheless be a disappointment to those who would prefer to see the more specific requirements of JCT 98 repeated in IFC 98.

An extension of time made under clause 2.3 will have the effect in relation to those 'neutral' events which are a contractor's risk of absolving the contractor from liability to pay or allow liquidated damages to the extent that the time for completion is thereby extended. Where the extension of time is made in respect of events which are the employer's responsibility, the effect of the extension is to keep the employer's entitlement to claim liquidated damages intact. See the general discussion on this earlier in this chapter (page 72).

It had been thought that JCT 63 had a potentially serious defect in its extension of time provisions, namely, that by its terms (clause 23 of that contract), it did not allow the architect to make an extension of time after the date for completion had gone by. If this contention was correct the effect would be to invalidate the liquidated damages provision.

Where the ground for an extension of time was what might be termed a neutral event, e.g. exceptionally adverse weather conditions, this might not in any event have attracted an extension of time as the contractor may well not have been delayed by the event but for the fact that he was in culpable delay at the time in question. Had he not been at fault, he would not have suffered delay from the neutral event. However, this would not inevitably be so, e.g. discovery of a defective water main which needs to be replaced by a statutory undertaker. At the end of the day it is a question of both causation and whether the facts are such that in the opinion of the architect it warrants a fair and reasonable extension of time. This particular issue was considered briefly by Mr Justice Coleman in *Balfour Beatty Building Ltd* v. *Chestermount Properties Ltd* (1993) when, in dealing with JCT 80, he said as follows:

> 'There may well be circumstances where a relevant event has an impact on the progress of the works during a period of culpable delay but where that event would have been wholly avoided had the contractor completed the works by the previously-fixed completion date. For example, a storm which floods the site during a period of culpable delay and interrupts progress of the works would have been avoided altogether if the contractor had not overrun the completion date. In such a case it is hard to see that it would be fair and rea- sonable to postpone the completion date to extend the contractor's time. Indeed, where the relevant event would not be an act of prevention it is hard to envisage any extension of time being fair and reasonable unless the contractor was able to establish that, even if he had not been in breach of overshooting the completion date, the particular relevant event would still have delayed the progress of the works at an earlier date. Such cases are not likely to be of common occurrence.'

However, if the delay was caused by the act of the employer or someone acting on his behalf, e.g. the issue of a variation instruction requiring extra work, the employer would have prevented the contractor from completing as soon as he would otherwise have done, and, it was argued, as there was no apparent power to make an extension of time, the liquidated damages provision would be invalidated. Under JCT 80 the position may arguably always have been otherwise as it envisaged the making of an extension of time after practical completion, presumably even where practical completion took place after the expiry of the contractual completion date, i.e. after a period of culpable delay during which the contractor was liable to liquidated and ascertained damages. JCT 80 also provided for the repayment of liquidated damages upon the fixing of a later completion date (see clauses 25.3.3 and 24.2.2 of that contract). However, the position was not, it is submitted, clarified beyond any doubt in JCT 80 (see particularly the words '…delayed thereby…' in JCT 80 clause 25.3.1.2), despite an apparent attempt to resolve the debate by an amendment introduced as part of Amendment 4 to JCT 80 in July 1987.

This debate now seems to have been settled by the judgment in the case of

Balfour Beatty Building Ltd v. *Chestermount Properties Ltd* (1993). The arguments, always somewhat technical, particularly so far as JCT 80 was concerned, were emphatically rejected by Mr Justice Coleman in the Commercial Court. In his judgment he said:

> 'The remarkable consequences of the application of the principle [that an act of prevention would disentitle the employer to liquidated damages] could therefore be that if, as in the present case, the contractor fell well behind the clock and overshot the completion date and was unlikely to achieve practical completion until far into the future, if the architect then gave an instruction of the most trivial variation, representing perhaps only a day's extra work, the employer would thereby lose all right to liquidated damages for the entire period of culpable delay up to practical completion or, at best, on the respondents' submission, the employer's right to liquidated damages would be confined to the period up to the act of prevention. For the rest of the delay he would have to establish unliquidated damages. What might be a trivial variation instruction would on this argument destroy the whole liquidated damages regime for all subsequent purposes.
>
> So extreme a consequence for the future operation of the contract could hardly reflect the common intention, particularly having regard to the very specific distribution of risk provisions which are agreed to be applicable in respect of relevant events occurring *before* the completion date. It is certainly a construction which is most improbable in the absence of some other express provision supporting it.'

Mr Justice Coleman made the point that to concede to such a technical argument would involve legal and commercial results which were so inconsistent with other express provisions and with the contractual risk distribution regime applicable to relevant delaying events that, in the absence of express words compelling such a construction, it could not be right.

In IFC 98, the second paragraph of clause 2.3 attempts to put the issue beyond question by expressly empowering the architect to make an extension of time upon the occurrence, after the expiry of the date for completion or any extended time previously fixed under the clause, of an event referred to in clauses 2.4.5, 2.4.6, 2.4.7, 2.4.8, 2.4.9, 2.4.12, 2.4.15, 2.4.17 or 2.4.18, all of which may be regarded as being in one way or another the responsibility of the employer. The intention behind this was to ensure that, despite such an event and the timing of it, the liquidated damages provision should remain effective. In the light of the decision in the *Balfour Beatty* case, such a provision is probably not now necessary. Indeed, it could pose a problem in relation to neutral events occurring during a period of contractor's culpable delay. It has been suggested above that an extension of time in respect of such an occurrence might be appropriate where the event would, at whatever time it occurred, have caused delay. This certainly appeared to be the case in JCT 80. However, including this second paragraph in clause 2.3 of IFC 98, may give rise to the argument that expressly permitting extensions of time for the events selected in that paragraph, means that an extension of time is not available for the other 'neutral' events. It is submitted that this was not an intended consequence of inserting this paragraph.

In passing, it is worth noting that, so far as IFC 84 containing Amendment 12

(April 1998) is concerned, the numbered events referred to in that version of IFC do not include certain acts of prevention on the part of the employer, namely in clause 2.4.17 (employer causing delay in connection with CDM Regulations), and clause 2.4.18 (contractor suspending due to non-payment by employer). This was presumably a drafting oversight. However, on a literal construction, it could pose real problems for the employer. As certain employer delays are expressly catered for, it could be argued that the intention is that those employer delays not referred to in the second paragraph do not attract an extension of time. As they are delays for which the employer is responsible and which prevent the contractor completing by the contractual completion date, the result could be that the completion date will cease to be binding, the contractor having a reasonable time in which to complete. The consequence of this could be that the employer will lose the right to claim any liquidated damages at all. This omission was corrected in producing IFC 98.

So far as the length of the extension is concerned, the words in the second paragraph, namely '... the length of the delay ... resulting from that event...' lend support to the contention that the extension given should be the net effect on the actual completion date which the event causes and not the gross period between the pre-existing contractual or extended completion date and the date when the delaying event has expended itself. On the other hand, the gross extension argument seems more logical than the net extension argument when considering the operation of clause 2.6. Where there has been a clause 2.6 certificate followed by a further extension of time involving the cancellation of the original clause 2.6 certificate and requiring the issue of a second clause 2.6 certificate to satisfy the condition precedent of an architect's certificate of non-completion before the employer can deduct any liquidated damages at all, the second certificate issued under clause 2.6 would, if the net extension argument is to prevail, have to contain a statement by the architect that the contractor has failed to complete by an extended completion date. This, by definition (as the contractor's culpable delay will not have been included within the extension of time given), must be a date earlier than that on which the delay for which the extension of time is granted expired, and possibly even before it commenced. When the employer then exercises his rights under clause 2.7 to deduct liquidated damages this will in fact cover a period of time during part of which the contractor was delayed by the employer. Even so, on balance it appears that the probable intention behind clauses 2.3 to 2.8 taken together favours the net extension approach.

This question of gross or net extension of time was also considered in the *Balfour Beatty* case in connection with similar provisions in JCT 80. The argument for the gross extension was roundly rejected in favour of the net extension approach, the court again adopting a commonsense and commercial approach to what was a highly technical, and many would say, unmeritorious argument. Mr Justice Coleman held that the gross extension approach was wrong in principle and that some support for this was to be found in the judgment of Lord Justice Denning (as he then was) in *Amalgamated Building Contract Ltd* v. *Waltham Holy Cross UDC* (1952). Mr Justice Coleman put it in this way:

'... the architect simply has to address the question what the completion date then is and whether the contractor had achieved practical completion by then.

There is no express or implied requirement that the completion date which should have been fixed or re-fixed having regard to what is fair and reasonable should be after the time of the fixing or re-fixing so as to make that date physically possible to be achieved. Because the function of adjusting the completion date is to delimit the contractual period of work, there is nothing in principle to stop the architect in an appropriate case from adjusting that date to a point of time which is both before the date of the exercise of that power and before the relevant event in question.'

The third paragraph of clause 2.3 provides for the architect, at any time up to 12 weeks from practical completion, to make an extension of time whether on reviewing a previous decision or otherwise, and whether or not the contractor has given the appropriate notice under the first paragraph of clause 2.3. This appears to give the architect considerable discretion where the contractor has failed to give the required notice. Any such extension must not reduce an extension of time previously made; any review can therefore only be to increase the period of extension of time. No doubt, however, if there are grounds for making an additional extension of time, but also factors which fairly reduce the extension required by the contractor, e.g. some variations requiring additional work and others omitting work, the architect can look at the net effect in order to determine the appropriate period for the extension, provided of course it does not have the effect of reducing the total of the extensions of time already given. If the architect fails to make a further extension of time when one is warranted, this will not invalidate the liquidated damages provisions in the contract, although of course the contractor could seek arbitration to secure an extension of time and recover any liquidated damages deducted: see *Temloc* v. *Errill Properties Ltd* (1987).

It must be at least arguable, though not as strongly as under JCT 98, that once the 12 weeks has expired the architect no longer has any power under the contract to make further extensions of time. He appears to be ex-officio in this regard.

Clauses 2.3 and 2.4 are basically traditional in character and scope and are for the most part consistent with other standard forms of contract issued by the Tribunal. They do, however, contain some additional features and these will be dealt with under the next heading and also under the notes which follow on page 97.

In relation to the grounds for making an extension of time, the contract adopts a listing approach rather than stating a general ground for an extension of time, e.g. 'for reasons beyond the control of the contractor'. This listing approach is to be preferred as the courts may be unwilling to construe a general provision for an extension of time as covering delays which are the responsibility of the employer, e.g. late instructions. This could be disastrous for the employer as his entitlement to liquidated damages would be lost: see *Peak Construction (Liverpool) Ltd* v. *McKinney Foundations Ltd* (1970).

If IFC 98 is compared with JCT 98, it can be seen that one of the relevant events in JCT 98 is not found in IFC 98, namely delay caused by the UK Government exercising any statutory power which directly affects the carrying out of the works by restricting availability of essential labour, goods or materials, fuel or energy in connection with the proper carrying out of the works (clause 25.4.9 of JCT 98). This provision was originally inserted into JCT 80 following the 'three day week' in the early 1970s. In certain circumstances such an event may be a *force majeure* falling within clause 2.4.1. All the other JCT 98 events are included in IFC 98, although some of them are worded differently.

THE EVENTS IN RESPECT OF WHICH AN EXTENSION OF TIME CAN BE MADE (CLAUSE 2.4)

Introduction

There are 18 clause 2.4 events listed although these break down into many more items. Those events in clauses 2.4.10, 2.4.11 and 2.4.14 do not apply unless stated to apply in the appendix.

If the employer intends to amend any of the events without this affecting the freezing of fluctuations it will be necessary to amend also, by deletion or otherwise, clause C4.7 and C4.8 of supplemental condition C or clause D13 of supplemental condition D as appropriate.

Some of the clause 2.4 events, namely those in clauses 2.4.5, 2.4.6, 2.4.7, 2.4.8, 2.4.9, 2.4.12, 2.4.14, 2.4.15, 2.4.17 and 2.4.18 have words which are similar or identical to some of the sub-clauses of clause 4.12 (or clause 4.11(a) in the case of 2.4.14) entitling the contractor to claim for loss and expense incurred by him as a result of one or more of those matters.

Although the wording may therefore be similar or identical, the clauses are not directly linked. If the architect makes an extension of time in respect of one of the clause 2.4 events, which has a similar or identical wording to any of the matters referred to in the sub-clauses of clause 4.12, this is evidence that the contractor has suffered delay as a result of a relevant event. This does not automatically mean that the delay has caused loss and expense. This has to be separately established: see *H. Fairweather & Co. Ltd* v. *London Borough of Wandsworth* (1987). On the other hand, the contractor may incur loss and expense by reason of one of these matters even though it does not delay the completion of the works, e.g. where plant or labour is idle or underproductive in relation to a non-critical part of the works, the delay and disruption to which does not affect the overall completion date.

The events themselves

Clause 2.4.1

Force majeure has been described as having reference to all circumstances independent of the will of man, and which it is not in his power to control. It has a meaning wider than *vis major* (act of God). Its interpretation will be conditioned to some degree by the nature and type of the other clause 2.4 events. It may well not cover an event, however catastrophic, which has been brought about by the negligence of the contractor: see *J. Lauritzen A/S* v. *Wijsmuller BV (the Super Servant Two)* (1989).

Clause 2.4.2

The weather must not be just adverse, it must be exceptionally adverse. This should render it a truly unusual if not rare occurrence. It is to be wondered, therefore, why architects appear so often to grant an extension of time under this head. Architects will often require evidence of exceptionally adverse weather conditions in the form of detailed meteorological records demonstrating the exceptional nature of the weather.

There are many debating points concerning this event. Examples are:

(1) Should the contractor have to provide meteorological evidence for the previous five, ten or fifteen years in order to demonstrate that the weather conditions were exceptionally adverse?

(2) To what extent, if any, should the architect have regard not only to the period of exceptionally adverse weather conditions but any exceptionally beneficial weather conditions occurring shortly before or after which may mean that *overall* progress has not been affected by such weather? For example, an exceptionally mild February which is particularly beneficial to the progress of the contractor's external works, during which month however there were two days of unprecedented snowfall which prevented progress being made on those two particular days.

It is submitted that it is for the architect to determine (reviewable in adjudication, arbitration or litigation) a fair and reasonable extension of time and that it is difficult, and often unhelpful, to try and lay down fixed rules or principles.

Clause 2.4.3

The specified perils are defined in clause 8.3 (Definitions). The list includes many of the insurable risks, e.g. fire, flood. It may be possible for the contractor to obtain an extension of time under this clause where the loss or damage was brought about by his own negligence or that of someone for whom he is responsible, even if this amounts to a breach of the express obligation of the contractor in clause 1.1 to carry out the works in a proper and workmanlike manner. It might have been thought that as the contractor must not benefit from his own breach of contract he might lose his entitlement to any extension of time. A contracting party cannot be seen to profit from his own breach of contract. However, in the case of *Surrey Heath Borough Council* v. *Lovell Construction Ltd* (1988) in which it was held at first instance, under a JCT 81 With Contractor's Design Contract, that the negligent destruction by fire by a subcontractor of part of the works was a breach by the contractor of an implied term similar to the express obligation in clause 1.1 of IFC 98 referred to earlier, it was nevertheless assumed by Judge Fox-Andrews QC, though without argument on the point, that an extension of time granted by the architect under a similar clause to that now under consideration was appropriate.

It is to be noted that the loss or damage is not restricted to the works. Any loss or damage caused by a specified peril which causes a delay to completion can fall within this event. This view is supported by the fact that the definition of

'Specified Perils' is to be found in clause 8.3 rather than in clause 6.3.2 dealing with insurance of the works; and by the scope of the insurance cover for lost liquidated damages set out in clause 6.3D.1 which includes loss or damage to temporary buildings, plant and equipment.

An argument that, in such circumstances, no extension of time need be made because the contractor has failed to use constantly his best endeavours to prevent delay is, it is submitted, untenable. The contractor's best endeavours relate to preventing *the delay* rather than preventing *the cause* of the delay.

There is provision for the employer to require the contractor to obtain insurance for the benefit of the employer for lost liquidated damages due to an extension of time for this event. This is discussed later in Chapter 9 (page 312).

Clause 2.4.4

It should be noted that the effect of such matters as civil commotion, strikes or lockouts etc. need not be direct, so long as the effect is on any of the trades employed on the works or any trade engaged in the preparation, manufacture or transportation of any of the goods or materials required for the works. It can therefore be of extremely wide application and virtually removes from the contractor all risks of delay (though not of course of costs which may result from any disruption) caused by industrial disputes and civil disobedience. However, it is submitted that the effect on any trade in the preparation, manufacture or transportation of any goods or materials must be of a national, or at any rate fairly general rather than local, nature as otherwise this sub-clause would appear to overlap 2.4.10 and 2.4.11.

Clause 2.4.5

See the specific clauses referred to for the instructions to which this clause relates.

Clause 2.4.6

This relevant event has to be considered together with clauses 3.12 and 3.13 (instructions as to inspection – tests and instructions following failure of work etc.) and clause 4.12.2 (loss and expense). So far as the application of this sub-clause is concerned, it matters not whether the instruction for opening up or testing etc. was issued under clause 3.12 or clause 3.13.1. In either case an extension of time can only be given where the inspection or test showed that the work, materials or goods were in accordance with the contract. However, oddly enough (see clause 3.13.2), it appears that an arbitrator or an adjudicator (see the commentary to clause 3.13.2) could award an extension of time in respect of an instruction given under clause 3.13.1 even though the result of the inspection or test etc. showed that the work, materials or goods were not in accordance with the contract.

Clause 2.4.7

Where clause 1.7.1 (information release schedule) applies, the architect must provide the information in accordance with that schedule. Where there is no such schedule or where, even though there is a schedule, information is necessary which is not contained in it, then the architect must provide the information, pursuant to clause 1.7.2, at such time as to enable the contractor to carry out and complete the works in accordance with the conditions, including enabling therefore the contractor to complete by the contractual completion date. However, if the contractor's progress suggests that he will overrun the contractual completion date the supply of information is linked to the contractor's actual progress. Reference should be made to the commentary on clause 1.7 (page 51).

Clause 2.4.8

This clause is to be read in conjunction with clause 3.11 (work not forming part of the contract). If it is intended at the outset that such other work shall be undertaken during the contract period, as much information as possible should be supplied to the contractor prior to his tendering. A certain amount of disturbance and disruption can then be allowed for by the contractor in his programming and tender price. In such a case it is only where the extent of the delay could not reasonably be foreseen by an experienced contractor that the architect should make an extension of time.

Sometimes local authorities and other public bodies employ their own direct labour organisations to carry out work. If that work is kept separate from the work for which the contractor is responsible under IFC 98, i.e. if the work does not form part of the contract, this sub-clause will apply. However, if their work forms part of the main contract work under IFC 98 (often the direct labour organisation is engaged by the contractor under a notional 'sub-contract'), this sub-clause has no application. In legal terms the public body concerned will be both the employer under the main contract for the works as a whole but also a sub-contractor to the contractor for the relevant sub-contracted part of the works.

If the public body, in breach of its sub-contract, delays the contractor, there appears to be no clause 2.4 event which readily covers the situation. Two possibilities arise. Firstly, if the contractor is thereby put in breach of the main contract and becomes liable for liquidated damages to the public body as employer, the contractor will have a right of recourse against the public body as sub-contractor. This circuity of action could be used to defeat the employer's claim for liquidated damages. Secondly, and more probably, the result will be that the public body as employer will have prevented (even if in its sub-contract role) the contractor from completing by the contractual completion date and as no appropriate event exists in clause 2.4 to cover the situation the employer will lose his right to claim liquidated damages.

Clause 2.4.9

The effect of a failure by the employer to supply or of a delay by the employer in supplying materials and goods is likely to have a significant impact on the contractor's programme – more so than delays by the employer or others on his behalf in relation to the execution of work. Subject to this important point, similar comments can be made on this clause as have been made in respect of clause 2.4.8 except that if it is a public body employer's direct service organisation which has entered into a notional 'supply' contract with the main contractor (in law it will be a supply contract between the public body and the contractor), this sub-clause will apply as it relates to the supply of materials or goods for the works themselves. Hence if it is a supply only rather than a supply and fix situation, an extension of time can be made and the liquidated damages provision will be preserved.

Clause 2.4.10

This sub-clause applies only where it is stated in the appendix to apply. Some employers are of the view that the risk of delay due to such an event as this is properly a contractor's commercial risk. Others may be of the opinion that including such a provision is more likely to produce comparable tenders as the need to price the risk is removed from the contractor, and if the event does not materialise the employer has not had to pay against the risk, whereas this might happen if the clause were deleted. Much may depend on the state of both the construction and labour markets.

Is the contractor expected to incur extra cost to overcome the problem, e.g. paying exceptionally high wages to attract skilled labour where there is a shortage of it? No doubt it is a matter which the architect will take into account in determining whether or not the contractor has used his best endeavours to prevent delay. It is submitted that the contractor would not be expected to incur considerable expenditure in order to overcome the problem.

What if the contractor obtains specialist sub-contract labour which produces defective work which no amount of supervision by the contractor could have avoided? Could the contractor contend that he has been unable 'for reasons beyond his control ... to secure such labour as is essential to the proper carrying out of the Works'? It would be surprising if such an argument were to succeed, as this sub-clause has generally been treated as relating to shortages of labour rather than to defective work. That said, it is by no means certain that the argument would fail: see the House of Lords case of *Scott Lithgow* v. *Secretary of State for Defence* (1989).

Clause 2.4.11

Similar comments can be made on this clause as have been made on clause 2.4.10.

Clause 2.4.12

It should be appreciated that this event has nothing to do with the employer giving possession of the site to the contractor as such. It is restricted to the means of access to, or egress from, the site over other connected or adjoining property or way in the possession and control of the employer.

Clause 2.4.13

Similar wording in JCT 63 has received some detailed judicial consideration: see *Henry Boot Construction Ltd* v. *Central Lancashire New Town Development Corporation* (1980) and note [19] on page 102.

Clause 2.4.14

This sub-clause applies only where stated in the appendix to apply. It should be read in conjunction with clause 2.2. It can only operate if the date of possession in the appendix is correctly completed either with a specific date inserted or alternatively, but less satisfactorily, a stated number of days or weeks from the date of acceptance of the tender. If the appendix carries some such phrase as 'on a date to be agreed' then it is vital that the date is agreed before the tender is accepted. To leave agreement until afterwards runs the risk of failure to agree a specific date with a possible invalidation of the liquidated damages provision.

This event is inserted very much for the employer's benefit in order to prevent an employer's delay from invalidating the liquidated damages provision. If it is known before the tender is accepted that there is going to be a delay in giving possession of the site to the contractor, then a new date for possession can be reached by agreement and there will be no place for this provision to operate. It applies only where, after a binding contract has been formed, the employer is unable to give possession of the site in accordance with clause 2.1. As this sub-clause refers to the deferment in giving possession of the site, it may well not include a deferment by the employer in giving possession of a part only of the site rather than a deferment in relation to the whole of it. If this is the case, then a deferment in relation to part of the site would amount to a breach of contract by the employer, at any rate where that part was required by the contractor in order for him to commence work on site. A delay occurring in such circumstances, as there is no appropriate extension of time event to cater for it, could invalidate the liquidated damages provision. This could also be the result of a failure by the employer to give possession of the whole of the site within or at the expiry of the deferred period stated in the appendix.

Clause 2.4.15

Where bills of quantities are a contract document, they are to be prepared in accordance with SMM 7 which recognises that bills drawn in accordance with the standard method can include approximate quantities which are the subject of a measure when the work is carried out. This is an exception to the lump sum principle of IFC 98, which does not itself require the quantities upon which the original unadjusted contract sum is based to be actually measured as completed work. This is considered again in the commentary to clause 3.7 (page 163).

If on measuring the work for which an approximate quantity is included in the bills, it is found that the approximate quantity stated is not a reasonably accurate forecast of the work as measured, this sub-clause applies. What is reasonably accurate will depend on all the circumstances.

Clause 2.4.16

This event is aimed at delay caused by terrorism. It is widely worded to extend not only to the direct consequences of a terrorist act but also to delays attributable to threats, e.g. a hoax telephone call. In addition it extends to delay caused by the actions of relevant authorities in dealing either with terrorism itself or the threat of it.

Clause 2.4.17

This event relates to clause 5.7.1 dealing with various obligations of the employer arising out of the Construction (Design and Management) Regulations 1994. It covers the situation where the employer's appointees under the Regulations – the planning supervisor or the principal contractor (but only where he is not also the contractor) – carry out or fail to carry out the statutory duties required and which by clause 5.7.1, the employer has agreed he will ensure are carried out. A typical example of its application would be where the planning supervisor has failed to comment on the safety implications of a variation, with the result that progress is delayed. Reference should also be made to the commentary on clause 5.7 (page 270).

Clause 2.4.18

This deals with delay caused by the contractor's exercise of his contractual right (and statutory right to suspend performance of his contractual obligations pursuant to section 112 of the Housing Grants, Construction and Regeneration Act 1996). The statutory right to suspend was considered earlier (see page 29). The Tribunal decided that despite there being a statutory provision for effectively extending the contractual time as a result of any such suspension, it would be sensible to provide a specific event in order to maintain a complete code in respect of delaying events.

   However, there appear to be differences between the statutory extension in section 112(4) and the entitlement under clause 2.4.18. It is arguable that the former relates the period of the extension to the contractual completion date directly to the period of the suspension. The period of suspension may cause a lesser or greater delay to completion than the period itself. On this basis the period of the statutory extension does not appear to depend on the delay caused by the suspension, whereas under clause 2.4.18 it is the delaying effect of the suspension which is taken into account. By way of example therefore a suspension of, say, 10 days after which the employer pays in full, might lead to a situation where an innocent sub-contractor, who was thereby prevented from working on the site, elected to remove his tower crane to relocate it to another contract in order to comply with his contractual obligations in relation to that other contract and which he would have been able to do by finishing his work had there been no such suspension. It may be that the sub-contractor cannot bring the tower crane back onto site for four weeks and it may not be practicable (even if contractually possible, which is doubtful) for the contractor to secure the services of another sub-contractor and tower crane in order to reduce the delay. The actual delay caused by the suspension could therefore be very much greater than the 10 day suspension itself.

The situation could be reversed, with the 10 day period of suspension resulting in no delay to completion. Under clause 2.4.18 the contractor would not be entitled to an extension of time whereas under the statutory provision, which would still apply, the contractor would be.

It is submitted that where the suspension is of such a period that the contractor acts reasonably in winding down the site, any delay attributable to having to gear up again will be a consequence of the suspension and therefore covered by the event.

CERTIFICATE OF NON-COMPLETION: PAYMENT AND REPAYMENT OF LIQUIDATED DAMAGES

Clause 2.6

The first paragraph makes it clear that the architect has a duty, not a discretion, as to the issue of a certificate of non-completion once the date for completion or any extended time has expired.

However, this certificate is automatically cancelled where, after its issue, a further extension of time is made whether or not the event occurred before the date stated in the certificate (e.g. on a reconsideration of a previous extension of time based on an estimate) or afterwards in respect of those events referred to in the second paragraph of clause 2.3. There is no provision for fixing an earlier date for completion than an existing one even where a subsequent event reduces the overall delay, e.g. a variation instruction requiring an omission. However, depending on the circumstances, it may be that if there are subsequent events causing delays as well as variations by way of omission then the architect would be acting reasonably in making an extension of time in respect of the net delay only.

If an extension of time is made after the certificate has been issued so that there is an automatic cancellation of the architect's certificate, and if the architect fails, if appropriate, to issue a further updated certificate in its place, any deduction of liquidated damages by the employer will be wrongful, and if limited to the actual period of culpable delay between the revised completion date and practical completion: see *A. Bell and Son (Paddington) Ltd* v. *C.B.F. Residential Care and Housing Association* (1989).

Clause 2.7

The benefit to the employer of a liquidated damages clause has been discussed elsewhere (page 70). On occasions, where the figure inserted in the appendix in respect of liquidated damages is less than the actual loss suffered by the employer as a result of the contractor's breach of contract in failing to meet the completion date or any extended time for completion, the liquidated damages clause in effect operates as a limitation of liability in favour of the contractor. This is not always appreciated.

The general effect of this clause can be summarised as follows. The architect must issue a certificate of non-completion under clause 2.6 and the employer must inform the contractor in writing prior to the date of the final certificate if he intends to operate the liquidated damages provision whether by way of requiring payment or by withholding or deducting from sums otherwise due to the contractor. Once the contractor has been so informed in writing, this remains effective even where the architect's certificate has been replaced by a further architect's certificate. It does not fall with the architect's certificate. This therefore makes it clear that the decision in *J. F. Finnegan Ltd* v. *Community Housing Association Ltd* (1995) and a decision to similar effect in the earlier case of *A. Bell & Son (Paddington)* v. *C.B.F. Residential Care and Housing Association* (1989) has no application under IFC 98.

The reason for the employer's notification prior to the date of the final certificate is on the basis that it is thought the contractor should know before the other financial aspects of the contract are concluded whether or not it is the employer's intention to deduct or recover liquidated damages. Whether this requirement can be met by employers stating in the tender documentation that they may exercise their right to claim liquidated damages, in order to guard against losing this right through oversight, clerical error or administrative incompetence, is debatable. It is certainly not within the spirit or intention of this clause and may well in any event be ineffective.

If the employer has met this requirement, he must in addition, if he requires the contractor to pay liquidated damages (on the basis that there are insufficient sums due to the contractor from which to withhold or deduct them), not later than five days before the final date for payment of the sum due under the final certificate, issue a written requirement to the contractor to pay liquidated damages. It is by no means clear if this further written requirement is a condition precedent to the right to make the claim. If it is not, the employer will be able to require payment of the liquidated damages subject only to the application of any relevant statutory limitation periods. It has to be said that the drafting in connection with this further written requirement is far from clear.

If the employer wishes to exercise his right to claim liquidated damages by way of withholding or deducting them, then again, not later than five days before the final date for payment of any sum due under the final certificate, he must give notice of intended deduction. This will have to comply with section 111 of the Housing Grants, Construction and Regeneration Act 1996 and these requirements are reflected in clause 4.2.3(b) in connection with interim certificates, clause 4.3(c) in relation to the interim payment on practical completion, and clause 4.6.1.3 in connection with the final certificate. These clauses set out the requirements for the contents of the notice. See notes 26 to 30 on page 104 *et seq* for further comments in relation to clause 2.7.

Clause 2.8

This clause provides for the repayment of liquidated damages where, subsequent to their deduction or recovery, the time for completion is extended or further extended. Can the contractor claim interest on the repayment of liquidated damages by the employer? The contract envisages the deduction followed by repayment where appropriate. The initial deduction cannot therefore subsequently be regarded as a breach of contract by the employer and the contract contains no express provision for interest in these circumstances. The provision in clause 4.2(a) entitling the contractor to interest in connection with late payment applies only where the employer fails to pay a sum due by its final date for payment. In connection with repayment of liquidated damages, it is likely that the due date for payment will be the date of the automatic cancellation of the relevant architect's certificate under clause 2.6 and the final date for payment is probably 17 days from that due date pursuant to paragraph 8(2) of Part II of the Scheme for Construction Contracts (England and Wales) Regulations 1998. Only after that date will the contractor be able to claim interest pursuant to express terms of the contract (see clause 4.2(a)). Even then, however, it is arguable that the interest provisions only apply in respect of late payment under certificates of the architect.

The decision of an Irish court in the case of *Department of Environment for Northern Ireland* v. *Farrans (Construction) Ltd* (1981) (a decision on JCT 63) may appear to indicate that interest is recoverable on repaid liquidated damages, but JCT 63, unlike IFC 98, does not expressly provide for repayment of liquidated damages. Furthermore, the decision in this case has received some critical comment (see the commentary by the learned editors of *Building Law Reports* at 19 BLR page 3 *et seq.*).

However, of more significance is the power of an arbitrator to award interest, compound or simple, under section 49 of the Arbitration Act 1996 in respect of any period up to the date of the award, not only in regard to the amount awarded but also in respect of any sum outstanding at the commencement of the arbitral proceedings but paid before the award has been made. This power is not dependent on establishing that the sum on which interest is ordered was being wrongfully withheld throughout the period for which interest is ordered. This was probably also the position before the 1996 Act came into force (see section 19A Arbitration Act 1979 and the case of *Amec Building Ltd* v. *Cadmus Investment Co. Ltd* (1996)). If of course repayment is made before arbitral proceedings are commenced, the only claim for interest will be that provided for by the express terms of the contract in clause 4.2(a), or where appropriate, under the Late Payments of Commercial Debts (Interest) Act 1998.

NOTES TO CLAUSES 2.3 TO 2.8

[1] '...progress of the Works is being or is likely to be...'
The contractor is required to give notice in respect of the progress of the works being delayed or of the likelihood of their being delayed and from whatever cause. This gives the architect the opportunity of taking action to negate or reduce the effect of the delay. It need not relate to a relevant event. It should be noted that a notice of the cause of the delay is required, not an application for an extension of time by the contractor as such. Also, it is delay to the progress of the works rather

than delay pushing the works beyond the date for completion which is the key to
the giving of the notice.

[2] '...forthwith...'
This must mean immediately progress of the works is or is likely to be delayed. Is
the requirement that the contractor's notice should be given forthwith a condition
precedent to the operation of the clause? It is suggested not, certainly in relation to
those clause 2.4 events which are inserted primarily for the benefit of the
employer. To construe the requirement for notice as a condition precedent would,
in the absence of due notice, prevent the architect from making an extension of
time and thus invalidate the liquidated damages provision. This cannot be the
intention of clause 2.3. Furthermore, the third paragraph of clause 2.3 appears to
give the architect a discretion to extend time in the absence of the notice.

In the case of *London Borough of Merton* v. *Stanley Hugh Leach Ltd* (1985) it was
held, under a JCT 63 contract, that the notice of delay required from the contractor
by that contract's extension of time clause (clause 23) was not a condition precedent to the contractor's entitlement to an extension of time. However, the failure
to give notice could well be a breach of contract by the contractor. If as a result
either the architect is prejudiced in his ability to make an accurate assessment of
the delay, e.g. if his opportunity to investigate the matter is now limited, or if the
architect has lost the chance to issue appropriate instructions, or to take other
measures which could have reduced the overall delay, then this will no doubt be
taken into account by the architect when assessing what amounts to a fair and
reasonable extension of time.

[3] '...is likely...'
The architect must grant a fair and reasonable extension of time for completion of
the works if he is of the opinion that completion has been or is likely to be delayed.
Clearly, when the delay is merely anticipated rather than actual, the assessment of
an appropriate extension of time may be somewhat speculative. Nevertheless, as
soon as he is able to make a reasonable estimate of the length of delay he must
make the extension of time. The estimate made can be reviewed under the provisions of the third paragraph of clause 2.3 during the progress of the works or at
any time up to 12 weeks after the date of practical completion.

[4] '...by any of the events...'
There is absolutely no power under this contract for the architect to make an
extension of time in respect of a cause of delay which does not fall within the
clause 2.4 events listed.

Where the same period of delay is attributable to both a clause 2.4 event and
some non-qualifying event, then it is reasonable, it is submitted, for the architect to
grant an extension of time. The reasoning for this is as follows: liquidated
damages clauses are generally construed strictly against the employer: see *Peak
Construction (Liverpool) Ltd* v. *McKinney Foundations Ltd* (1970). It is for the
employer to claim his entitlement to liquidated damages. In order to do so he
must, it is submitted, be able to demonstrate that, but for the contractor's culpable
delay (i.e. a delay for which no extension of time event is applicable), the employer
would have got his building on time. If there is a genuinely concurrent delaying

event for which the contract provides an extension of time, running alongside the contractor's culpable delay, it follows that even if the contractor had not been at fault the employer would still not have got his building on time due to a delay for which the contractor is entitled to an extension of time and thus to relief from liability to pay liquidated damages. This clearly ought to be the position where the concurrent delay is one of those events in clause 2.4 for which the employer is responsible. It seems likely that it should also be the position where the concurrent delay is of the 'neutral' kind, although the argument here is not as strong as the employer will not have contributed to the delay.

The situation could well be reversed in relation to claims by the contractor for reimbursement of direct loss or expense where there are concurrent causes, one of which is a clause 4.12 matter and the other a matter for which the contractor takes the commercial risk, as it is the contractor in this instance who is in the role of claimant and has to prove his case. Those readers particularly interested in this difficult area may wish to read *Keating on Building Contracts*, 6th edition, page 209 *et seq.*, where the general conclusion (page 213) appears to be that the so-called 'dominant cause approach' is or should be the correct approach rather than what might be called the 'burden of proof approach' referred to in this paragraph.

Having said all this, there is some authority for saying that neither delay should dominate the other and that some sort of apportionment is required: see *H. Fairweather & Co. Ltd v. London Borough of Wandsworth* (1987). In practice it will often be impossible for the architect to do this on any rational or certain basis. Nevertheless, this in itself is no reason for rejecting such an approach which may produce the fairest result on a case by case basis.

[5] '. . . the Architect/the Contract Administrator shall . . .'
Failure to do so if wilful could well prevent the employer from being able to operate clause 2.7 to deduct liquidated damages. Further, the freezing effect of the fluctuations provisions at the date for completion or any extended time for completion will not operate (see clause 4.9 and supplemental condition C (clause C4.7 and C4.8) and supplemental condition D (clause D13)).

If the architect's failure is not wilful it is possible that the employer's right to deduct liquidated damages may still remain on the basis that the architect's failure to make the assessment *'so soon as he is able . . .'* relates to a matter of administration rather than of substance: see *Temloc Ltd v. Errill Properties Ltd* (1987) where however it was, exceptionally, the employer who was unsuccessful in claiming that the right to liquidated damages was lost due to the architect's failure to comply with the timetable for determining extensions of time under a JCT 80 contract. If it had been the contractor so claiming, the decision might just have gone the other way. Of course, if the contractor is aggrieved that an extension of time to which he is entitled has not been granted, this can be the subject of a challenge in adjudication or arbitration proceedings, or if the parties choose not to arbitrate then in the courts: see *Beaufort Developments (NI) Ltd v. Gilbert-Ash NI Ltd and Others* (1998).

[6] '. . . in writing . . .'
The architect should make it absolutely clear that he is making an extension of time and should state the extended time for completion of the works.

[7] '...fair and reasonable...'
The decision of the architect is reviewable by an adjudicator or arbitrator or, if the parties agree not to operate the arbitration clause, then the courts: see *Beaufort Developments (NI) Ltd* v. *Gilbert-Ash N I Ltd and Others* (1998). The architect must take into account the proviso in the fourth paragraph of clause 2.3, that the contractor shall constantly use his best endeavours to prevent delay.

[8] '...At any time up to twelve weeks...'
Presumably, by implication, the architect has no power to operate this paragraph after the 12-week period has expired, although it appears that under a similar though not identical clause in JCT 80, clause 25.3.3, the court found otherwise: see *Temloc* v. *Errill Properties Ltd* (1987).

[9] '...may...'
The exercise by the architect of the power to make further extensions of time under this paragraph of clause 2.3 is thus not mandatory. This is to be compared with the somewhat similar provision in JCT 98, clause 25.3.3, which uses the word 'shall'.

[10] '...or otherwise...'
These words do not mean that the contractor is entitled to an extension of time for delaying events which occur between the date for completion or any extended time for completion and the date of practical completion, unless such events fall within those listed in the second paragraph to clause 2.3. The events there listed are of course inserted for the benefit of the employer.

[11] '...best endeavours...'
This is often a difficult phrase to interpret in a given context. What is the position, for instance, if the contractor could take certain very expensive measures to prevent delay, e.g. importing labour to avoid the effect of a widespread strike in a particular trade. Is he bound to do so? It is submitted that there is no requirement for the contractor to commit himself to significant expenditure even though the word used is 'best' and it is nowhere qualified in terms of the cost involved.

If the contractor fails to use his best endeavours to prevent the delay, then he will be liable to pay or allow to the employer liquidated damages.

If the clause 2.4 event was one which also materially affected the regular progress of the work and so could form the basis of a claim by the contractor for reimbursement of direct loss and expense under clauses 4.11 and 4.12, could any of the extra cost incurred by the contractor in using his best endeavours to prevent delay becoming part of that claim for reimbursement? There seems no reason in principle why not, if the steps taken were reasonable.

It is to be hoped that the words 'best endeavours' will be construed relatively. It would have been preferable from the contractor's point of view if the words 'reasonable endeavours' had been used.

[12] 'Provided always that the Contractor shall use constantly his best endeavours to prevent delay and shall do all that may be reasonably required to the satisfaction of the Architect/the Contract Administrator to proceed with the Works'
This paragraph contains the normal type of proviso to be found in extension of time provisions. In the event of delay occurring, it is clear that the contractor is not

permitted to sit back and let the delay take its natural course. He must take such positive steps as are available to prevent delay. In any event this will usually be in his own best interests, especially in those situations where any disruption costs will have to be borne by him.

The proviso has two stings to it. Firstly, the constant use by the contractor of his best endeavours: this is independent of any instructions of the architect. Secondly, the doing of all that may be reasonably required to the satisfaction of the architect to proceed with the works: this enables the architect to issue instructions as to how the delay should be overcome or reduced. This is qualified by the use of the word 'reasonably'.

[13] *'The Contractor shall provide such information required by the Architect/the Contract Administrator as is reasonably necessary for the purposes of clause 2.3'*

This requirement for the contractor to provide information is worded very generally. Clearly it will cover matters such as an assessment of the length of any delay already suffered, together with an estimate of the extent of any expected delay. It is submitted that the contractor will also have to give information as to the expected effects of the delay where this is reasonably required by the architect in order that he can determine what may be reasonably required of the contractor to proceed with the works. It may also be reasonably necessary for the architect to have this information in order to determine whether or not the contractor has used his best endeavours to prevent delay.

As to the extent and detail of the information required from the contractor, the case of *London Borough of Merton* v. *Stanley Hugh Leach* (1985) is instructive. Though considering the requirement for a contractor to provide information relating to a claim for reimbursement of direct loss and expense rather than an extension of time, the summary of Mr Justice Vinelott in this case is still apposite when he said at 32 BLR page 97/98:

'But in considering whether the contractor had acted reasonably and with reasonable expedition it must be borne in mind that the architect is not a stranger to the work and may in some cases have a very detailed knowledge of the progress of the work and of the contractor's planning. Moreover it is always open to the architect to call for further information either before or in the course of investigating a claim. It is possible to imagine circumstances where the briefest and most uninformative notification of a claim would suffice: a case, for instance, where the architect was well aware of the contractor's plans and of a delay in progress caused by a requirement that the works be opened up for inspection but where a dispute whether the contractor has suffered direct loss or expense in consequence of the delay had already emerged. In such a case the contractor might give a purely formal notice solely in order to ensure that the issue would in due course be determined by an arbitrator when the discretion would be exercised by the arbitrator in place of the architect.'

[14] *'...adverse...'*

This can include hot and dry as well as cold and wet weather conditions.

[15] *'...except ... for the expenditure of a provisional sum for defined work...'*

This is because if the provisional sum is for defined work it is deemed to have been taken into account by the contractor in programming and planning the works so as to complete them by no later than the contractual completion date: see SMM 7 General Rule 10.4. General Rule 10.4 is set out as part of the footnote to clause 8.3 Definitions (see page 371). See also the discussion on this matter earlier on page 49.

[16] '...*in due time*...'
If a time is agreed or clearly indicated on an agreed contractor's programme this will no doubt be the due time, other things being equal. If no precise time is agreed in advance, then access or egress must be given in reasonable time taking into account the overall situation including, among other matters, the employer's position.

[17] '...*in the possession and control*...'
   It is submitted that the employer must have not only the right to possession but also the *de facto* control of the land, buildings, way or passage. Presumably some form of shared control is sufficient, e.g. a joint tenancy.

[18] '...*as otherwise agreed*...'
The question of access and egress may be dealt with in the contract documents or otherwise agreed. If dealt with in the contract documents, the agreement is clearly directly between the employer and the contractor. If otherwise agreed, the clause suggests that the agreement is between the architect and the contractor with the former clearly acting as the agent of the employer. The architect should therefore check his own authority to make sure that he is permitted to bind the employer in this fashion. As the contract clearly bestows such power on the architect, the contractor need not concern himself with checking the authority of the architect unless perhaps he has knowledge that the architect has limited actual authority.

[19] '...*its statutory obligations in relation to the Works*...'
In reality, the carrying out by a statutory body of work strictly in pursuance of its statutory obligations will often be very limited. Cases can arise where part of the work is carried out pursuant to statutory obligations and part not. If the statutory body carries out work otherwise than pursuant to its statutory obligations, this will sometimes take the work outside the contract so that clause 2.4.8 will apply. On other occasions the work will remain part of the contract works and will be carried out under a sub-contract between the main contractor and the statutory body concerned. If the work is carried out by way of a domestic sub-contract, then the contractor takes the risk of delay in respect of non-statutory work. However, where such work is covered by a provisional sum item, then although it forms part of the contract in that the contract sum includes the value of provisional items, it was nevertheless held in the case of *Henry Boot Construction Ltd* v. *Central Lancashire New Town Development Corporation* (1980) that such work did not form part of the contract works for which the contractor was responsible under JCT 63 as the employer entered into separate contracts with the statutory undertakers. In such cases the contractor will be given an extension of time under clause 2.4.8 and will be entitled to reimbursement of loss and expense under clause 4.12.3 if the regular progress of the works has been materially affected. Bearing in mind the increasing privatisation of former statutory bodies, it is difficult to see how this

clause can be applied in many situations, especially where a privatised body carries out a statutory duty or function. It may be time for a reconsideration of the wording of this particular event.

[20] '... the deferment ...'
This must relate to the delay in giving the initial possession of the site. The wording is not appropriate to cover interruptions in possession of the site or parts of it once the contractor has obtained possession. Such interruptions should be covered wherever possible by an instruction as to postponement of work pursuant to clause 3.15. However, if in truth there is a failure of the employer to give possession of part of the site during the course of the contract rather than a postponement of work, the effect could be to invalidate the liquidated damages provision (see also commentary on clause 2.4.14 on page 93).

[21] '... possession of the site ...'
These words are used also in clause 2.1. In the context of clause 2.1 they appear to refer to the whole of the site and cannot easily be read to mean part only of the site. There seems no reason to give the words any different meaning in this sub-clause. Accordingly the deferment cannot relate to part only of the site.

[22] '... terrorism ...'
If Amendment TC/94 issued in April 1994 is incorporated into the IFC 98 contract, clause 6.3.2 provides a definition of terrorism: '... any act of any person acting on behalf of or in connection with any organisation with activities directed towards the overthrowing or influencing of any government *de jure* or *de facto* by force or violence.'

No doubt even without Amendment TC/94 being incorporated, terrorism would be similarly defined.

[23] '... compliance ...'
The event covers not only delay as a result of the employer's failure to ensure that the planning supervisor or principal contractor carry out their duties but also any delay attributable to the employer making sure that they carry out such duties. Some might argue that this is a risk which should fall on the contractor rather than the employer but it is for the employer under clause 5.7.1 to ensure compliance and if in so doing delay is caused then it is appropriate that an extension of time should be given.

[24] '... performance of his obligations ...'
Note that this does not just cover suspending construction of the works but also the suspending of any contractual obligations, e.g. protection of the works, security of the site, ordering of goods or materials, renewal of insurances.

[25] '... as may be necessary'
If the extension of time which has led to the automatic cancellation of the clause 2.6 certificate brings the extended time for completion in line with the date of practical completion, there will be no further certificate. If the extension of time referred to pushes the extended contract completion date into the future, then

again no further certificate will be necessary at that time, though it may be later on, should the date of practical completion be later than the extended time for completion.

[26] *'Provided the Architect/the Contract Administrator has issued a certificate under clause 2.6...'*
This is a condition precedent to the right to deduct liquidated damages (see *Amalgamated Building Contractors Ltd* v. *Waltham Holy Cross UDC* (1952); *Token Construction Co. Ltd* v. *Charlton Estates Ltd* (1973)). The right to deduct may be lost if the architect wilfully fails to perform his functions in relation to the proper consideration of clause 2.4 events and the making of extensions of time as this would affect the validity of the certificate. However, the fact that the architect is guilty only of tardiness in his consideration of possible extensions of time will not of itself cause an existing clause 2.6 certificate to be invalid: see the cases of *Temloc Ltd* v. *Errill Properties Ltd* (1987) and *J. F. Finnegan Ltd* v. *Community Housing Association Ltd* (1993) at first instance (95 BLR 103).

[27] *'Provided the Employer has informed the Contractor in writing...'*
This is another condition precedent to the employer's right to claim liquidated damages under this clause. If, following the employer's requirement in writing that liquidated damages are to be deducted or claimed, the architect makes a further extension of time, the employer's written requirement will continue to be effective as a result of the last paragraph of clause 2.7.2.

[28] *'...give a notice pursuant to clause 4.2.3(b)...'*
Quite apart from the employer's written requirement or notice under the first paragraph of clause 2.7, it is also necessary for the employer to provide a further notice under clause 4.2.3(b) relating to deductions from interim payment certificates, not later than five days before the final date for payment in respect of the certificate from which it is intended to make the deduction. This notice which is a requirement under section 111 of the Housing Grants, Construction and Regeneration Act 1996, will be adequate in respect of not only the certificate from which the initial deduction is intended to be made, but also deductions from later payment certificates where earlier certificates are insufficient to cover the notified amount of liquidated damages. In such a case the same notice could also satisfy the requirements of clause 4.6.1.3 relating to deductions from the final certificate. If the employer's requirement in writing under the first paragraph of clause 2.7 states the amount intended to be deducted, which it need not and probably should not do (see comments under note [30] below), both notices could be combined. It is unwise to attempt to do this as it would produce difficulties of application where at one point in time the notice may satisfy both sets of requirements, whereas at some later point in time it may only satisfy one set of requirements.

[29] *'...at the rate stated in the Appendix...'*
In the case of *Temloc Ltd* v. *Errill Properties Ltd* (1987) under a JCT 80 contract the appendix entry was '£nil'. This was held to be fully binding on the employer and was treated as exhaustive of the employer's remedies against the contractor for delay in completion.

[30] '...shall remain effective...'

This makes it clear that the employer's written requirement or notice given to the contractor remains effective despite any cancellation of the architect's clause 2.6 certificate, so that it will continue to serve its purpose in respect of subsequent clause 2.6 certificates. The fact that the first written requirement or notice of the employer remains effective must also mean that the contents of the employer's notice need only be of the most general kind and need not, and indeed should not on this basis, be made up of a detailed calculation of the liquidated damages intended to be deducted. It is a statement of general intent. Therefore, the decision in *J. F. Finnegan Ltd* v. *Community Housing Association Ltd* (1995) that the notice should state the sum being deducted, does not apply here. In any event, the additional notice requirements set out in clauses 2.7.1 and 2.7.2 effectively require the amount being claimed or deducted to be stated.

Clauses 2.9 and 2.10

Practical Completion

2.9 When in the opinion of the Architect/the Contract Administrator Practical Completion of the Works is achieved and the Contractor has complied sufficiently with clause 5.7.4 he shall forthwith issue a certificate to that effect. Practical Completion of the Works shall be deemed for all the purposes of this Contract[1] to have taken place on the day named in such certificate[2].

Defects liability

2.10 Any defects, shrinkages or other faults which appear[3] and are notified by the Architect/the Contract Administrator to the Contractor not later than 14 days[4] after the expiry of the defects liability period named in the Appendix from the date of Practical Completion, and which are due to materials or workmanship[5] not in accordance with the Contract or frost occurring before Practical Completion[6], shall be made good by the Contractor at no cost to the Employer unless the Architect/the Contract Administrator with the consent of the Employer[7] shall otherwise instruct; and if the Architect/the Contract Administrator does so otherwise instruct then an appropriate deduction[8] in respect of any such defects, shrinkages or other faults not made good shall be made from the Contract Sum. The Architect/the Contract Administrator shall, when in his opinion the Contractor's obligations under this clause 2.10 have been discharged, issue a certificate to that effect[9].

COMMENTARY ON CLAUSES 2.9 AND 2.10

For a summary of the general law in relation to practical completion see page 73.

Clause 2.9 does not expressly deal with those outstanding items of work or minor defects which exist but which the architect regards as insufficient to hold up the issue of a certificate of practical completion. In practice, the contractor will be asked to deal with such matters during the defects liability period. As it could be argued by a contractor that clause 2.10 does not cover items of outstanding or defective work which were known to exist at the time of practical completion, it is advisable, even though such an argument would probably not succeed (see earlier page 73 considering the case of *Emson Eastern (in Receivership)* v. *E. M. E. Developments Ltd* (1991)), for the architect in such cases to refer to such items by way of list or other suitable record and to obtain the contractor's confirmation as to how they will be dealt with before actually issuing the certificate of practical completion (but see [3] below under notes to clauses 2.9 and 2.10).

This clause in no way affects the employer's right to recover under the general law the consequential losses flowing from defective work. Furthermore, even after the expiry of the defects liability period, the contractor still has a legal liability in respect of defective work, subject to the effects of the issue of a final certificate (see clause 4.7).

In the case of IFC 98, practical completion does not only involve completion of the building work etc. but also the additional requirement that the contractor must have complied sufficiently with clause 5.7.4 which provides that within the time reasonably required by the planning supervisor the contractor is to provide, and ensure that the sub-contractors provide, such information as the planning supervisor reasonably requires in order to prepare the health and safety file as required by the Construction (Design and Management) Regulations 1994. This is a very significant requirement falling on the contractor and is often likely, in practice, to delay the issue of the certificate of practical completion and involve the contractor in liability for liquidated damages. On the other hand, if the employer, as will very often be the case, wishes to occupy and use the works for their intended purpose, the architect may be prevailed on to issue a 'certificate of practical completion' without this requirement having been met. This could result in the employer as a client, and the architect if also the planning supervisor, falling foul of the Regulations.

The health and safety file is essentially an enlarged maintenance manual to inform those responsible for the works after handover as to risks that must be managed when the works and any associated plant is maintained, repaired, renovated or demolished. Though it is based on risks to health and safety, it will still very often require the provision of quite detailed maintenance and operation manuals. These traditionally are often difficult for the architect (and therefore will be difficult for the planning supervisor) to obtain. Making their provision a condition precedent to achieving practical completion under the contract can be expected to create some change to the custom and practice in the industry of providing such manuals at a later date.

The requirement to 'comply sufficiently' with clause 5.7.4 is far from precise. It could lead to arguments about the scope of the CDM Regulations and in particular 14(d), 14(e) and 14(f). More is said about what these Regulations provide and about the content of the health and safety file in the commentary on clause 5.7 (see page 273).

Clause 2.10 deals with the contractor's obligations during the defects liability period. It requires the contractor to make good any defect, shrinkages or other faults which appear and are notified by the architect to the contractor not later than 14 days after the expiry of the defects liability period named in the appendix. If no period is stated in the appendix then it is to be treated as being six months from the day named in the certificate of practical completion.

NOTES TO CLAUSES 2.9 AND 2.10

[1] '...for all the purposes of this Contract...'
The achieving of practical completion has some very important contractual consequences:

(1) Within 14 days the architect must issue a certificate including 97.5 per cent of the value of work as against a previous certified payment of 95 per cent, which thus effectively releases one moiety of retention. See clause 4.3(a).

(2) The defects liability period commences on the day named in the certificate in accordance with clause 2.10.

(3) The procedures leading up to the issue of a final certificate commence. See clauses 4.5 and 4.6.

(4) The liquidated damages payable for delay in completion in accordance with clause 2.7 will stop on the day named in the certificate of practical completion.

(5) The contractor's risk in respect of loss or damage to the works will switch to the employer on this date. Up until practical completion the contractor will generally be responsible for the risk of damage or loss to the works by virtue of his general obligation to complete in clause 1.1. However, in respect of certain risks, the contractor will not be liable if clause 6.3B or clause 6.3C has been chosen whereby the risk of loss or damage to the works for the events described will fall on the employer. The contractor's obligation to insure, where clause 6.3A is selected, will usually cease at this date.

(6) This date marks the time beyond which variation instructions may not be validly given (except possibly in relation to items of outstanding work).

(7) This date is usually the last date on which time will run under the Limitation Acts for actions for breach of contract (except in relation to outstanding items completed or defects remedied during the defects liability period, in which case the date of their completion or remedy will be the appropriate date).

[2] '... the day named in such certificate'
It is the day named in the certificate which is relevant and not the date of the certificate itself. There is no need therefore for the architect to issue a certificate bearing a retrospective date.

[3] '... which appear ...'
This clause does not as it might have done (see for example JCT 98 clause 17.2) go on to refer to defects etc. which appear during the defects liability period, i.e. after practical completion. Even under the JCT 98 wording (clause 17.2) the accepted view is that defects known to exist prior to practical completion will be picked up: see for example *Keating on Building Contracts*, 6th edition, page 592. Further, under the JCT Minor Works Form 1980 it has been held that defects appearing before practical completion are covered where the words used, 'which appear within three months of the date of practical completion' are far more restrictive than in IFC 98: see *William Tomkinson & Sons Ltd* v. *The Parochial Church Council of St Michael and Others* (1990). It is open to the employer therefore to contend that defects etc. which are known to exist at the time of practical completion must be remedied by the contractor under this clause.

[4] '... not later than 14 days ...'
The notification can be at any time from the date of practical completion to 14 days after the expiry of the defects liability period. It can be notification from time to time during this period. There is no express provision for a schedule of defects

although in practice one will often be prepared at the end of the defects liability period. The architect clearly has an important duty to his client to ensure that the works are thoroughly examined and tested as appropriate before expiry of the defects liability period (literally up to 14 days after that date) and that any defects are notified to the contractor.

[5] '... materials or workmanship ...'
It should be noted that there is no reference to defects in design. These are not of course the contractor's responsibility. The remedying of such defects falls outside this clause and poses problems generally, e.g. a variation instruction for remedial works owing to defective design cannot be given after the date of practical completion.

[6] '... occurring before Practical Completion ...'
In relation to damage caused by frost, this must relate to injury or potential injury suffered during the contract period but which reveals itself during the defects liability period. Frost damage caused after practical completion will be at the employer's risk.

[7] '... with the consent of the Employer ...'
After practical completion the architect has no inherent power under this clause to accept work not in accordance with the contract or to omit work required by the contract documents. If the employer is to accept something different from that for which he contracted, then he must consent to this. No doubt this provision will prove very useful where the defect is either irremediable or would involve unreasonable cost to put right, e.g. if it would involve the undoing and redoing of a considerable amount of good work.

[8] '... an appropriate deduction ...'
It was thought in some of the JCT contracts, e.g. early versions of JCT 80, that the architect in exercising his power to 'otherwise instruct' could produce a result whereby the defect could be made good at the employer's expense. It is unlikely that this was ever intended and this clause now makes it clear beyond doubt that it is only an appropriate deduction from the contract sum which is permitted in cases where the contractor is relieved from his obligation to make good the defects etc.

An appropriate deduction should generally, it is submitted, be based on what would have been the cost to the contractor of remedying the defects etc., excluding for this purpose any further saving to the contractor as a result of his not having to incur related costs, e.g. removal of good work and its replacement which is necessary in order to remedy defective work, or meeting the employer's dislocation costs. There should be no ransom element in the valuation of an appropriate deduction. The valuation of the appropriate deduction is taken into account by way of an adjustment to the contract sum, thus ensuring that it is the quantity surveyor rather than the employer who determines what it should be – see clause 4.5.

[9] '... issue a certificate to that effect'
For the relationship between this certificate and the release of the last moiety of retention and the issue of the final certificate, see clause 4.6.1.

Clause 2.11

Partial Possession by the Employer

2.11 If at any time or times before the date of issue by the Architect/the Contract Administrator of the certificate of Practical Completion the Employer wishes to take possession of any part or parts of the Works and the consent of the Contractor (which consent shall not be unreasonably delayed or withheld) has been obtained then, notwithstanding anything expressed or implied elsewhere in this Contract, the Employer may take possession thereof. The Architect/The Contract Administrator shall thereupon issue to the Contractor on behalf of the Employer a written statement identifying the part or parts of the Works taken into possession and giving the date when the Employer took possession (in clauses 2.11, 6.1.4 and 6.3C.1 referred to as 'the relevant part' and the 'relevant date' respectively); and

- for the purpose of clause 2.10 (*Defects liability*) and 4.3 (*Interim payment on Practical Completion*) practical completion of the relevant part shall be deemed to have occurred and the defects liability period in respect of the relevant part shall be deemed to have commenced on the relevant date;
- when in the opinion of the Architect/the Contract Administrator any defects, shrinkages or other faults in the relevant part which he may have required to be made good under clause 2.10 shall have been made good he shall issue a certificate to that effect;
- as from the relevant date the obligation of the Contractor under clause 6.3A or of the Employer under clause 6.3B.1 or clause 6.3C.2 whichever is applicable to insure shall terminate in respect of the relevant part but not further or otherwise; and where clause 6.3C applies the obligation of the Employer to insure under clause 6.3C.1 shall from the relevant date include the relevant part;
- in lieu of any sum to be paid by the Contractor or withheld or deducted by the Employer under clause 2.7 (*Liquidated damages for non-completion*) in respect of any period during which the Works may remain incomplete occurring after the relevant date there shall be paid such sum as bears the same ratio to the sum which would be paid apart from the provisions of clause 2.11 as the Contract Sum less the amount contained therein in respect of the said relevant part bears to the Contract Sum; or the Employer may give a notice pursuant to clause 4.3(c) that he will deduct such sum from monies due to the Contractor.

COMMENTARY ON CLAUSE 2.11

This clause is for use where the employer with the contractor's consent wishes to take possession of part or parts of the works before completion of the whole.

Where it is intended before the contract is entered into that completion of parts of the works is required before other parts, this cannot be satisfactorily achieved by using clause 2.11. The IFC 98 Sectional Completion Supplement should be used in such a case (this is dealt with under the next heading, page 111).

Clause 2.11 enables the employer, with the contractor's consent, to take possession of part only of the works which part, once taken over, must be made the subject of a written statement issued by the architect on behalf of the employer identifying the part taken over ('the relevant part') and specifying the date when it was taken over ('the relevant date').

The contractor's consent must be obtained and must not be unreasonably withheld. In most instances the contractor will be prepared to consent as this has certain advantages as will be seen below when looking at the contractual effects of partial possession. However, occasionally there may be reasonable grounds on which the contractor can withhold consent, e.g. where consenting to partial possession would result in difficulties of access or working which would inevitably cause the contractor to overrun on the remainder of the contract into a period of time when his resources were to be fully committed elsewhere.

The three key elements are the identification of the relevant part, the valuation of the relevant part, and the relevant date.

In deciding whether to seek partial possession the employer and the architect must carefully consider the practicability of it. While it may pose few problems in some situations, e.g. individual housing or industrial units, it may raise very considerable difficulties in others. Of particular importance are common services such as light, heat and power where to take over part of a building may require taking over the common services for the whole. Examples where there could be potential difficulties in relation to common parts are central staircases or roof structures in relation to blocks of flats. Even in relation to individual units there could be common roads, accesses etc. to be considered. If only part of a common service is to be taken over there may be difficulties both in defining and valuing it.

Once taken into possession the relevant part is deemed to be practically completed. Great care is therefore required to identify the relevant part precisely. As it is a deemed practical completion there is no certificate as such and clause 2.9 requiring the architect to issue a certificate of practical completion does not apply. Instead the architect issues a written statement identifying the part and the date when the employer took possession. Practical completion of the relevant part is then deemed to have occurred on that date. Care should be taken to ensure that there has been sufficient compliance by the contractor with clause 5.7.4 dealing with the provision of relevant information for the health and safety file under the Construction (Design and Management) Regulations 1994 prior to the employer taking possession.

The architect should consider the possibility in appropriate circumstances of utilising clause 2.1 where the requirement is for storage or very limited use or occupation as, if clause 2.1 is used, this will not be deemed to amount to practical completion.

The precise identity of the relevant part can be vital as the defects liability period under clause 2.10 operates in respect of the relevant part from the relevant date. For example, the whole of the heating system may have to be taken over to obtain beneficial use of part. All of the relevant part will be subject to the defects liability period. This could prejudice the employer if it turns out that there is a defect in that part of the heating system not utilised.

The architect must issue a certificate of making good defects for the relevant part when in his opinion any defects, shrinkages or other faults have been made good pursuant to clause 2.10.

The valuation of the relevant part and the fixing of the relevant date are important for a number of reasons including:

(1) One half of the retention is released for the relevant part as it is deemed to be practically completed.
(2) The insurance of the works obligations under clause 6.3A, 6.3B or 6.3C ceases. Where clause 6.3C operates, the employer's obligation to insure existing structures will extend to the relevant part as from the relevant date. Strangely enough, where clause 6.3B applies and the employer takes part into possession there is no requirement for him to insure it against specified perils.

(3) The liquidated damages figure stated in the appendix is reduced by the proportion which the value of the relevant part bears to the unadjusted contract sum. The operation of this reduction or apportionment formula makes no allowance for the different effects in terms of financial loss which any delay in completion of outstanding parts may have. For instance, a retail unit may be taken over in part with only, say, £20,000 worth of landscaping outstanding. Here any delay in the outstanding part of the works may cause little or no loss in terms of the unit opening for trade and maintaining it. On the other hand, such a unit may be taken into possession for the employer to arrange to have it fitted out ready for trade, leaving the public access roads still to be finished off by the original contractor at a cost of, say, £15,000. Here any delay in finishing off the balance of work beyond the employer's fitting out period, could prevent trading and the estimated loss could be just as much as if no part had been taken over at all and yet the liquidated damages rate will have been significantly reduced. The operation of this formula is therefore inherently unlikely to produce a liquidated damages figure which is based on a genuine pre-estimate. As a result it has been suggested by some that a contract with this type of clause in it may cause the liquidated damages provision to be a penalty. However, if in general terms it remains an attempt, in the circumstances, to achieve a genuine pre-estimate then it is likely to be upheld even if it is lacking in precision: see for example *Philips (Hong Kong) Ltd* v. *Attorney General of Hong Kong* (1993).

SECTIONAL COMPLETION SUPPLEMENT FOR IFC 98

The Tribunal has issued for use with IFC 98 a Sectional Completion Supplement (1998). It provides a series of amendments to the IFC 98 conditions to enable completion of the works by sections. Article 7 of the articles of agreement to IFC 98 provides for the sectional completion supplement to apply by means of a deletion of option B.

Where it is known in advance of the contract being entered into that either possession of parts of the site is to be given at different times or that completion and use by the employer of some parts of the works is required before others, the sectional completion supplement should be used. To rely on the optional clause 2.11 partial possession provision is not appropriate in such circumstances. Further, to try and achieve a similar result by including somewhere in the contract documentation a schedule of handovers and hand-backs, e.g. for a housing refurbishment scheme, is very likely to fail in its purpose. At best it is unlikely to deal adequately with such matters as practical completion of the part handed back, release of retention, insurance, apportioned liquidated damages and separate extension of time provisions for each part. At worst it will not even get off the ground and could well invalidate the liquidated damages provisions of the contract. Further, by virtue of clause 1.3 of IFC 98, seeking to achieve sectional completion by using contract documents other than the conditions is likely to fail. Clause 1.3 provides that:

'Nothing contained in the Specification/Schedules of Work/Contract Bills shall override or modify the application or interpretation of that which is contained in the Articles, Conditions, Supplemental Conditions or Appendix.'

Examples of unsuccessful attempts are to be found in such cases as *Gleeson* v. *Hillingdon Borough Council* (1970) and *Bramall and Ogden Ltd* v. *Sheffield City Council* (1983). However, in the Scottish case of *Barry D. Trentham Ltd* v. *Robert McNeil* (1995) it was held that an attempt to introduce sectional completion through provisions in the bills was successful where the bills contained the appendix to the conditions and the appendix referred to sections of the works. The court held that the appendix remained a contract document even though it was in the bills and as such the appendix could be considered alongside the conditions, including clause 1.3. If there was a conflict between them then the appendix was to prevail. See also the earlier discussion of this case (page 47).

Where this supplement is to be used the works must be divided into sections. This must be clearly done in the contract documents. The sections should be numbered and the numbers inserted into the amended appendix entries which will also state the date of possession, completion, rate of liquidated damages, defects liability period and value for each section.

Each section has its own start and completion date, its own certificate of practical completion and its own defects liability period. Each section has its own liquidated damages and extension of time provision applicable to it.

Care must be taken to ensure that where a section is to be completed and taken into possession for use by the employer, any necessary enabling services are considered, as it may be necessary to take over part or the whole of such services as part of the first section to be completed. Much will depend on whether the common elements, e.g. heating or power, can be divided into a number of sections.

There is only one final certificate for the whole of the works even though sectional completion applies. Accordingly its issue will usually be dependent on the issue of a certificate of making good defects for the last section in which defects are made good.

Where possession of one section is dependent on progress on any other section, the appendix entries for related sections need to be considered with great care in order to keep the liquidated damages clause intact. The solution with dependent sections is to give them fixed completion dates by reference to the calendar but to fix the date for possession of the section by reference to a critical date for an earlier section, such as the achieving of practical completion. For example, a date could be inserted for possession of section 2 which is 'ten days after practical completion of section 1'.

Finally, it sometimes happens in practice that while sectional completion is required, it is not possible at the time of contracting to identify the section easily, e.g. in a housing 'decant' scheme where the particular batch of houses to be handed over is dependent on when tenants are prepared to vacate to alternative temporary accommodation. There may be no easy answer to this problem. It may be possible to identify a section within the contract documents by reference to, say, any four adjoining houses. Even so difficulties can easily arise, for instance in identifying and valuing the individual sections which the amended appendix to IFC 98 requires to be done at the outset.

Chapter 5
Control of the works (Part I)

CONTENT

This chapter looks at clauses 3.1 to 3.3 of section 3 of IFC 98 which deal with matters relating to assignment and sub-contracting, including consideration of named persons as sub-contractors.

SUMMARY OF GENERAL LAW

(A) Assignment

Assignment in relation to a contract means the transfer of contractual rights or benefits by a party to the contract to some other person who was not originally a party to it, i.e. a stranger to the contract.

A contract is made up in part of benefits and in part of burdens. What amounts to a benefit to one party will very often be a burden to the other, e.g. under a building contract the employer obtains the benefit of a building and the contractor assumes the burden of it or the liability for building it. The law treats the transfer of benefits and burdens differently.

The benefits

Subject to what a contract may expressly provide, a contracting party may assign the benefits due to him under the contract, e.g. the contractor under a building contract may assign the benefit of the retention money or the employer may assign the partly constructed building together with the right to have it completed. If a party to a contract assigns any benefit under it, he should give notice of that fact to the other party. Once notified the other party must, if the benefit being assigned represents a burden to him, discharge that burden in favour of the new beneficiary, i.e. the assignee.

Certain benefits of a truly personal nature cannot be assigned, e.g. the right to litigate where the cause of action is purely personal.

The burdens

The law does not permit the unilateral assignment of contractual liabilities. The burdens under a contract can only be transferred with the consent of the other contracting party. Once that consent is forthcoming the original party can be released from the burdens. The person to whom the burdens are transferred becomes liable in the place of the person who transferred them. This in law is called a novation. It must be distinguished from vicarious performance of contractual liabilities, i.e. someone else actually performing the obligation whilst the contracting party remains fully liable under the contract. Vicarious performance is dealt with below under the next heading.

The law in relation to assignment stems essentially from simple contracts such as money debts or simple contracts of sale where discussion of benefits and burdens is straightforward. However, in relation to building contracts and other complicated contracts, these rather elementary concepts are not always suitable. In particular a distinction can sometimes be drawn between an assignment of rights under the contract such as a right to claim damages or to payment on the one hand and the right to have the contract performed on the other. Depending on the wording of any particular assignment clause in the contract, the former may be assignable while the latter is not. However, when a court comes to consider a clause prohibiting assignment, it will require very clear words before it construes the clause as allowing assignment of the fruits of the contract such as damages or payment on the one hand while prohibiting assignment of the right to performance of the contract on the other: see *Linden Gardens Trust Ltd* v. *Linesta Sludge Disposals Ltd* (1993). As Lord Browne-Wilkinson said in this case:

> 'In my view they (the parties) cannot have contemplated a position in which the right to future performance and the right to benefits accrued under the contract should become vested in two separate people.'

The likely result therefore is that the prohibition against assignment will be total rather than partial.

Many building contracts, including IFC 98, purport expressly to prohibit assignment, even of benefits, without the consent of the other contracting party.

(B) Sub-contracting

Sub-contracting of contractual obligations, i.e. vicarious performance of the actual work, is permitted under the general law unless the obligations are personal in nature (see next paragraph). In many instances, so long as the original party remains responsible for any failure to properly and fully perform the burden or obligation, it matters not who actually carries it out. For example, a main contractor may of his own choice arrange for a sub-contractor to carry out certain of the work. Subject to what the express terms of the contract may say, the main contractor will be permitted to do this while remaining fully liable to the employer for any failure of the sub-contractor to adequately fulfil his contractual obligations where this leads to a similar failure of the main contractor *vis-à-vis* the employer. The employer will be able to claim from the main contractor who must in turn claim from the sub-contractor. The relationship between the main contractor and the sub-contractor is of no concern to the employer. This arrangement is known in the construction industry as domestic sub-contracting to distinguish it from nominated or named sub-contracting. If it is the employer who selects a particular sub-contractor to the main contractor then different considerations apply. This topic is dealt with under the next heading.

If the employer can demonstrate that he is placing reliance on the particular skill or expertise of a given contractor then even a building contract may be regarded as being of a personal nature, thus preventing the sub-contracting of any of the work. A good example of this is to be found in the case of *Southway Group Ltd* v. *Wolff* (1991). This case concerned a contract between a developer and the owners of a warehouse under which the developer undertook to have the warehouse refurbished prior to its resale. The developer was closely involved in many decisions concerning the refurbishment, such as the number of windows, the location of staircases and entrance and the internal sub-division of walls, such that it was clear that the expertise of the developer, and in particular its principal shareholder, was being heavily relied on. It was held that this made the contract personal to the extent that it could not be vicariously performed. This case shows that the personality of the contractor in regard to control and co-ordination of a construction project may be sufficiently important to prevent that part of the work being vicariously performed by sub-contractors, even if physical work can be sub-contracted.

(C) Selection by employer of a sub-contractor to the main contractor

The general principle of English law that a main contractor will be responsible to the employer for the defaults and breaches of contract of a sub-contractor in respect of workmanship and materials can be significantly qualified by the parties' contractual arrangements.

(i) *Full nomination*

It may be that by the terms of the main contract the employer reserves to himself the right, through his architect or other agent named under the contract, to nominate, name or otherwise select a sub-contractor with whom the main contractor is obliged to enter into contract for part of the contract works. The main contract often achieves this by the use of a provisional or prime cost sum expended on the instruction of the architect. A typical building contract will probably go on to say that this provisional or prime cost sum is adjustable in line with the actual sub-contract final account for the work done, barring only such increases in the sub-contract sum as are due to the default of the main contractor. Thus, the actual cost of such work will be met by the employer, the main contractor not being required or even permitted to price the work for himself.

Where there is full nomination by the use of a provisional or prime cost sum, then in most standard forms of construction contract there is likely to be some restriction placed on the liability of the main contractor for the nominated sub-contractor's defaults, even though this could leave the employer without a remedy in respect of losses suffered due to unfinished work.

Where the nominated sub-contractor completes the sub-contract work required of him but that work is defective, the main contractor will be responsible for it to the employer. However, if the nominated sub-contractor fails to complete the sub-contract works, the employer will often have to re-nominate and the cost of completing the outstanding work will fall on the employer. The question of whether the contractor is responsible for a nominated sub-contractor's defective workmanship and materials prior to any re-nomination can be a difficult question. Clearly the terms of the contract have to be carefully considered. While the terms of the particular contract may prohibit the main contractor from himself carrying out work covered by a prime cost item, this may not absolve him from the responsibility of meeting the cost of remedying defects in that work. On the other hand, it may. For example, in the case of *Fairclough Building Ltd* v. *Rhuddlan Borough Council* (1985) the contractor under a JCT 63 contract was held not liable for part of the cost of remedial works required when a nominated sub-contractor failed to complete his sub-contract work, leaving behind defective work.

While the employer may have to meet the cost of having the unfinished work of the nominated sub-contractor completed, the contractor will also no doubt suffer losses which he must bear if he cannot recover them from the nominated sub-contractor.

If the employer has no direct contract between himself and the nominated sub-contractor, he will have no contractual rights at all to recover this extra cost of completing the work. The main contractor will not be liable (he generally cannot be made to finish off the work at his own cost if it is a prime cost item) to the employer for the nominated sub-contractor's breach in failing to complete and the employer will have no contract with the nominated sub-contractor.

In recent years on a number of occasions the courts have been called on to consider the legal position following defaults by nominated suppliers or nominated sub-contractors. Three particularly important decisions are in the cases of *Young and Marten Ltd* v. *McManus Childs* (1969); *Gloucester County Council* v. *Richardson* (1969); and *North-West Metropolitan Regional Hospital Board* v. *T.A. Bickerton & Sons Ltd* (1970).

In the case of *North-West Metropolitan Hospital Board* v. *T.A. Bickerton & Sons Ltd* (1970) the contracts before the court were JCT 63 and the standard 'green' form of sub-contract for nominated sub-contractors. The nominated heating sub-contractor went into liquidation before commencing work. The main contractor contended that the employer was bound to nominate a second sub-contractor and to pay the main contractor the amount of the second sub-contractor's account. The employer on the other hand contended that there was no duty to re-nominate or to pay more than would have been payable had the original sub-contractor not defaulted. It was held that the wording of the contract meant that the contractor was forbidden to carry out works for which a prime cost item was provided in the contract and, accordingly, the contractor's argument was accepted.

It is clear that the particular contractual arrangements made between the parties will be of the essence and no general statement can cover the multifarious situations which arise. The contract may seek to deal expressly with the sub-contractor's failure to complete, as does JCT 98 in clause 35, or it may leave the situation open, in which case it is tentatively submitted that, if the sub-contractor fails to complete for whatever reason, then in most instances where a prime cost item is used, a re-nomination will be required and the employer will generally have to meet the subsequent nominated sub-contractor's account.

However, the important House of Lords' case of *Percy Bilton Ltd* v. *Greater London Council* (1982) (dealt with in detail on page 131) demonstrates that the point has not yet been reached whereby the employer warrants to the main contractor that a nominated sub-contractor will complete his sub-contract work without interruption.

What is very clear from the present state of the law in relation to defaulting nominated sub-contractors is that the employer needs to create a direct contract between himself and the nominated sub-contractor in order that he can claim directly against him in respect of losses brought about by a failure of the nominated sub-contractor to perform his contractual obligations. While there is always a possibility of seeking a remedy directly against the defaulting nominated sub-contractor by virtue of establishing a duty of care in the tort of negligence – see *Junior Books Ltd* v. *The Veitchi Co. Ltd* (1982) – the present state of the law suggests that this is by no means an easy remedy to establish: see *D & F Estates Ltd* v. *Church Commissioners* (1988) and *Murphy* v. *Brentwood District Council* (1991).

There may in the right circumstances be the possibility of establishing a collateral warranty between the nominated sub-contractor and the employer: see *Shanklin Pier Co. Ltd* v. *Detel Products* (1951). A collateral warranty might also be found to exist between the employer and a nominated or uniquely specified supplier: see *Greater London Council* v. *Ryarsh Brick Co. Ltd and Others* (1985).

(ii) Less than full nomination

There are a number of ways in which the employer can be involved in the selection of a sub-contractor to the main contractor without the full machinery of nomination being involved. For instance, the employer may provide a list of sub-contractors from whom the main contractor can choose one. The work will be fully described in the contract documents and will be priced by the contractor. In such circumstances the resulting sub-contract may be very close to a domestic sub-contract. For an example of this see JCT 98 clause 19.3.

Alternatively, the employer may, without using a provisional or prime cost sum, identify a named sub-contractor while providing the main contractor with sufficient information to enable him to price the work himself, even though it will be carried out by the named sub-contractor. The consequences of this will depend on the wording of the contract, although it may be said that if the main contractor has full or fairly full information before tendering as to the identity, sub-contract conditions, sub-contract price, programme, attendances etc., so far as these can be finalised before a sub-contract is entered into, it is the more likely that the main contractor will carry some responsibility for the sub-contractor's failure to perform.

While the conditions of the sub-contract cannot in themselves affect any plain express terms in the main contract as to the main contractor's responsibility for the work of a nominated or named sub-contractor, it is clear that such conditions may be relevant when considering the liabilities of the various parties.

The position in relation to responsibility for defective workmanship or materials of named sub-contractors under IFC 98 is dealt with later in this chapter (see particularly the commentary on clause 3.3.4 (see page 139)).

CONSIDERATION OF THE RELEVANT CLAUSES OF IFC 98

Clause 3.1

Assignment
3.1   Neither the Employer nor the Contractor shall, without the written consent of the other, assign this Contract.

COMMENTARY ON CLAUSE 3.1

For a summary of the general law see page 113. This clause is in a typical form for a building contract. Even without such a clause, neither the employer nor contractor could transfer the burden or liabilities arising under this contract without the consent of the other party. However, benefits can generally be assigned unless the contract prohibits it without consent, as does this one.

The House of Lords in *Linden Gardens Trust Ltd* v. *Linesta Sludge Disposals Ltd and Others*, and in *St Martin's Property Corporation Ltd and Another* v. *Sir Robert McAlpine & Sons Ltd* (1993) had to consider a JCT contract containing identical wording to clause 3.1 of IFC 98. It was held that such wording prohibited assignment not only of the benefit of the contract but also of the fruits of performance of the contract (see earlier page 114). Furthermore such a prohibition on assignment was not void as being contrary to public policy on the alleged grounds that such a prohibition rendered a chose in action inalienable.

The consent under clause 3.1 must be in writing.

Clause 3.2

Sub-contracting
3.2   The Contractor shall not sub-contract any part of the Works other than in accordance with clause 3.3 without the written consent of the Architect/the Contract Administrator whose consent shall not be unreasonably delayed or withheld. The Contractor shall remain wholly responsible for carrying out and completing the Works in all respects in accordance with clause 1.1 notwithstanding the sub-contracting of any part of the Works.
It shall be a condition in any sub-contract to which clause 3.2 refers that:

3.2.1   the employment of the sub-contractor under the sub-contract shall determine immediately upon the determination (for any reason) of the Contractor's employment under this Contract;
and

3.2.2   the sub-contract shall provide that:
(a) subject to clause 1.10 of these Conditions (in clauses 3.2.2(b) to (d) called 'the Main Contract Conditions'), unfixed materials and goods delivered to, placed on or adjacent to the Works by the Sub-Contractor and intended therefor shall not be removed except for use on the Works unless the Contractor has consented in writing to such removal, which consent shall not be unreasonably delayed or withheld;
(b) where, in accordance with clause 4.2.1(b) (*Materials and goods delivered to Works*) of the Main Contract conditions, the value of any such materials or goods shall have been included in any interim payment certificate under which the amount properly due to the Contractor shall have been paid[1] by the Employer to the Contractor, such materials or goods shall be and become the property of the Employer and the Sub-Contractor shall not deny that such materials or goods are and have become the property of the Employer;

(c) provided that if the Contractor shall pay the Sub-Contractor for any such materials or goods before the value therefor has, in accordance with clause 4.2.1(b) (*Materials and goods delivered to Works*) of the Main Contract Conditions, been included in any interim payment certificate under which the amount properly due to the Contractor has been paid[1] by the Employer to the Contractor, such materials or goods shall upon such payment by the Contractor be and become the property of the Contractor;

(d) the operation of sub-clauses (a) to (c) hereof shall be without prejudice to any property in any materials or goods passing to the Employer as provided in clause 1.11 (*Off-site materials or goods: passing of property, etc.*) of the Main Contract Conditions.

3.2.3   The sub-contract shall provide that if the Contractor fails properly to pay the amount, or any part thereof, due to the Sub-Contractor by the final date for its payment stated in the sub-contract, the Contractor shall pay to the Sub-Contractor in addition to the amount not properly paid simple interest thereon for the period until such payment is made; that the payment of such simple interest shall be treated as a debt due to the Sub-Contractor by the Contractor; that the rate of interest payable shall be five per cent (5%) over the Base Rate of the Bank of England which is current at the date the payment by the Contractor became overdue; and that any payment of simple interest shall not in any circumstances be construed as a waiver by the Sub-Contractor of his right to proper payment of the principal amount due from the Contractor to the Sub-Contractor in accordance with, and within the time stated in, the sub-contract or of any rights of the Sub-Contractor under the sub-contract in regard to suspension of the performance of his obligations to the Contractor under the sub-contract or determination of his employment for the failure by the Contractor properly to pay any amount due under the sub-contract to the Sub-Contractor.

COMMENTARY ON CLAUSE 3.2

For a summary of the general law see page 114. Clause 3.2 permits the contractor to enter into domestic sub-contracts only with the consent of the employer. However, such consent must not be unreasonably delayed or withheld. The test of unreasonableness is likely to be objective rather than subjective. However, if an employer chooses a particular contractor because of his skill and judgment in a certain field it may be that a withholding of consent to sub-contracting would be a reasonable stance to take. Where the skill and expertise relied upon relates in particular to management or organisation skills rather than to physical construction work, a withholding of consent is likely to be upheld. See for example *Southway Group Ltd* v. *Wolff and Another* (1991) (discussed earlier at page 115).

Where it is the physical work which is intended to be sub-contracted, it is more likely that sub-contractors will be more specialist than the main contractor and therefore it is less likely that the particular skill and judgment of the main contractor is being relied on. Withholding of consent in such a situation is therefore less likely to be reasonable. Provided that the contractor will remain as fully liable to the employer for the sub-contractor's defaults as if he had carried out the work himself, the reason for withholding consent would have to be compelling to prevent it being unreasonable. It is submitted that the correct position is that the contractor is fully liable to the employer in such circumstances despite apparent aberrations from this fundamental principle implicit in some decisions, e.g. *John Jarvis Ltd* v. *Rockdale Housing Association* (1987); *Greater Nottingham Co-Operative Society Ltd* v. *Cementation Piling and Foundations Ltd* (1988); *Scott Lithgow* v. *Secretary of State for Defence* (1989). This principle is expressly made part of the contract in the last paragraph of clause 3.2 by which the contractor remains wholly responsible notwithstanding any sub-contracting.

It is submitted that the reason for withholding consent must rest on the ability of the sub-contractor to perform the work competently, efficiently and on time and it would be a very liberal interpretation of the clause for an employer to validly withhold consent, as some public sector employers apparently do, to a sub-contract with anyone who does not appear on an approved list of contractors.

Any consent which is given is conditional on the sub-contract containing, as a minimum, certain conditions:

(1) Automatic determination of sub-contract

The sub-contract must contain a condition that the employment of the sub-contractor shall determine immediately on a determination, for any reason, of the employment of the contractor under the main contract.

If the sub-contractor's employment is determined in this fashion, the question arises as to whether or not this determination is a breach of the sub-contract by the contractor? In itself, it will not be. However, if the determination of the employment of the main contractor under the main contract came about as a result of the contractor's breach of that main contract, it is at least arguable that the contractor is in breach of an implied term in the sub-contract that he will not give the employer grounds for determining his employment under the main contract. If such is the case a main contractor will be liable in damages for breach of contract.

Additionally, the sub-contract itself may well make provision for what is to happen following such determination as does the recommended form of sub-contract for use in conjunction with IFC 98, i.e. IN/SC which provides for the contractor to recompense the sub-contractor.

If the determination of the sub-contractor's employment under IN/SC is due to a proper determination by the main contractor of his own employment under IFC 98, then this expenditure will be recoverable by the contractor from the employer under IFC 98 – see clause 7.11.3.

It will be seen therefore that the main contractor will, at any rate initially, and even though himself innocent, suffer financial consequences similar to those which would flow from a serious breach of contract by him. The reason for this is that it results eventually in the financial losses being met by the party at fault, i.e. the employer in this particular example. The innocent sub-contractor claims from the innocent main contractor who in turn claims from the employer who is at fault. The sub-contract conditions for use between the named sub-contractor and the main contractor in relation to contracts under IFC 98 are on the whole similar in this respect to the recommended domestic sub-contract conditions IN/SC.

If there were no provision in the sub-contract for an automatic determination of the sub-contractor's employment under the sub-contract, the effect on the sub-contract of a determination of the main contractor's employment under IFC 98 would fall to be considered under the common law and this could produce uncertainty. For instance, if, through no fault of the main contractor, he lost possession of the site, due to the employer's wrongly retaking it, it might be argued that the sub-contract was thereby frustrated. It is better therefore to have an immediate determination and the implications of it dealt with expressly in the sub-contract conditions.

(2) Materials and goods

By clause 3.2.2, the sub-contract must contain certain terms relating to materials and goods.

Firstly, there must be a term in the sub-contract that any unfixed materials and goods delivered to, placed on or adjacent to the works by the sub-contractor shall not be removed without the consent in writing of the contractor, which consent shall not be unreasonably delayed or withheld. This provision is made expressly subject to clause 1.10 which means that, so far as the contractor is concerned, he must obtain the consent of the architect before giving his consent to the sub-contractor.

Secondly, there must be a term in the sub-contract to the effect that where the value of any materials or goods has been included in any interim payment certificate under which the amount properly due to the contractor has been paid by the employer, such materials or goods shall become the property of the employer and the sub-contractor shall not deny that this is so. This provision is included to deal, so far as the contract terms can do so, with the type of problem experienced by the employer in the case of *Dawber Williamson Roofing Ltd* v. *Humberside County Council* (1979).

Facts:

The main contract was based on JCT 63. There was a domestic sub-contractor and his contract was based on the standard form of domestic sub-contract (the 'blue' form). By clause 14 of the main contract, it was stated that any unfixed materials or goods delivered to, and placed on or into the works should not be removed without consent, and that when the value of those goods had been included in a certificate under which the contractor had received payment, such materials or goods should become the employer's property. By clause 1 of the domestic sub-contract the sub-contractor was deemed to have notice of all the provisions of the main contract (other than prices). (It should be noted that there was no express provision in the sub-contract as to when, if at all, the property in the sub-contractor's materials or goods was to pass to the main contractor.)

The sub-contractor delivered 16 tons of roof slates to the site and submitted invoices to the main contractor. Under the main contract an interim certificate was issued including the value of the slates. The employer paid the appropriate sum to the main contractor. According therefore to the main contract, as the amount had been certified and paid, the ownership in the slates would vest in the employer. The main contractor did not pay the domestic sub-contractor and went into liquidation. The sub-contractor claimed that he was still the owner of the slates and therefore entitled to possession of them.

Held:

The slates were still owned by the sub-contractor. They were never at any time owned by the main contractor so he could not pass the title in them to the employer.

This case has been criticised as not being in line with sound commercial practice in that an employer had to pay for goods without ownership in them thereby being transferred to him, even though in the overall scheme of main contract and sub-contract it may have been thought that this was the intention. There is thus a gap in JCT 63 as there was also in JCT 80 until Amendment 1.1984 dealt with the situation in a similar way to IFC 98.

Thirdly, the sub-contract must contain a term whereby, if the contractor has paid the sub-contractor for any such materials or goods *before* their value has been included in a payment certificate and the amount paid by the employer, then such materials or goods shall, on payment by the contractor, become his property.

By virtue of these provisions appearing in the sub-contract, the sub-contractor cannot in such circumstances challenge the title either of the employer or the contractor, as the case may be, to the materials or goods in question. While this will provide some protection for the employer where he has discharged the amount due to the contractor in respect of the materials and goods, albeit the contractor has not paid the sub-contractor, this provision will not assist the employer where the sub-contractor is not himself the owner of the materials or goods, e.g. where there is a retention of title clause in his contract with the supplier of the materials or goods. (For a summary on the general law relating to retention of title see the commentary to clause 1.10 on page 55).

The third provision referred to above fills a gap which existed in the JCT forms and related sub-contract documents, e.g. DOM/1, which did not expressly provide that materials or goods should become the property of the main contractor even where he had paid for them. The provision which is now required to be contained in the sub-contract should prevent the sub-contractor from claiming that ownership remains with him even though he has been paid, e.g. under a retention of title arrangement covering all contracts generally between that sub-contractor and main contractor.

Of course, once the materials or goods are fully and effectively incorporated into the works, they will become the property of the landowner in any event.

Fourthly, the operation of the three provisions referred to above is expressly stated to be without prejudice to any property in any materials or goods passing to the employer as provided in clause 1.11 of IFC 98.

Finally, clause 3.2.3 provides that the sub-contract must include a provision that if the contractor fails to properly pay the amount or any part thereof, due to the sub-contractor by the final date for payment, the contractor shall pay simple interest at 5% over the Bank of England Base Rate current at the date when the payment became overdue. The sum is expressly treated as a debt and any claim for such interest is not to be treated as a waiver of the sub-contractor's right to proper payment of the principal amount at the appropriate time and does not in any way interfere with a sub-contractor's contractual right to suspend performance or to determine his own employment due to the failure to pay by the contractor. This provision of course mirrors that contained in clause 4.2(a) of IFC 98.

The standard domestic sub-contract form for use with IFC 98, IN/SC, contains all of the provisions required to be included by clause 3.2.2.

By virtue of clause 7.2.1(d) (see page 338) any failure by the contractor to comply with the provisions of clause 3.2 by reason of a default on his part is a ground for determination of the contractor's employment by the employer, subject however to the proviso in clause 7.2.4 that the notice of determination shall not be given unreasonably or vexatiously.

NOTES TO CLAUSE 3.2

[1] '...has been paid...'
See note [2] to clause 1.11 (page 59).

NAMED PERSONS AS SUB-CONTRACTORS

Introduction

Clauses 3.3.1 to 3.3.8 inclusive provide for a system of selection by the employer of a sub-contractor to carry out part of the contract work. While this involves the selection by the employer of a named person to carry out work as a sub-contractor to the main contractor it is significantly different to the traditional system of nomination.

The system is structured so as to provide two methods of naming:

(1) The person who it is intended shall be the sub-contractor is named in the main contract tender documents and the work to be done is described in detail there. The main contractor puts in his own tender price for the work, after being informed of the sub-contract price and most of the relevant conditions relating to the sub-contract works, e.g. conditions of contract, attendances and possibly programme details of the proposed sub-contractor. The contractor is therefore in a position to price the sub-contract works as he thinks fit but with due regard to the information which has been made available to him. The detailed sub-contract information is obtained by the employer through the architect obtaining from the sub-contractor whom it is proposed to name, the completed Parts I and II of a sub-contract tender document known as Form of Tender and Agreement NAM/T. Once the named sub-contract is duly entered into, the main contractor accepts a greater degree of responsibility for delays and defaults by the named sub-contractor than does the contractor in respect of a nominated sub-contractor under say, JCT 98. There is no liability on the contractor for design failures where the named sub-contractor has design obligations and the contractual provisions for naming expressly exclude design responsibility (see clause 3.3.7). The architect is thus not concerned with questions of payment or delay on the part of a named sub-contractor. These will be matters between the named sub-contractor and the main contractor.

However, following on a determination of the employment of the named sub-contractor by the contractor, under clauses 27.2, 27.3 or clause 27.4 of NAM/SC, the extent to which the contractor is responsible for completing the outstanding work and the financial implications flowing therefrom will depend on which of the three options listed in clause 3.3.3 is chosen by the architect. (Clauses 27.2, 27.3 or 27.4 of NAM/SC basically provide for the contractor to determine the employment of the named sub-contractor due to the latter's default or bankruptcy etc. The provisions of clauses 27.2, 27.3 and 27.4 are considered in more detail later when considering the operation of clause 3.3.4 of IFC 98.) The consequences of each instruction are set out in clause 3.3.4.

If the named sub-contractor's employment is determined either as a result of the contractor's default or by the contractor accepting a repudiation of the sub-contract following a repudiatory breach by the named sub-contractor, the architect will be required to choose one of the three options listed in clause 3.3.3, but the costs associated with this will be borne by the contractor. In such circumstances the financial consequences in clause 3.3.4 apply only to the extent that they result in a reduction in the contract sum, e.g. where there has been an omission of work under clause 3.3.3(c) or where the cost of completing the outstanding balance of work is less than it would have been had there been no default by the first named sub-contractor. Furthermore, the provision for an extension of time and for loss and expense will not apply – see clause 3.3.6(a). This is discussed in more detail in the commentary to clause 3.3.6 (see page 146).

Experience suggests that where the architect is fully conversant with this first method of naming and where the works are in an advanced state of design and specification before the main contract tender is let, this system works well for employer, contractor and sub-contractor. However, the industry remains plagued with ignorance and/or apathy in this respect and all too often the procedure is ignored or flouted leading to uncertainty and complications as the building project unfolds.

(2) A named sub-contractor can be introduced by means of an instruction as to the expenditure of a provisional sum after the main contract has been entered into. Here, the contractor has a right of reasonable objection. In the event of the contractor determining the employment of the named sub-contractor under clauses 27.2, 27.3 or 27.4 of NAM/SC, the subsequent instruction of the architect under clause 3.3.3 is to be regarded as a further instruction issued under the provisional sum. The provisional sum will therefore generally be adjusted as appropriate to take into account any variation in the cost of completing the work. The second system resembles more closely therefore the traditional form of nomination under JCT 98. However, should the sub-contractor's employment be determined in any other manner than under clauses 27.2, 27.3 or 27.4 of NAM/SC, the instruction under clause 3.3.3 will not be regarded as a further instruction under the provisional sum unless it results in a reduction in the contract sum – see clause 3.3.6(a).

Of the two methods, naming in the main contract tender document will generally be more advantageous to the employer but the use of this method presupposes that it is possible for the architect to name the proposed sub-contractor at that stage and this will not always be the case.

Provided the named sub-contractor's employment is determined under clauses 27.2, 27.3 or 27.4 of NAM/SC, the main contractor will be entitled, depending on which of the three types of instruction is issued under clause 3.3.3, to payment from the employer of the cost of completing the work and/or an extension of time and/or direct loss and expense. The employer therefore incurs extra costs and suffers losses. Clause 3.3.6(b) attempts to provide an indirect method whereby the employer recovers such increased cost and losses from the named sub-contractor through the main contractor. This is discussed in detail later in the commentary on this particular sub-clause.

On a determination of the employment of the named sub-contractor by the contractor, express provision is made in clause 3.3.3(a) for the architect to name another person to finish off the outstanding work. However, as will be seen, when the detailed drafting is considered later, the possibility of a second named sub-contractor's employment being determined does not appear to have been considered so far as work described in the contract documents for execution by a named person is concerned. Certainly, the opening paragraph of clause 3.3.4 restricts the operation of the financial consequences flowing from a re-naming to the situation where the instruction to re-name arises 'in respect of a person named in the [contract documents]...'. The second named person is not a person so named.

The system of naming adopted in IFC 98 is represented in Fig. 5.1 and Tables 5.1 and 5.2.

Clause 3.3.1

Named persons as sub-contractors

3.3.1 Where it is stated in the Specification/Schedules of Work/Contract Bills that work described therein for pricing by the Contractor[1] is to be executed by a named person who is to be employed by the Contractor as a sub-contractor the Contractor shall not later than 21 days after entering into this Contract[2] enter into a sub-contract with the named person using Section III of the Form of Tender and Agreement NAM/T referred to in the First recital.

If the Contractor is unable so to enter into a sub-contract in accordance with the particulars given in the Contract Documents[3] he shall immediately inform the Architect/the Contract Administrator and specify which of the particulars have prevented the execution of such sub-contract. Provided the Architect/the Contract Administrator is reasonably satisfied that the particulars so specified have prevented such execution the Architect/the Contract Administrator shall issue an instruction which may

(a) change the particulars so as to remove the impediment to such execution; or
(b) omit the work; or
(c) omit the work from the Contract Documents and substitute a provisional sum.

An instruction under clause 3.3.1(a) or 3.3.1(b) shall be regarded as an instruction under clause 3.6 requiring a Variation which shall be valued under clause 3.7[4] and the provisions of clauses 2.3 (*Extension of time*) and 4.11 (*Disturbance of progress*) as relevant shall apply. Where the instruction is under clause 3.3.1(b) the Employer may, subject to the terms of clause 3.11, have the work omitted executed by a person to whom clause 3.11 refers. An instruction under clause 3.3.1(c) shall be dealt with in accordance with clause 3.3.2.

The Contractor shall notify the Architect/the Contract Administrator of the date when he has entered into the sub-contract with the named person. The Architect/the Contract Administrator may, but, subject to clause 3.3.3, not after the date so notified[5], issue an instruction that the work to which this clause 3.3.1 refers is to be carried out by a person other than the person named in the Specification/Schedules of Work/Contract Bills. Such instruction shall omit the work from the Contract Documents and substitute a provisional sum which shall be dealt with in accordance with clause 3.3.2.

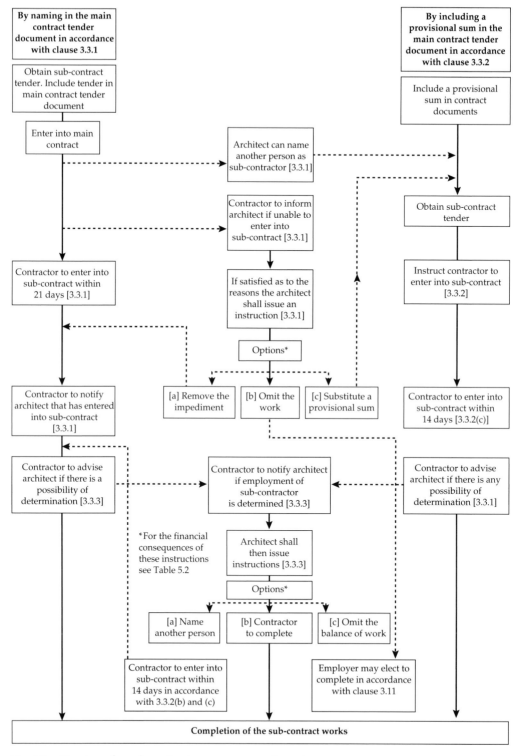

Fig. 5.1 Naming of a sub-contractor – sequence of events

Table 5.1 Clause 3.3.1 – Consequences of the architect's instruction following the failure of the contractor to enter into a sub-contract with the person named in the contract documents.

Clause 3.3.1 Architect's instructions – Options	(a) Remove the impediment by changing the particulars	(b) Omit the work (see note below)	(c) Omit the work and substitute a provisional sum
Consequences of instruction	Clause 3.3.1 — Valued as a variation – clause 3.7 — Extension of time – clause 2.3 — Disturbance of progress – clause 4.11		Any subsequent instruction as to the expenditure of a provisional sum treated as follows: — Valued as a variation – clause 3.7 — Extension of time – clause 2.3 — Disturbance of progress – clause 4.11

Table 5.2 Clause 3.3.3 – Consequences of the architect's instruction following the contractor's determination of the employment of the named sub-contractor in accordance with the provisions of clause 27.2, 27.3 or 27.4 of NAM/SC.

Clause 3.3.3 Architect's instructions – Options	(a) Name another person to execute the outstanding balance of work	(b) Contractor to complete (can sub-contract with consent – clause 3.2)	(c) Omit the outstanding balance of work (see note below)
Consequences of the instruction where the sub-contractor is named in the contract documents	Clause 3.3.4(a) — Contract sum adjusted by net difference in price of first and second named sub-contractor for the remaining work (discounting any cost included for rectification of defects in the work of the first named sub-contractor) — Extension of time – clause 2.3 N.B. Clause 4.11 (disturbance of progress) does not apply	Clause 3.3.4(b) — Valued as a variation – clause 3.7 — Extension of time – clause 2.3 — Disturbance of progress – clause 4.11	
Consequences of the instruction where the sub-contractor is named in an instruction as to the expenditure of a provisional sum	Clause 3.3.5 — Valued as a variation – clause 3.7 — Extension of time – clause 2.3 — Disturbance of progress – clause 4.11		
Recovery of monies from first named sub-contractor	Clause 3.3.6(b) Contractor to take action to recover from the named sub-contractor the additional amounts payable to the contractor by the employer together with employer's lost liquidated damages		
Where the employment of the named person is determined otherwise than in accordance with clause 27.2, 27.3 or 27.4 of NAM/SC	Clause 3.3.6(a) The provisions of clause 3.3.4(a) or 3.3.4(b) or 3.3.5 as applicable (see above) shall apply in any adjustment arising out of the instruction but only to the extent that they result in a reduction of the contract sum N.B. Clause 2.3 (extension of time) and clause 4.1.1 (disturbance of progress) do not apply		

Note: Where the architect's instruction omits work then the employer can complete under the provisions of clause 3.11

COMMENTARY ON CLAUSE 3.3.1

This clause deals with one of only two available methods which this contract provides for the employer to select a sub-contractor to carry out part of the work, the other method being described in clause 3.3.2.

The employer must provide a detailed description of the work in the specification/schedules of work/contract bills. This description of the work, together with other necessary information, is communicated to the proposed sub-contractor by use of a standard form of tender known as Tender and Agreement NAM/T. Section I of NAM/T is the invitation to tender and it will contain such matters as a list of those potential main contractors invited to tender and the main contract information, if available, e.g. the appendix entries from the proposed main contract (and departures from this for the purposes of the sub-contract where these are known). It will contain the expected commencement dates and periods of sub-contract work and other relevant information. The sub-contractor whom it is proposed to name will, in section II, provide information as to his requirements.

Once the selection has been made by the employer, details of the sub-contract price and all the other relevant information are included in the main contract tender documents to form part of the information on which the contractor tenders. All the documents on which the sub-contract tender is based should be provided to the main contractor.

These will include, as appropriate, drawings, bills of quantities and a specification or schedule of works (see the last paragraph of the first recital to IFC 98). If a specification is included on its own and is not priced, the sub-contractor's tender being based on drawings and a specification only, the sub-contractor should supply a sub-contract sum analysis or a schedule of rates in support of his tender. Wherever possible of course, the sub-contract documentation should mirror that used for the main contract.

The contractor prices the sub-contract work himself and it is this price which forms part of the contract sum of the main contract. If the main contractor and named sub-contractor enter into a sub-contract and the named sub-contractor completes the sub-contract works, the named sub-contractor's price will be of little concern to the employer. However, should the employment of the named sub-contractor be determined, the sub-contract price could, as will be seen later, become of considerable significance to the employer. Once the main contract has been entered into, the contractor has just 21 days in which to enter into a sub-contract with the named person on the basis of sections I and II of NAM/T and this is achieved by completing the Agreement in Part III by which the standard sub-contract for named persons, known as NAM/SC, is incorporated. The contractor must notify the architect of the date when the named sub-contract has been entered into.

This rather short period of 21 days is likely to put time at a premium. It may well be therefore that on occasions the proposed main contractor will make contact with the person who has been named in the main contract tender documents during the main contractor's own tender period in order to clarify any uncertain areas and also, if the main contract is being let some considerable time after the completion of sections I and II of NAM/T, to confirm that the details contained therein are still applicable. Such a practice may well be regarded unfavourably by the sub-contracting side of the industry.

It is advisable for the architect, just before accepting the tender of a main contractor, to check with the person whom it is proposed to name as a sub-contractor, that there is no departure from the basis on which he has agreed to carry out the work, or that if there is, that this can be quickly resolved. This should reduce to a minimum the chances of a disagreement arising between the main contractor and proposed named sub-contractor during the 21 day period, particularly in those cases where there has been no contact between main contractor and named sub-contractor before the commencement of that period.

Since the contractor tenders on the basis of the particulars given in the contract documents in relation to the sub-contract works, it is to be hoped that those particulars will remain unchanged and form the basis of the sub-contract. However, should any difficulty occur in the proposed named sub-contractor and contractor agreeing the particulars, which is not resolved between them during the 21 day period, the contractor must immediately inform the architect of this fact, specifying which of the particulars in the contract documents has caused the impasse.

If the proposed named sub-contractor withdraws his offer completely during the 21 day period for reasons unconnected with any problem concerning the particulars, e.g. a liquidation of the sub-contractor, the architect is able to issue an instruction requiring that the work be carried out by some other person. This will be done by omitting the work from the contract documents and substituting a provisional sum to be dealt with in accordance with clause 3.3.2 – see last paragraph of clause 3.3.1.

Contractor's inability to enter into a sub-contract in accordance with the particulars given in the contract documents

The second and third paragraphs of clause 3.3.1 deal with the situation where the contractor and sub-contractor cannot agree to enter into a sub-contract based on the particulars given in the contract documents. The drafting of these paragraphs probably proceeds on the assumption that the contractor and proposed sub-contractor cannot agree with one another as to the particulars. What if they are in agreement that the particulars should change, e.g. programme times for the sub-contract works being varied from that in the particulars given in NAM/T? There is no impasse in such a situation. The problem is that such a change is strictly speaking a variation and must therefore be the subject of an instruction from the architect. No doubt this can be done by the architect issuing an instruction under clause 3.3.1(a) but this does appear to be using a sledgehammer to crack a nut. Provided the terms of the main contract remain unaffected, particularly in terms of the contractor's price for the named sub-contract work, it would have been sensible to build into clause 3.3 a facility for the contractor and proposed sub-contractor to agree to change the particulars subject to approval by the architect, not to be unreasonably withheld, and for this approval not to amount to a variation requiring a formal instruction.

Provided the architect is reasonably satisfied that the particulars specified by the contractor as preventing him from entering into a sub-contract have in fact caused the impasse, he may issue an instruction which may:

(a) Change the particulars so as to remove the impediment
(b) Omit the work
(c) Omit the work from the contract documents and substitute a provisional sum.

These three possibilities are considered here.

(a) Change the particulars so as to remove the impediment

Such a change will very often have financial implications. So far as the contractor is concerned, the instruction changing the particulars is to be regarded as a variation and will therefore be valued accordingly and clauses 2.3 (extension of time) and 4.11 (loss and expense) will apply where appropriate. The extent to which they will apply is not easy to discern.

To take an example, suppose there is a failure to agree between main contractor and proposed named sub-contractor on the periods required for the sub-contract works in relation to a proposed sub-contract for a small amount of piling work to be carried out in the very early stages of the contract. The employer may have been late in going out to tender and what was originally envisaged by the proposed named sub-contractor as a summer contract is now a winter contract so that the proposed named sub-contractor requires further time in which to carry out the sub-contract works and further money in respect of it. If the architect now changes the particulars in order to accommodate the proposed named sub-contractor's longer period during which to carry out the works and increased sub-contract price, what is the contractor's position? Do the changed particulars have to be agreed by both the proposed named sub-contractor and the contractor in order 'to remove the impediment' or must the contractor accept the changed particulars even though they will have a direct bearing on the overall completion date and the figure at which he priced the sub-contract work? Presumably, if he must accept these changed particulars, whether he likes them or not, his recompense will be found in the valuation of the instruction, the extension of time and the loss and expense provisions.

(i) Valuation of the instruction

It is highly likely that the valuation will be made under clause 3.7.5. This will often be based on the new sub-contract price with perhaps a percentage added for the contractor's pricing of the work. By clause 3.7.8 no loss and expense can be included in this valuation although it can be recovered, if appropriate, under clause 4.11.

(ii) Extension of time

It is only a delay in the date for completion due to compliance with an instruction removing the impediment which will be allowed. As clause 2.4.5 is expressly restricted to *compliance* with instructions under clause 3.3, delays prior to the instruction cannot, of course, be attributable to compliance with it.

(iii) Loss and expense

The loss and expense must arise as a result of the regular progress of the works being materially affected due to the instruction. If the changed particulars affect the contractor's programme then it is quite probable that his regular progress will be disturbed and he will accordingly be entitled to reimbursement of any loss and expense incurred. Again, the period up to the issue of the instruction changing the particulars so as to remove the impediment is not covered – see clause 4.12.7.

Delays and disturbances to regular progress occurring between the contractor notifying the architect that a sub-contract cannot be entered into and the architect issuing an instruction changing the particulars etc.

It is submitted that, provided the architect acts within a reasonable time in issuing the instruction changing the particulars etc., there can be no claim for an extension of time or for loss and expense. To the extent that any delay is unreasonable there is a possibility that the contractor can claim an extension of time under clause 2.4.7 (late instructions etc.) and loss and expense under clause 4.12.1 (late instructions etc.). Support for this view, it is suggested, is to be found in the important House of Lords' case of *Percy Bilton Ltd* v. *Greater London Council* (1982), dealing with a form of contract based on JCT 63.

Facts:

The contract was on the GLC's own conditions of contract but based on JCT 63. It involved the question of extensions of time and liquidated damages following non-performance by a nominated sub-contractor. The nominated sub-contractor for mechanical services went into liquidation. The main contractor, relying on the principle laid down in the case of *North-West Metropolitan Regional Hospital Board* v. *T. A. Bickerton & Sons Ltd* (1970), called on the employer through his architect to make a fresh nomination. However, there was a considerable delay in making the new nomination due mainly to the fact that the firm first proposed for the renomination withdrew without ever starting work. Another sub-contractor was eventually nominated some four months after the repudiation by the first nominated sub-contractor. There were thus two parts to the delay: the first part which was regarded as a reasonable time and the second part which was regarded as an unreasonable time between repudiation and renomination.

Held:

The employer had a major responsibility for the unreasonable delay and under clause 23(f) (extension of time for late instructions under JCT 63) the contractor should have had a reasonable extension of time in respect of this part of the delay. In relation to the first part of the delay, i.e. that which was a reasonable period after the repudiation by the first nominated sub-contractor, there was nothing in JCT 63 which imposed what would, in effect, be a warranty on the employer that the nominated sub-contractor would invariably carry on work continuously. This would place an unreasonable burden on the employer. It was the duty of the employer, acting through his architect, to give instructions for a renomination within a reasonable time. The contractor was therefore not entitled to an extension of time for the first part of the delay, and if this meant that the contractor was late in completing then he would suffer liquidated damages.

It is submitted that there is nothing in IFC 98 to suggest that any principles other than those stated in this case would apply.

(b) Omit the work

An instruction omitting the work is to be regarded as a variation. It will be valued as appropriate under clause 3.7.3 and will result in an adjustment downwards of the contract sum. Again the extension of time and loss and expense provisions apply but it is unlikely that any delay or disruption will be caused by a straightforward omission of work.

Where an instruction requiring such an omission is given, the omitted work may be carried out by another contractor engaged directly by the employer. The contractor's consent (which shall not be unreasonably withheld) will be required to this course of action by virtue of clause 3.11.

(c) Omit the work from the contract documents and substitute a provisional sum

It may not be possible for the proposed named sub-contractor and contractor to reach an agreement which results in a concluded sub-contract. The architect cannot simply instruct the contractor to enter into a sub-contract with another named sub-contractor, e.g. the next lowest sub-contracting tenderer, as perhaps the architect may do in nominating a sub-contractor under JCT 98. The contractor will not have priced the work on this changed basis. The option open to the architect is to omit the sub-contract work and then to substitute a provisional sum giving instructions naming a sub-contractor in accordance with clause 3.3.2 (dealt with below).

This instruction may follow protracted negotiations which may have caused delay and disruption. It is submitted that, provided the architect acts with reasonable expedition, there can be no claim by the contractor for an extension of time or loss and expense: see *Percy Bilton Ltd* v. *Greater London Council* (1982). But there could be other problems. For example, the delays running prior to the instruction could produce a result whereby whoever is named under the provisional sum cannot complete the sub-contract work within sufficient time to enable the main contractor to complete by the contract completion date. If this is so, and if no extension of time can be given under clause 2.4.5 (clauses 3.8 or 3.3) because in truth the effective delay was already suffered before the instruction was issued so that *compliance* with it did not as such *cause* any delay, the contractor can most probably object under clause 3.3.2(c) within 14 days of the date of the issue of the instruction.

Independently of any failure to agree between the main contractor and the proposed named sub-contractor, at any time before a sub-contract with a named person has been entered into, the architect may issue an instruction omitting the work from the contract documents and substituting a provisional sum to which clause 3.3.2 will apply.

NOTES TO CLAUSE 3.3.1

[1] '...for pricing by the Contractor...'
It is clear that for the purposes of this contract the employer will generally only be concerned with the contractor's price and not that of the named sub-contractor. Provided the named sub-contractor enters into a sub-contract and subsequently completes the work, the employer has no interest in the price submitted by the named sub-contractor. It is only in the event of the sub-contract not being entered into or of the employment of the named sub-contractor being determined that the sub-contractor's price becomes of interest to the employer. The extent of the employer's interest in the named sub-contractor's price is dealt with in the commentary to clause 3.3.3.

[2] '...not later than 21 days after entering into this Contract...'
This could be a tight timetable and it is envisaged that, prior to the main contract being let, the contractor, sub-contractor and architect may have liaised to ensure that the basic agreement holds good in relation to essential terms and particulars in the main contract tender documents.

[3] '...in accordance with the particulars given in the Contract Documents...'
The particulars are those in the tender of the named person contained in sections I and II of the Form of Tender and Agreement NAM/T (see first recital to IFC 98). The particulars required are extensive, covering not only main contract information but also the sub-contractor's price, dates of anticipated commencement on site and of the periods required to carry out the work, periods of notice required, periods required for the submission of sub-contractor's drawings, insurances, fluctuations etc.

[4] '...which shall be valued under clause 3.7...'
If an instruction is given under clause 3.3.1(a) then no doubt the valuation will often be carried out under clause 3.7.5.

[5] '...not after the date so notified...'
In practice, whether or not the architect has been so notified, if the sub-contract has been entered into, it will be too late for the architect to instruct that the work is

to be carried out by some other person. The architect has no power of course to instruct the contractor to breach the sub-contract conditions.

Clause 3.3.2

3.3.2   (a) In an instruction as to the expenditure of a provisional sum under clause 3.8 the Architect/the Contract Administrator may require work to be executed by a named person who is to be employed by the Contractor as a sub-contractor.
(b) Any such instruction shall incorporate a description of the work and all particulars of the tender of the named person for that work in a Form of Tender and Agreement NAM/T with Sections I and II completed together with the Numbered Documents referred to therein.
(c) Unless the Contractor shall have made reasonable objection[1] to entering into a sub-contract with the named person within 14 days of the date of issue of the instruction he shall enter into a sub-contract with him using Section III of the Form of Tender and Agreement NAM/T for the execution of the said work.

COMMENTARY ON CLAUSE 3.3.2

By the use of this clause, the architect may, by issuing an instruction as to the expenditure of a provisional sum after the contract has been entered into, select a named person to carry out sub-contract work. The instruction must provide the detailed information contained in sections I and II of NAM/T. However, if the work concerned is crucial to the contractor's programme, it is the better practice to name the sub-contractor when the main contract tenders are being invited and allow the contractor to price for the work in his own tender.

If compliance with the instruction involves the contractor in delay or disturbance to regular progress, e.g. where the named sub-contractor requires a period for completing the sub-contract works which will necessarily cause delay and disruption to the contractor, the contractor will be entitled to an extension of time under clauses 2.3 and 2.4.5 and loss and expense under clauses 4.11 and 4.12.7.

Once the sub-contract has been entered into, any delay or disturbance suffered by the contractor due to the activity or inactivity of the named sub-contractors will not attract an extension of time or loss and expense as it will not be due directly to compliance with the instruction under the provisional sum.

Many of the risks associated with a failure by the named sub-contractor to complete the sub-contract works fall squarely on the employer (see clause 3.3.5).

There is no provision in IFC 98 either for the contractor to make his own allowance for attendance and profit on the expenditure of provisional sums in favour of named sub-contractors or for payment thereof.

In practice some suitable system is often devised to cater for this. One method is to include an item for the contractor to price, but bearing in mind the domestic nature of the subsequent sub-contract, it may be unwise to describe the item as being an allowance for attendance and profit. However, the inclusion of a priceable item has the merit of providing a firm basis for subsequent adjustment and of making the amount inserted by the contractor subject to competition.

If nothing is provided it is difficult to see how the contractor can recover anything for attendance and profit.

NOTES TO CLAUSE 3.3.2

[1] 'Unless the Contractor shall have made reasonable objection . . .'
Under clause 3.3.1 if the main contractor is dissatisfied with the selection arrangements or conditions, he can always elect not to submit a tender for the main contract work. This opportunity does not arise under clause 3.3.2. Accordingly, the contractor has a right of reasonable objection which, if he intends to exercise it, must be exercised within 14 days of the date of the issue of the instruction. This period does not give the contractor long to assess the particulars given in NAM/T. If any of the particulars are not acceptable, or if any outstanding items are not settled within the 14 day period, the contractor should report this fact to the architect within the period.

If the right to reasonable objection is not exercised, it might be felt by some employers that thereafter the contractor should take the financial and other risks of a failure to perform by the sub-contractor. However, this is not the case in IFC 98, for although the contractor does take the risk of delays and disruption caused by the named sub-contractor up to the point of the determination of the named sub-contractor's employment, thereafter he is entitled to seek an extension of time and loss and expense, if any, suffered as a result of the instruction given by the architect under clause 3.3.5. Further, any increased cost of completing the sub-contract works will itself fall on the employer.

It is submitted that, in considering what is a reasonable ground for objection, factors other than the sub-contract particulars may be considered, e.g. the contractor's previous experience with the proposed named sub-contractor by which he can demonstrate that the sub-contractor concerned has in the past had difficulty in achieving a reasonable standard of work or in achieving reasonable progress.

Clause 3.3.3

> 3.3.3 The Contractor shall not determine the employment of the named person otherwise than under clause 27.2 or 27.3 or 27.4 of the Sub-Contract Conditions NAM/SC, nor let that employment be determined by accepting repudiation[1] of the sub-contract.
> The Contractor shall advise the Architect/the Contract Administrator as soon as is reasonably practicable of any events which are likely[2] to lead to any determination of the employment of the named person howsoever arising[3].
> Whether or not such advice has been given if, before completion[4] of the sub-contract work, the employment of the named person is howsoever determined, the Contractor shall notify the Architect/the Contract Administrator in writing stating the circumstances. The Architect/The Contract Administrator shall issue instructions as may be necessary in which he shall:
> (a) name another person[5] to execute the work or the outstanding balance of work in accordance with clause 3.3.2(b) and subject to clause 3.3.2(c), or
> (b) instruct the Contractor to make his own arrangements for the execution of the work or the outstanding balance of the work, in which case the Contractor may sub-contract the work in accordance with clause 3.2, or
> (c) omit the work or the outstanding balance of work.

COMMENTARY ON CLAUSE 3.3.3

As it is the employer who has selected the named person to be a sub-contractor to the contractor, in the event that the sub-contract works are not completed by the named sub-contractor, this clause requires the employer through the architect to give appropriate instructions as to the unfinished work. Certain financial consequences flow depending on which option contained within this clause is chosen and the method by which the named sub-contractor's employment is determined. These financial consequences operate as a safeguard to the contractor in the event of the employment of the named sub-contractor being determined by the contractor under clauses 27.2, 27.3 or 27.4 of NAM/SC. Looked at another way, it is the cost to the employer in exercising the privilege of selection of a sub-contractor.

The first paragraph of clause 3.3.3 provides that the contractor shall not determine the employment of the named sub-contractor otherwise than under clauses 27.2, 27.3 or 27.4 of NAM/SC. Nor must he let that employment be determined by accepting a repudiation by the named sub-contractor of the sub-contract itself. The reasons for this are tied in with clause 3.3.6(b) regarding the obligation on the contractor to pursue the sub-contractor for the extra costs incurred by the employer following the architect's issuing an instruction under clause 3.3.3 in relation to the unfinished work. It will be seen later that there are certain financial consequences flowing from an instruction under this clause, and depending on the type of instruction issued, the employer will have to pay for the increased costs of having the work finished and is likely to be deprived of liquidated damages. Furthermore, he may have to pay the contractor loss and expense.

The method of recovery is dealt with in the commentary to clause 3.3.6(b). However it is worth noting at this point that the main contractor is required to seek recovery of the employer's losses etc. from the named sub-contractor under clause 27.6.4 of NAM/SC which imposes a contractual obligation on the named sub-contractor to pay these sums to the main contractor who then holds any sums recovered to the account of the employer. However, as the claim by the main contractor will be one under, and in accordance with, the contract rather than for breach of it, it is essential that when the named sub-contractor's employment is determined, the sub-contract conditions themselves remain intact in order that clause 27.6.4 can be relied on by the main contractor. If the sub-contract itself is brought to an end by the contractor accepting a repudiation of it by the named sub-contractor, the effect will be that with exceptions, e.g. the arbitration and adjudication clauses, the sub-contract will fall to the ground and with it the contractor's right to recover the employer's losses. It is for this reason that the contractor must not determine the employment of the named sub-contractor otherwise than under clauses 27.2, 27.3 or 27.4 of NAM/SC. It will be seen in the commentary to clause 3.3.4 and 3.3.5 that, should the named sub-contractor's employment be determined in any way other than under clauses 27.2, 27.3 or 27.4 of NAM/SC, the financial consequences to the contractor can be very severe indeed.

Furthermore, the restriction as to the manner in which the contractor can bring the named sub-contractor's employment to an end can pose serious problems for the contractor.

Firstly, it may be that the factual situation does not easily lend itself to a system of notices which is required in order for the contractor to rely on the provisions of clauses 27.2, 27.3 or 27.4 of NAM/SC. The relationship between named sub-contractor and contractor may have deteriorated to a level whereby the named sub-contractor has left the site.

The main contractor may well be keen to prevent any delays occurring which could affect the overall completion date. However, in such a situation he must still serve the appropriate notices and wait the appropriate time before he can determine the employment of the named sub-contractor.

Secondly, it may be that for good reason the contractor may wish to keep his options open in relation to the determination of the named sub-contractor's employment. He may wish to purport to determine not only in accordance with the provisions of clauses 27.2, 27.3 or 27.4 but also, if it turns out that for some reason the sub-contractor's employment was not validly determined, on the basis of accepting a repudiation of the contract on the part of the named sub-contractor.

Thirdly, the contractor could inadvertently fail to follow the correct procedures for serving the appropriate notices (see clause 27.1 of NAM/SC). This may not become apparent until some considerable time after the notices have been served and the named sub-contractor has left the site. If the determination was invalid because of a defective notice then the contractor will have failed to determine the named sub-contractor's employment in accordance with clauses 27.2, 27.3 or 27.4 and potentially grave financial consequences follow for the contractor.

Fourthly, there may be situations where the contractor simply cannot determine the named sub-contractor's employment under clauses 27.2, 27.3 or 27.4, e.g. the death of a named sub-contractor who is an individual or the frustration in law of the sub-contract. These situations are likely to be extremely rare and presumably in such a case it would not amount to a breach by the contractor of clause 3.3.3. It is uncertain in such a case who will have to meet the increased costs etc. of having the work completed. By a literal reading of clause 3.3.6(a) the costs would appear to fall on the contractor. This might be regarded by some as harsh.

There is no doubt that the restriction imposed on the contractor as to the method by which the named sub-contractor's employment may be determined represents a radical step which could well be a cause of considerable concern to contractors.

The second paragraph of clause 3.3.3 requires the contractor to advise the architect as soon as reasonably practicable of any events which are likely to lead to a determination of the employment of the named person, howsoever that determination may arise. This will enable the architect and employer to consider or take appropriate steps, perhaps to contain or improve a deteriorating situation. Further, the third paragraph of clause 3.3.3 requires the contractor to notify the architect in writing of the determination of the employment of the named sub-contractor. The written notice must state the circumstances in which the determination has come about.

Once notified of the determination of the named sub-contractor's employment, the architect must issue any necessary instructions which must include one of the three options set out in this clause. He must:

(a) Name another person to execute the work or the outstanding balance of work in accordance with clause 3.3.2(b) and subject to clause 3.3.2(c); *or*

(b) Instruct the contractor to make his own arrangements for the execution of the work, in which case the contractor may sub-contract the work in accordance with clause 3.2; *or*

(c) Omit the work requiring to be completed.

Although these alternatives are separated by what appears to be a disjunctive 'or', there seems no good reason why more than one should not be used for different elements of the outstanding work, e.g. the naming of a specialist under (a) to test and commission some specialist plant, with the contractor himself being instructed under (b) to finish off other associated work. However, it must be said that the words 'the work' used in (a), (b) and (c) of this clause do not facilitate this interpretation.

The financial consequences flowing from an instruction under this clause are dealt with under clauses 3.3.4 and 3.3.6 where the naming of a sub-contractor arises under clause 3.3.1, and in clauses 3.3.5 and 3.3.6 where a provisional sum has been used under clause 3.3.2.

NOTES TO CLAUSE 3.3

[1] '...nor let that employment be determined by accepting repudiation...'
It is clear that the sub-contractor's employment can come to an end either by the operation of clauses 27.2, 27.3 or 27.4 or by the contractor (albeit wrongly under this clause) accepting a repudiation of the contract on the part of the named sub-contractor. Without these words it might have been arguable whether the acceptance of a repudiation by the contractor which brought the sub-contract itself to an end involved a 'determination of the employment' of the named sub-contractor as that phrase is used throughout this contract.

[2] '...which are likely...'
This advice could be useful to the architect in relation to his general administration and control of the works. There will no doubt be a temptation to mediate between the contractor and named sub-contractor which could in certain circumstances be advantageous, provided it is done with extreme care so as not to either directly influence the contractor's decisions or prejudice the employer's position. Whatever else he does, the architect should familiarise himself with the problem. A failure by the contractor to so advise the architect could lead to the contractor being the author of his own misfortune. For instance, advance warning to the architect of a possible determination of the named sub-contractor's employment may enable the architect to make some preliminary investigations or enquiries in advance of the possible issue of an instruction under clause 3.3.3(a), (b) or (c) and, should a determination in fact occur, such advance investigations or enquiries could reduce the time reasonably taken by the architect between the determination of the named sub-contractor's employment and the issue of the instruction in relation to the unfinished work, thereby reducing what would otherwise be the delay for which the contractor, and not the employer, is responsible: see *Percy Bilton Ltd* v. *Greater London Council* (1982).

[3] '...howsoever arising'.
i.e. whether under clauses 27.2, 27.3 or 27.4 of NAM/SC or in any other way.

[4] '...before completion...'
This means 'apparent completion'. See the case of *City of Westminster* v. *Jarvis* (1970) in which the House of Lords held that completion meant completed ready to hand over, even if in reality, but unknown to the parties, there was defective work. If the named sub-contractor's work is apparently completed, i.e. it has

achieved practical completion within the meaning of clause 15 of NAM/SC, then it is suggested that it is completed for the purposes of this clause even if this is before practical completion of the works as a whole. If this is so and the named sub-contractor's employment is determined after such apparent completion but before practical completion of the whole of the works, then the provisions of this clause do not apply so that for example any failure by the named sub-contractor to remove or remedy work found to be defective during this period will (unless it is a question of defective design for which the contractor is not responsible) be solely the contractor's responsibility and cannot be dealt with by an instruction under this clause.

[5] '... name another person ...'
This appears to preclude the possibility of renaming the same person. This might be desirable, for example where a receiver is appointed to manage the business of a named sub-contractor and the contractor decides to determine the named sub-contractor's employment on this ground (see clause 27.3 of NAM/SC), the named sub-contractor may be in a position, with the agreement of the receiver, to complete the sub-contract works in accordance with the sub-contract as to time and price. In such a case the employer might very much like to rename the sub-contractor. However, in such circumstances, the employer does have the option of issuing an instruction under clause 3.3.3(c) to omit the outstanding balance of work which can then form the basis of a direct contract between the employer and the sub-contractor (see last paragraph of clause 3.3.4).

Clause 3.3.4

3.3.4 The following provisions of this clause 3.3.4 shall apply where an instruction referred to in clause 3.3.3(a) to (c) arises in respect of a person named[1] in the Specification/ Schedules of Work/Contract Bills under clause 3.3.1 whose employment has been determined under clause 27.2 or 27.3 or 27.4 of the Sub-Contract Conditions NAM/SC:
(a) such an instruction under clause 3.3.3(a) shall be regarded as an event to which clause 2.3 (*Extension of time*) applies, but not as a matter to which clause 4.11 (*Disturbance of progress*) applies, and the Contract Sum shall be adjusted by the amount of the increase or the reduction in the price of the second named sub-contractor for the work not carried out by the first named sub-contractor when compared with the price of the first named sub-contractor for that work[2]; provided that in the foregoing adjustment there shall be excluded from the price of the second named sub-contractor any amount included therein for the repair of defects[3] in the work of the first named sub-contractor;
(b) such an instruction under clause 3.3.3(b) or (c) shall be regarded as one requiring a Variation which shall be valued under clause 3.7 and as an event to which clause 2.3 (*Extension of time*) applies and as a matter to which clause 4.11 (*Disturbance of progress*) applies.
Where the instruction is under clause 3.3.3(c) the Employer may, subject to the terms of clause 3.11, have the omitted work executed by a person to whom clause 3.11 refers.

COMMENTARY ON CLAUSE 3.3.4

This clause deals with the financial consequences which arise when the employment of a sub-contractor named under clause 3.3.1 has been determined by the contractor under clauses 27.2, 27.3 or 27.4 of NAM/SC, and the architect issues instructions with regard to the unfinished sub-contract work under clause 3.3.3.

Clauses 27.2, 27.3 and 27.4 of NAM/SC

The operation of clause 3.3.4, dealing as it does with payment by the employer to the contractor in respect of the costs of, and associated with the completion of the outstanding work etc., is of vital importance to the contractor. However the provisions of clause 3.3.4 operate only where the named sub-contractor's employment has been determined by the contractor under clauses 27.2, 27.3 or 27.4 of NAM/SC. It is of the utmost importance therefore that the contractor takes great care to ensure that any determination is in accordance with those clauses.

Clause 27.2 provides that the contractor may determine the employment of the sub-contractor if the latter defaults in any of the following ways:

(1)   Without reasonable cause wholly or substantially suspends the carrying out of the works
(2)   Without reasonable cause fails to proceed regularly and diligently with the sub-contract works
(3)   Refuses or neglects to comply with a written direction from the contractor requiring him to remove any work, material or goods not in accordance with the sub-contract and by such refusal or neglect the works are materially affected
(4)   Fails to obtain consent to an assignment or sub-letting
(5)   Fails to comply with the requirements of the CDM Regulations.

Clause 27.2 provides for the issue of a notice specifying the default. It must be given by actual delivery or sent by special delivery or recorded delivery. If the default then continues for 14 days after receipt of the notice (clause 27.2.2) or at any time thereafter is repeated (clause 27.2.3), the contractor may within 10 days of that continuance, or within a reasonable time after repetition, again by notice given by actual delivery or sent by special delivery or recorded delivery, determine the employment of the sub-contractor. Such determination shall take effect on the date of receipt of the notice. Such notice shall not however be given unreasonably or vexatiously. The question of the service of notices is discussed in some detail in Chapter 10 dealing with the determination provisions of IFC 98.

Clause 27.3 of NAM/SC provides for determination of the sub-contractor's employment under NAM/SC where certain insolvency events occur. If the insolvency event is the appointment of a provisional liquidator or trustee in bankruptcy or if a winding-up order is made or if the sub-contractor passes a resolution for voluntary winding-up (except for the purposes of amalgamation or reconstruction), then the sub-contractor's employment is forthwith automatically determined though it may be reinstated if the contractor and sub-contractor both agree. If the insolvency event is any other type, e.g. composition or arrangement with creditors, proposal for a voluntary arrangement for a composition of debts or scheme of arrangement, appointment of an administrator or an administrative receiver, there is no automatic determination of the sub-contractor's employment.

The contractor is entitled by notice to determine the employment of the sub-contractor with the determination taking effect on the date of receipt of the notice.

The financial consequences set out in clause 3.3.4 depend on which of the three options contained in clause 3.3.3 is chosen by the architect. The position is as follows.

Case (A): The instruction names another person to execute the work or the outstanding balance of work

The instruction does not amount to a variation. Express provision is made for the adjustment of the contract sum in relation to the cost of completing the outstanding work and also for an extension of time in relation to any delays due to compliance with the instruction. It is also expressly stated that clause 4.11 (dealing with reimbursement of loss and expense) shall not apply, although these words are probably inserted *ex abundanti cautela* (out of an abundance of caution) as nothing in the wording of clause 4.11 or 4.12 would suggest that loss and expense could be claimed arising out of an instruction under clause 3.3.3(a). If therefore the architect's instruction causes the regular progress of the works to be materially affected, any loss and expense in which the contractor is thereby involved cannot be recovered from the employer. No doubt the contractor will look to the first named sub-contractor for recompense. Of course, if the architect takes an unreasonable time in issuing the instruction then clause 4.12.1.2 (late instructions etc.) may apply.

Adjustment of contract sum

The contract sum is to be adjusted by adding to, or subtracting from it, the amount by which the new sub-contract price for the outstanding work exceeds or is below the price, as the case may be, of the original named sub-contract for that work. It should be noted that, in making the adjustment, it is the price of the first named sub-contractor which must be used in the valuation and not the price for that work included by the contractor in his own tender. Generally, it will involve an addition to the contract sum. It is provided in clause 3.3.4(a) that any amount included in the second named sub-contractor's price for the repair of defects in the first-named sub-contractor's work is to be excluded from the adjustment as the main contractor is responsible for the defective workmanship executed and defective materials supplied by the named sub-contractor. These words put the matter beyond doubt although it is submitted that the wording in clause 3.3.4(a), '... the price of the second named sub-contractor for the work not carried out by the first named sub-contractor ...', would in any case be sufficient to enable the architect or quantity surveyor in calculating the additional cost, if any, due to the contractor to discount any part of the second named sub-contractor's price which relates to the cost of remedying defects in the first named sub-contractor's work. This position is in marked contrast to that on the renomination of a sub-contractor under JCT 63 and JCT 80 (and now JCT 98): see *Fairclough Building Ltd* v. *Rhuddlan Borough Council* (1985).

It will be essential for the architect to obtain from the second named sub-contractor a separate price for the repair of defects so that the adjustment to the contract sum can be made. Even then, the architect or quantity surveyor on his behalf has, it is suggested, considerable problems of evaluation to cope with. A valuation must be made of that part of the first named sub-contractor's price which relates to unfinished work. This is no simple task.

It can be seen that the employer is therefore concerned with the sub-contract price of the named sub-contractor in this instance rather than the contractor's own price for the work.

The Form of Tender and Agreement NAM/T does not in parts readily lend itself to use on renaming, although it is required by clause 3.3.3(a) to be used. For example, there is no express provision requiring the price of the second named sub-contractor to be split between outstanding work and repairing defects, though no doubt this can be required in the 'Numbered Documents' (see NAM/T section I).

The situation where the sub-contractor's employment is determined otherwise than under clauses 27.2, 27.3 or 27.4 of NAM/SC is dealt with in clause 3.3.6(a). Briefly, where the determination is otherwise than under these two clauses, the contract sum cannot be adjusted upwards but only downwards, e.g. where a second named sub-contractor's cost for completing the outstanding balance of work results in a saving. Furthermore, the extension of time provisions will not apply.

The financial benefits to the contractor flowing from clause 3.3.4 on the determination of the named sub-contractor's employment are likely to be of immense importance and for them to be lost, perhaps because of a technical defect in the serving of an appropriate notice or because the factual situation is such that the contractor is obliged to accept the named sub-contractor's repudiation of the sub-contract, seems indefensible.

Extension of time

As clause 2.3 (extension of time) applies, the architect must make an extension of time for completion of the works if such completion has been or is likely to be delayed owing to the contractor's compliance with the instruction (see clause 2.4.5). Any such delay must therefore occur after the instruction is issued. The operation of clause 2.4.5 does not relate to any delay occurring before the instruction, e.g. delay on the part of the first named sub-contractor. Once the architect is notified of the determination of the employment of the first named sub-contractor it will of necessity take some time to reach a position whereby an instruction to rename can be issued. The time taken could cause delay and disruption to the contractor's programme. It is submitted that provided the architect acts with reasonable expedition the risks associated with this delay, i.e. the imposition of liquidated damages, must be borne by the contractor. However, if entering into a second named sub-contract would inevitably result in the contractor being unable to complete by his contract completion date where the real cause of the delays occurred prior to the instruction (so that no extension of time event under clause 2.4 is appropriate), the contractor could raise a reasonable objection pursuant to clause 3.3.2(c). The implications of the House of Lords' decision in *Percy Bilton Ltd* v. *Greater London Council* (1982) (dealt with earlier on page 131) are again important here.

Difficulties in assessing extensions of time are bound to arise, e.g. where compliance by the contractor with the instruction involves the second named sub-contractor completing unfinished work sufficiently late or out of sequence to cause the contractor to complete the total works late.

The delay may be due to one or more of the following factors:

(1) Delay on the part of the first named sub-contractor
(2) The time taken by the architect between determination of the employment of the first named sub-contractor and the instruction naming a second sub-contractor
(3) Delay on the part of the second named sub-contractor.

Whichever one of these delays is appropriate, it is not compliance with the instruction which caused a delay in completion of the works as a whole. However, if the second named sub-contractor's programme for the unfinished work, to be carried out under similar conditions, shows a longer period for completion of the unfinished work than did that of the first named sub-contractor, this would, if it delayed completion of the work as a whole, entitle the contractor to an appropriate extension of time; similarly, if the sequence of working of the second named sub-contractor differed from that of the first in such a manner that the main contractor was delayed in completing the works.

It is submitted that the delay caused to completion of the works brought about by the time taken by the second named sub-contractor repairing defects in the first named sub-contractor's work will not entitle the contractor to an extension of time.

Where the determination of the employment of the named sub-contractor is not in accordance with the provisions of clauses 27.2, 27.3 or 27.4 of NAM/SC, no extension of time can be given (see clause 3.3.6(a)). For a comment on the potential injustice of this see the last paragraph under the previous heading (page 142).

Case (B): The contractor is instructed to make his own arrangements for the execution of the work or the outstanding work is omitted

This instruction is to be regarded as a variation and will accordingly be valued under clause 3.7. Clause 2.3 (extension of time) and clause 4.11 (loss and expense) apply.

Valuation of variations

No doubt the contractor's rates and prices will form the basis of any valuation. Any changes in the conditions under which any work is to be carried out must be considered. However, there may well be some significant limitations to this. For instance, if prior to the determination of the employment of the named sub-contractor he was guilty of delay which resulted in the outstanding work having to be carried out in a winter period rather than in the summer period for which it was programmed and priced, can that factor be taken into account in the valuation or is it a matter for which the contractor under the contract is wholly responsible? It is submitted that under the contract the main contractor is responsible for the delays of his named sub-contractor and he cannot benefit from his own default so that such factors cannot be taken into account. This view is reinforced by the terms of clause 3.3.9 which provide that unless otherwise expressly stated in the conditions, the contractor is to remain wholly responsible for carrying out and completing the work notwithstanding the naming of a sub-contractor.

Where the architect's instruction omits the outstanding balance of work it will be valued under clause 3.7.3.

Where the employment of the named sub-contractor is determined otherwise than under clauses 27.2, 27.3 or 27.4 of NAM/SC the variation can only be valued to the extent that it reduces the contract sum – see clause 3.3.6(a). The potential injustice of this provision has already been commented on earlier on page 142.

Extension of time

Clause 3.3.4(b) expressly provides that clause 2.3 (extensions of time) applies to an instruction issued under clause 3.3.3(b) or (c). The implications of the decision of the House of Lords in *Percy Bilton* v. *Greater London Council* (1982) dealt with earlier on page 131 are again relevant here. It may be difficult in a given situation to determine whether or not a late finish by the contractor was as a result of complying with an instruction to finish off the sub-contractor's work or was in truth due to the time taken by the architect in issuing the instruction in the first place. The comments under the heading of 'Extension of time' under case (A) above are also relevant here.

Where the determination of the sub-contractor's employment was not brought about by the contractor under the provisions of clauses 27.2, 27.3 or 27.4 of NAM/ SC, the extension of time provisions do not apply (see clause 3.3.6(a)). This particular aspect of the clause has already been discussed earlier on page 142.

Loss and expense

Clause 3.3.4(b) expressly provides that clause 4.11 (disturbance of progress) shall apply to an instruction issued under clause 3.3.3(b) or (c).

Only such loss and expense as is due to the issue of the instruction is reimbursable. Difficulties in establishing the effective cause of the disturbance are bound to arise and the example instanced under the heading 'Extension of time' under case (A) above should be considered here.

Once again, the implications of the House of Lords' decision in *Percy Bilton Ltd* v. *Greater London Council* (1982) are relevant.

The loss and expense provision does not apply where the employment of the named sub-contractor is determined otherwise than under clauses 27.2, 27.3 or 27.4 of NAM/SC.

Where an instruction under clause 3.3.3(c) is issued, the employer may employ another direct contractor to finish off the work or he may even finish it off himself provided the contractor's consent is obtained in accordance with clause 3.11.

NOTES TO CLAUSE 3.3.4

[1] '. . . arises in respect of a person named . . .'
Surely the instruction should relate to the work of the person named and not to the person. The effect of this inappropriate language could prevent this clause applying when it is desired to issue an instruction involving a further naming where it is a second named person whose employment has been determined. This is only compounded by the reference in clause 3.3.4(a) to the second named sub-contractor.

[2] '. . . the price of the first named sub-contractor for that work . . .'
The architect, or where appointed the quantity surveyor on his behalf, must endeavour to value the outstanding work. Where there are bills of quantities as a sub-contract document there is unlikely to be much difficulty. If there are no bills then the task is likely to be more difficult. In practice the quantity surveyor is likely to try and agree this value with the contractor if at all possible.

[3] '. . . for the repair of defects . . .'
Does repair include complete replacement? The normal meaning of the word repair would tend to suggest something less than complete replacement. Else-where in the contract, e.g. clause 2.10 (defects liability), reference is made to making good defects, which would cover replacement if this was necessary to make good the defect. The choice of word is perhaps unfortunate. It is the defective work rather than the defect itself which is repaired. It is with some diffidence therefore that it is suggested that, in order to make sense of the risk sharing between employer and contractor in the event of the employment of a named sub-contractor being determined, the words used should be construed to extend to replacement as well as mere repair of defective work. There appears to be no ulterior reason for construing the words in a narrow sense. It is further suggested that the words used can extend even to carrying out of work not done by the first named sub-contractor where the defective work is defective because of an omission.

Clause 3.3.5

3.3.5   Where an instruction referred to in clause 3.3.3(a) to (c) arises in respect of a person named in an instruction as to the expenditure of a provisional sum under clause 3.3.2 whose employment has been determined under clause 27.2 or 27.3 or 27.4 of the Sub-Contract Conditions NAM/SC, such instruction shall be regarded as a further instruction issued under the provisional sum.

COMMENTARY ON CLAUSE 3.3.5

Where the employment of a named person following an instruction as to the expenditure of a provisional sum under clause 3.3.2 is determined, the instruction under clause 3.3.3 will be regarded as a further instruction issued under the provisional sum. The financial consequences set out in clause 3.3.4 do not apply as they are applicable only to the situation where the determination is of the employment of a sub-contractor whose work was described as being required to be executed by a person named in the specification/schedules of work/contract bills. The further instruction as to the expenditure of a provisional sum issued under clause 3.8 will be valued under clause 3.7 and can give rise to an extension of time and reimbursement of loss and expense – clause 2.4.5 and clause 4.12.7.

Where the sub-contractor's employment is determined otherwise than under clauses 27.2, 27.3 or 27.4 of NAM/SC the valuation will only apply to the extent that the contract sum is adjusted downwards. The extension of time and loss and expense provisions will not apply. This may prove in some situations to be unfair to the contractor (see earlier page 142).

The wording of clause 3.3.2, 3.3.3 and this clause does not appear to rule out the possibility of a determination of the employment of a second or further named person followed by a further instruction. Also, as it remains an instruction as to the expenditure of a provisional sum under clause 3.3.2, the contractor retains a right of reasonable objection as provided therein.

Clause 3.3.6

3.3.6   (a) Where the employment of the named person is determined otherwise than under clause 27.2 or 27.3 or 27.4 of the Sub-Contract Conditions NAM/SC in respect of the instructions under clause 3.3.3(a) or (b) or (c) the provisions of clause 3.3.4(a) or (b) or 3.3.5 as appropriate shall apply but only to the extent that they result in a reduction in the Contract Sum and the instruction shall not be regarded as an event to which clause 2.3 (*Extension of time*) applies or a matter to which clause 4.11 (*Disturbance of progress*) applies.
(b) The following provisions of this clause 3.3.6(b) shall apply where the employment of the named person is determined under clause 27.2 or 27.3 or 27.4 of the Sub-Contract Conditions NAM/SC:
– the Contractor shall take such reasonable action as is necessary to recover from the named sub-contractor under clause 27.6.4 of the Sub-Contract Conditions NAM/SC any additional amounts payable to the Contractor by the Employer as a result of the application of clause 3.3.4(a) or (b) or 3.3.5 together with an amount equal to any liquidated damages that would have been payable or allowable by the Contractor to the Employer under clause 2.7 but for the application of clause 3.3.4(a) or (b) or 3.3.5;
– the Contractor shall account to the Employer[1] for any amounts so recovered;
– in taking such action the Contractor shall not be required to invoke the procedure under this Contract[2] relevant to the resolution of disputes or differences unless the Employer shall have agreed to indemnify the Contractor against any legal costs reasonably incurred in relation thereto;
– if the Contractor has failed to comply with this provision[3] he shall repay to the Employer any additional amounts paid as a result of the application of clause 3.3.4(a) or (b) or 3.3.5 and shall pay or allow an amount equal to the liquidated damages referred to herein.

COMMENTARY ON CLAUSE 3.3.6

By virtue of clause 3.3.6(a) if the named sub-contractor's employment is determined otherwise than under clauses 27.2, 27.3 or 27.4 of NAM/SC, the obvious financial benefits to the contractor from the operation of clause 3.3.4 and clause 3.3.5 are not to apply. Clauses 3.3.4 and 3.3.5 will only operate to the extent that their application would in any event result in a reduction in the contract sum, e.g. an instruction under clause 3.3.3(c) omitting the outstanding balance of work. The opening words of clause 3.3.3 forbid the contractor from determining a named sub-contractor's employment otherwise than under clauses 27.2, 27.3 or 27.4. These three clauses and also the possible difficulties which could be experienced by a contractor faced with a defaulting sub-contractor have been discussed earlier on pages 140 and 142.

No doubt the purpose of the first paragraph of clause 3.3.3 and clause 3.3.6(a) is to try to ensure that the named sub-contractor's employment is determined under clauses 27.2, 27.3, or 27.4 of NAM/SC in order that the main contractor can pursue the sub-contractor under clause 27.6.4 of NAM/SC to recover on behalf of the employer the extra cost incurred by the employer by the operation of the financial provisions in clauses 3.3.4 and 3.3.5. These of course only come into operation when the named sub-contractor is in default. A determination of the named sub-contractor's employment by any other method is likely to prevent recovery under NAM/SC by the contractor of the employer's extra costs (see below). IFC 98 in such a situation therefore deprives the contractor of the financial benefits contained in clause 3.3.4, leaving the contractor to seek recovery from the named sub-contractor of the extra costs incurred by the contractor in complying with the architect's instructions issued under clause 3.3.3(a), (b) or (c).

Clause 3.3.6(b) imposes on the contractor an obligation to recover from the named sub-contractor under clause 27.6.4 of NAM/SC any additional amounts payable to the contractor by the employer as a result of the application of clauses 3.3.4(a) or (b) or clause 3.3.5, together with an amount equal to any liquidated damages which the employer would have deducted or recovered but for the operation of the extension of time provisions.

Clause 27.6.4 of NAM/SC cross refers to clause 3.3.6(b) and imposes an undertaking on the named sub-contractor not to contend that the contractor has suffered no loss or that the sub-contractor's obligation to pay the contractor should in any way be reduced or extinguished by reason of the operation of clause 3.3.4(a) or (b) or clause 3.3.5 of IFC 98.

It can be seen therefore that in truth the contractor receives the financial benefits provided in clause 3.3.4 but nevertheless has an obligation to obtain a sum equivalent to these benefits from the named sub-contractor, which must then be handed over to the employer.

Bearing in mind that the main contractor does not suffer any loss in this regard, it might conceivably be argued that clause 27.6.4 of NAM/SC represents something akin to a penalty provision in that it requires the payment of a sum of money from the named sub-contractor to the contractor which bears no relationship whatsoever to any actual loss suffered or incurred by the contractor. See *Campbell Discount Co. Ltd* v. *Bridge* (1962), particularly the judgments of Lords Denning and Devlin.

However, if the overall scheme is considered, in particular the absence of a direct contractual relationship between employer and named sub-contractor under which the employer can obtain redress for the increased expenditure incurred as a result of the defaults of the named sub-contractor, it is clear that the provisions are reasonable and it would involve a strict application of the doctrine of privity of contract to deprive these provisions of their intended effect. It is hoped that a court or an arbitrator may not adopt too strict an approach based on privity of contract alone, as the overall commercial as well as contractual scheme of things should be considered.

Some, albeit limited, support for the argument that this type of provision is not a penalty is to be found in the Australian case of *Corporation of Adelaide* v. *Jennings* (1985) CLJ 1984-85 Vol. 1 No. 3 at page 205 where Wilson J in the Australian High Court said:

> 'A sub-contractor who refuses or fails to complete his sub-contract does not escape liability for the reasonable cost of doing the sub-contract work under a plea that a third party (the proprietor) will become liable under the head contract to pay an amount equal to that cost. The natural consequence of his failure to complete is the incurring of the reasonable cost of completing the sub-contract less any unpaid balance of the sub-contract price. Those damages are recoverable under the first branch of the rule in *Hadley* v. *Baxendale* (1854) and it is immaterial that the builder has no claim under the second branch. The builder's right under the head contract to recoup from the proprietor any increase in the cost of completing the sub-contract does not relieve the sub-contractor from liability to the builder; the measure of that liability depends, of course, on the sub-contract. But if a proprietor pays the builder a sum representing an increase in the contract price occasioned by the sub-contractor's default, the proprietor is subrogated to such rights as the builder may have to recover that sum from the sub-contractor.'

Admittedly this judgment was given in relation to a case based on an Australian standard form of building contract known as the RAIA Form of Contract. Even so the principle is sound.

These potential difficulties could have been avoided if a different method of recovery of the employer's increased costs had been employed, e.g. either a direct contract between employer and named sub-contractor dealing with the question of the named sub-contractor's default, or for the contractor to suffer the increased costs as a result of the failure of the named sub-contractor to complete the sub-contract works, supported perhaps by an indemnity provision on the part of the employer in favour of the contractor should the contractor be unable, after taking reasonable steps, to obtain recompense from the defaulting named sub-contractor. Of course, the latter alternative has from the contractor's point of view the distinct disadvantage, when compared with the actual scheme of recovery in IFC 98 and NAM/SC, that the purse strings would be held by the employer and not by the contractor.

Clause 3.3.6(b) requires the contractor to account to the employer for any amounts recovered from the named sub-contractor. The contractor is not required to invoke the contractual dispute procedures to pursue the sub-contractor unless the employer provides an indemnity against legal costs reasonably incurred. If the contractor fails to comply with clause 3.3.6(b) he is required to repay the employer any additional amounts paid by the employer as a result of the application of clause 3.3.4(a) or (b) or clause 3.3.5, together with an amount equal to the lost liquidated damages because of the application of those provisions.

These provisions are complicated and cumbersome and may well prove rather difficult to operate in practice.

NOTES TO CLAUSE 3.3.6

[1] '...account to the Employer...'
These words are probably sufficient to impose a trust or fiduciary status on the amount so recovered. In such a case, provided the amounts so recovered are in a separate and identifiable fund in the contractor's hands, the employer will have a right to the sum recovered in priority to the claim of any trustee in bankruptcy, liquidator, or administrative receiver of the contractor.

[2] '...under this Contract...'
The reference to 'this Contract' must be a drafting slip. It is the dispute resolution procedures under the sub-contract and not the main contract which are relevant.

[3] '...with this provision...'
This presumably refers to the whole of clause 3.3.6(b). Having regard to the significance of failing to comply, the drafting could do with being made clearer, e.g. by replacing these words with '...*with the provisions of this clause 3.3.6(b)*...'.

Clause 3.3.7

3.3.7 Whether or not a person who has been named as a sub-contractor under any of clauses 3.3.1 to 3.3.5 is responsible to the Employer for exercising reasonable care and skill in:
– the design of the sub-contract works insofar as the sub-contract works have been or will be designed by the named person;
– the selection of the kinds of materials and goods for the sub-contract works insofar as such materials and goods have been or will be selected by the named person; or
– the satisfaction of any performance specification or requirement relating to the sub-contract works,
the Contractor shall not be responsible to the Employer under this Contract for anything to which the above terms relate, nor, through the Contractor, shall the person so named or any other sub-contractor be so responsible[1]; provided that this shall not be construed so as to affect the obligations of the Contractor or any sub-contractor in regard to the supply of goods and materials and workmanship.
The provisions of this clause 3.3.7 shall apply notwithstanding that the Sub-Contract Sum stated in article 2 of Section III of the Tender and Agreement NAM/T referred to in clause 3.3.1 or 3.3.2 included for the supply of any design, selection or satisfaction as referred to herein, and that such Sub-Contract Sum is included for within the Contract Sum or the Contract Sum as finally adjusted.

COMMENTARY ON CLAUSE 3.3.7

IFC 98 is a work and materials contract. It is not a design contract and the contractor is not responsible under this contract for design. Therefore, if a named sub-contractor has a design element to perform this clause attempts to make it clear that the contractor is not responsible for failures in relation to such design. The contractor's liability is only in relation to the quality of goods and materials supplied and the standards of workmanship.

The three insets all relate to aspects of design or fitness for their purpose of the sub-contract works. It is perhaps a criticism of this clause that it may cover all matters of design however trivial or insignificant. Even if the named sub-contractor has a certain level of discretion in relation to quite minor matters which could be classified as design, e.g. the type of nails to be used or the type of joint in woodwork, this clause may effectively exclude any liability of the contractor in respect of failures in relation to such matters. Such examples are arguably matters of design or fitness for purpose.

It is submitted that the contractor, even under IFC 98, will ordinarily be responsible for the exercise of his discretion in relation to such minor details. Why therefore should he not be responsible where such minor details are left to the discretion of a named sub-contractor? It may be possible for the employer to successfully contend that such minor matters are part of the contractor's obligations in regard to the supply of goods and materials and workmanship and will therefore be caught by the proviso at the end of the first paragraph of this clause. If this is not the case, the employer must seek a direct design warranty from the named sub-contractor in every case, even where there does not appear to be any element of traditional design in the named sub-contract. A form of design agreement has been produced for use between employer and named sub-contractor by two of the constituent bodies on the former Joint Contracts Tribunal: the Royal Institute of British Architects and the Specialist Engineering Contractors' Group. It is known as ESA/1 and is discussed briefly later (page 152).

The last paragraph of clause 3.3.7 attempts to make it clear that the contractor has no liability in connection with the design matters referred to in the first paragraph of the clause, even where the contract sum contains, as it is likely to, the design fee of the named sub-contractor within the main contractor's pricing of the work.

NOTES TO CLAUSE 3.3.7

[1] *'... nor, through the Contractor, shall the person so named or any other sub-contractor be so responsible ...'*
These words prevent the traditional legal chain of liability being established, i.e. the employer looking to the contractor, who in turn would look to the sub-contractor, who may look to a sub-sub-contractor etc. through a contractual link down to the point where the ultimate responsibility rests with the person by whom it should properly be borne (though the clause would probably have this effect in any event).

Clause 3.3.8

> 3.3.8   Clauses 3.3.1 to 3.3.7 shall not apply to the execution of part of the Works by a local authority or a statutory undertaker executing such work solely in pursuance of its statutory rights or obligations.

COMMENTARY ON CLAUSE 3.3.8

This is an important clause which states that local authorities or statutory undertakers who execute part of the works cannot become named sub-contractors where their work is carried out in pursuance of statutory obligations, i.e. work which they must do, or statutory rights, i.e. work which they are not bound to do but which they have the right to do. The word 'rights' is difficult to construe in this context. Does it mean that if a certain kind of work is to be carried out, the local authority or statutory undertaker can insist that they and no one else may do it? Or does it mean work which they have a power to do but only by agreement with the employer? Perhaps the word 'powers' would have been more appropriate.

If work relating to part of the contract works falls to be carried out by a local authority or statutory undertaker in pursuance of its statutory obligations or rights, it may be carried out by a direct arrangement with the employer. If included as a provisional sum for the purpose of placing the responsibility for delays or failures of the local authority or statutory undertaker on the contractor, it will probably not work: see *Henry Boot Construction Ltd* v. *Central Lancashire New Town Development Corporation* (1980).

Clause 3.3.9

> 3.3.9 Save as otherwise expressed in the Conditions the Contractor shall remain wholly responsible for carrying out and completing the Works in all respects in accordance with clause 1.1 notwithstanding the naming of a sub-contractor for the execution of work described in the Specification/Schedules of Work/Contract Bills.

COMMENTARY ON CLAUSE 3.3.9

This makes it clear that unless clauses 3.3.1 to 3.3.7 in relation to named sub-contractors expressly provide otherwise, the contractor is to remain wholly responsible for carrying out and completing the works (including work carried out by a named sub-contractor) in all respects in accordance with clause 1.1; in other words in a proper and workmanlike manner and in accordance with the contract documents. The clause expresses what is in any event probably implicit when considering the conditions as a whole. However, it puts the matter beyond doubt.

Contractor's failure to comply with the provisions of clause 3.3

A failure on the part of the contractor to comply with any of the provisions of clause 3.3 due to his default is a ground for the determination of his employment by the employer, subject however to the proviso that the notice of determination must not be given unreasonably or vexatiously – see clauses 7.2.1(d) and 7.2.4.

Standard documents for use in connection with the naming of a sub-contractor

Where clause 3.3.1 (named persons to carry out sub-contract work described in the specification/schedules of work/contract bills) is to be used, the contractor in tendering for the main contract must price the named sub-contractor's work as described in the tender documents and, unlike the situation regarding traditional nomination under JCT 98, the contract sum will not subsequently be adjusted to take account of changes in the sub-contract sum. For this reason it is important that contractors are fully aware of the basis on which the proposed named sub-contractor has tendered so that a proper assessment of the proposed named sub-contractor's price can be made by the contractor in building up his own price for the work.

For example, he needs to know as much as possible about the period or periods required for the completion of the sub-contract work, special attendances, main contract appendix details, as well as the proposed sub-contract price. The contractor takes a significant proportion of the financial risks of poor performance by the proposed named sub-contractor. It is only right therefore that as much information as possible is provided. It is clearly better that this be provided in a standard form and it is to be found in Parts I and II of the Form of Tender and Agreement NAM/T. It is also obviously desirable that there be a standard set of sub-contract conditions for named sub-contractors and this is provided in NAM/SC.

Form of Tender and Agreement NAM/T

This is the form of tender on which tenders will be invited by the architect from persons who may subsequently be named to carry out sub-contract works, whether the works are described in the specification/schedules of work/contract bills or under an instruction as to the expenditure of a provisional sum. The form currently consists of 15 pages and comprises three sections. It has been amended on a number of occasions (currently up to Amendment 8 issued April 1998). It has now been consolidated into a 1998 edition. Section I is the invitation to tender for completion by the architect and contains main contract information and sub-contract information, there being 19 items in all. Section II comprises the tender by the sub-contractor in which there are six items for completion. Finally, section III comprises the articles of agreement which is the actual agreement for execution incorporating the sub-contract documents including the Tender and Agreement NAM/T, the Sub-Contract Conditions NAM/SC and the numbered documents which are listed on the first page of section I of NAM/T.

The Sub-Contract Conditions NAM/SC

These are the standard conditions of sub-contract between the named sub-contractor and the contractor. They are considerably detailed and are of course designed to be compatible with IFC 98. The conditions are based to a large extent on the standard domestic form of contract associated with JCT 98, i.e. DOM/1. There are some differences to take account of the fact that a named sub-contractor is not purely domestic.

ESA/1

This is an agreement prepared outside the Tribunal by two of the constituent bodies of the former Tribunal: the Royal Institute of British Architects and the

Specialist Engineering Contractors' Group. It deals with design responsibilities undertaken by named sub-contractors in relation to the sub-contract works and accordingly, as the main contractor will generally have no responsibility for design under IFC 98, this agreement is between the named sub-contractor and the employer only. The design obligation is expressed in similar, though not identical, terms to that in the JCT Standard Form of Employer/Nominated Sub-Contractor Agreement (NSC/W) for use between employer and nominated sub-contractor under JCT 98 (see clause 2.1 of NSC/W).

While IFC 98 is designed for works of a simple nature not involving complicated electrical or mechanical services, nevertheless in practice named sub-contractors will undoubtedly be used whose work will carry within it a design element, often of a specialist nature. Clause 3.3.7 of IFC 98 expressly relieves the contractor from responsibility to the employer for any defects in the design of a named sub-contractor's work. It is absolutely imperative therefore that in every contract in which a named sub-contractor is used, the employer or those advising him consider carefully the question of whether or not a named sub-contractor is providing any design service, and if so the separate agreement ESA/1 or some other suitable agreement should be used.

ESA/1 has existed with only minor amendments since 1984. At the time of writing it has not been updated. In its current form it still seems to be applicable for use with IFC 98. However it does need generally to be modernised. It may be that the Tribunal will itself produce a suitable form of warranty in line with its declared intention to produce a comprehensive suite of interlinking contract documents.

If such an agreement is not used the employer will find himself without a contractual remedy in respect of defective design by a named sub-contractor. While the employer may in appropriate situations be able to establish a cause of action against the named sub-contractor for breach of a duty of care or by virtue of some collateral warranty, this will not often be the case, and the absence of a suitable direct agreement between the employer and the named sub-contractor in such a situation will leave a serious gap in the employer's panoply of remedies in respect of defective work.

Chapter 6
Control of the works (Part II)

CONTENT

This chapter looks at clauses 3.4 to 3.15 of section 3 of IFC 98 which deals with the following matters:

Contractor's person-in-charge
Architect's instructions
Variations (including their valuation)
Instructions to expend provisional sums
Levels and setting out
Clerk of works
Work not forming part of the contract
Instructions as to inspection – tests
Instructions as to the removal of work etc.
Instructions as to postponement

SUMMARY OF GENERAL LAW

(A) Contract administrator's instructions

Building contracts which provide for an architect will generally empower him to issue instructions. While as between the employer and the contractor the validity of the architect's instruction will depend on the terms of the building contract, as between the architect and the employer the authority of the architect to issue certain instructions, e.g. as to variations in the contract works, will depend on his conditions of engagement with the employer. These may impose restrictions on his power to issue certain instructions, particularly where this would add to the contract sum.

The building contract will often require the architect's instructions to be in writing. Such instructions can and very often do cover a wide variety of matters, e.g. variations to the contract work or to the conditions under which they are carried out, the removal of defective work, testing, expenditure of provisional sums, postponement, resolution of inconsistencies in the requirements laid down in the contract documents and re-nomination of sub-contractors. Compliance with an instruction of the architect may be a good ground for the contractor to seek an extension of time for completion where it causes delay, and monetary compensation where it causes financial loss.

The contract will often provide a mechanism for the contractor to challenge the validity of an instruction and also an express power for the employer to have any work required by an instruction carried out by someone other than the contractor if he fails to comply with it, the additional cost being borne by the contractor.

(B) Variations

When the word 'variation' is used in relation to building contracts, it generally refers to a variation in the contract works or occasionally to the conditions (not contract conditions) under which the contract work is to be carried out, e.g working hours, working space, access to site and suchlike. It does not relate to a variation in the terms of the contract itself. It cannot be stressed too often that the architect has no inherent power to vary the terms or conditions of the contract. If the architect agrees with the contractor that the terms of the contract will in some way be varied, the contractor will be taking a risk as the employer will not be bound by the agreement. The contractor should therefore check the actual authority of the architect to agree to such a variation.

The building contract between employer and contractor will bestow on the architect the power to order variations in relation to the contract works themselves. To the extent therefore that the instruction requiring a variation under the main contract is in accordance with its terms, the contractor need not check with the employer that the architect is empowered by the employer to order such a variation. The employer, in appointing the architect to act under the contract, is bestowing on him the power and authority which the main contract gives to the architect. The contract can of course expressly contain limits as to the contract administrator's authority in this connection, e.g. see clause 2 and Part II of the FIDIC contract (4th edition). Often a restriction is placed on the architect's freedom to issue instructions ordering variations where the contract sum is exceeded by more than a given percentage.

It is imperative that the architect, when issuing instructions requiring a variation to the contract works, does so strictly in accordance with the terms of the contract giving him that authority. A failure to do so can cause severe difficulties with the possibility of the contractor being disentitled to payment for such unauthorised, though ordered, variations, and the architect running a risk of personal liability in such circumstances for losses suffered by the contractor.

In the absence of an express power for variations in the contract works, there is no implied right of the employer to have variations carried out by the contractor. The drawings, specification etc. would in fact be frozen at the date that the contract was entered into, and if the employer required any departure from the requirements of these documents, this could only be achieved by a separately negotiated agreement between himself and the contractor.

Most contractual variation clauses contain a power not only to add or vary work but also to omit it. It would not be a proper exercise of this power to omit work in circumstances where the employer intended to have the omitted work carried out by a third party: see *Amec Building Ltd* v. *Cadmus Investment Company Ltd* (1996).

Work indispensably necessary and standard methods of measurement

As a general principle, an obligation to do specified work includes an obligation to do all necessary ancillary work. Even where a detailed bill of quantities is prepared in accordance with a standard method of measurement, it may not descend to every detail of construction. There may be some items or some work processes which are not fully described in the standard method of measurement, and it is at least arguable that this work will not amount to a variation and will have to be carried out within the overall contract sum without adjustment therefor.

CONSIDERATION OF THE RELEVANT CLAUSES OF IFC 84

Clause 3.4

Contractor's person-in-charge

3.4   The Contractor shall at all reasonable times keep upon the Works a competent person-in-charge and any written instructions given to him by the Architect/the Contract Administrator shall be deemed to have been issued to the Contractor.

COMMENTARY ON CLAUSE 3.4

Clearly, the contractor should appoint a responsible person in this position. Any instruction of the architect given to the person-in-charge is deemed to have been given to the contractor. This is especially important wherever time limits are related to an instruction, e.g. clauses 3.3.2 (entering into a named sub-contract), 3.13.1 (instructions on opening up and testing), 3.5 (compliance with the architect's instructions generally) and 3.7.1.2 A4.2 (time for acceptance of amended price statement).

Clauses 3.5.1 and 3.5.2

Architect's/Contract Administrator's instructions

3.5.1   All instructions of the Architect/the Contract Administrator shall be in writing. The Contractor shall forthwith comply with such instructions issued to him which the Architect/the Contract Administrator is empowered by the Conditions to issue; save that where such instruction is one requiring a Variation within the meaning of clause 3.6.2 the Contractor need not comply to the extent that he makes reasonable objection in writing to the Architect/the Contract Administrator to such compliance.
If within 7 days after receipt of a written notice from the Architect/the Contract Administrator requiring compliance with an instruction the Contractor does not comply therewith then the Employer may employ and pay other persons to execute any work whatsoever which may be necessary to give effect to such instruction and all costs incurred thereby may be deducted by him from any monies due or to become due to the Contractor under this Contract or shall be recoverable from the Contractor by the Employer as a debt.

3.5.2   Upon receipt[1] of what purports to be an instruction issued to him by the Architect/the Contract Administrator the Contractor may request the Architect/the Contract Administrator to specify in writing the provision of the Conditions which empowers the issue of the said instruction. The Architect/the Contract Administrator shall forthwith comply with any such request, and if the Contractor shall thereafter comply with the said instruction (neither Party before such compliance having invoked the procedures under this Contract relevant to the resolution of disputes or differences in order that it may be decided whether the provision specified by the Architect/the Contractor Administrator empowers the issue of the said instruction), then the issue of the same shall be deemed for all the purposes of this Contract to have been empowered by the provision of the

Conditions specified by the Architect/the Contract Administrator in answer to the Contractor's request.

COMMENTARY ON CLAUSE 3.5

For a summary of the general law see page 154.

All the architect's instructions must be in writing. This contract makes no provision whatsoever for oral instructions even if the contractor confirms them in writing and they are not dissented from by the architect. However, no doubt compliance by the contractor with an oral instruction varying the works, followed by a written ratification by the architect, will produce an effective variation from the date of the oral instruction (see clause 3.6 first paragraph). Nevertheless, the contractor should be very cautious indeed about acting on a verbal instruction.

Having received the written instruction (and this includes receipt by the person-in-charge as referred to in clause 3.4), the contractor is bound to comply *forthwith*, except that if the instruction is one requiring a variation under clause 3.6.2 (imposition of or changes from the original contract documents in relation to obligations or restrictions in regard to access, working space, working hours or sequence of work), the contractor need not comply to the extent that he makes reasonable objection in writing to such compliance. What would be a reasonable objection in these circumstances is difficult to define, particularly as an instruction under clause 3.6.2 can give rise to reimbursement of direct loss and expense under clause 4.12.7 and an extension of time under clause 2.4.5, as well as involving a fair valuation of the variation itself. Presumably therefore a reasonable objection must generally relate to a matter for which money is insufficient recompense.

If within seven days of receipt by the contractor of a written notice from the architect requiring him to comply with an instruction the contractor does not comply therewith, the architect can employ others to do the work and the costs involved can be deducted from money due to the contractor or can be recovered from him as a debt. Where it is intended to withhold or deduct such costs from sums due or to become due to the contractor, it will be necessary for the employer to serve an appropriate notice of intended set-off in accordance with section 111 of the Housing Grants, Construction and Regeneration Act 1996. This requirement is embodied in clause 4.2.3(b) in respect of interim payments before practical completion; clause 4.3(c) in respect of the interim payment on practical completion; and clause 4.6.1.3 in respect of payment under the final certificate. Reference should be made to those clauses for details of the notice requirements in terms of both timing and content.

Placing work with another contractor under this provision is an extreme remedy which can cause serious practical problems on site and hinder the smooth running of the contract.

It should be remembered that a failure by the contractor to implement the instruction *forthwith* is in itself a breach of contract. The second paragraph of this clause relating to the seven day time limit does not have the effect of giving the contractor a further seven days in which to comply with the instruction. If he fails to comply forthwith then he will be liable to the employer for breach of contract even for the period before as well as within the seven day period and will become responsible for any foreseeable losses thereby suffered by the employer.

Instructions can only be issued so far as they are empowered under the conditions of contract. Clause 3.5.2 enables the contractor to test the validity of the instruction by requesting the architect to specify in writing the provision of the conditions which empowers the issue of the instruction in question. The architect must thereupon comply with the request. If the contractor thereafter complies with the instruction without having firstly invoked the relevant dispute resolution procedures under the contract, in order to test the validity of the instruction, such instruction is deemed to have been empowered by the provisions of the conditions as specified by the architect in his response to the contractor's request.

If dissatisfied, the contractor would generally be well advised to comply with the instruction but may preserve the right to challenge the architect's answer by invoking an appropriate dispute procedure before compliance. Of course if the instruction is demonstrably not a valid one, the contractor would not be in breach of contract in refusing to comply with it. However, before refusing, the contractor should be very clear as to his ground for refusal. The speed within which an adjudicator's decision can be obtained (see clause 9A), will hopefully lead to the rapid resolution of such disputes without too much dislocation being caused to the progress of the works.

If the contractor complies with the instruction, having first taken the precaution of serving the appropriate notice, what is his position if in the result, the instruction is not a valid one? The factual position in each case would need to be carefully considered. If the circumstances were such that the architect was, despite issuing an invalid instruction, acting as agent of the employer, then the employer will be liable to pay a reasonable sum in respect of any benefit which the employer has obtained. Alternatively, if the architect is not acting as the employer's agent, the architect may find himself with personal liability if it can be established that he is in breach of warranty of authority, i.e. if he has held himself out as being authorised to issue such an instruction. Hopefully the situation will be rare, although it cannot be ruled out, in which the contractor, having formed the view that the instruction is invalid, but having nevertheless carried it out, will find himself without any legal recourse to recover the costs incurred in having done so.

If the architect delays in giving an answer to the contractor's request, it is difficult to see what remedy the contractor has. The contractor should comply with the instructions forthwith and the fact that he has invoked the provisions of clause 3.5.2 does not affect this, so that any delay on the architect's part in responding to the contractor's clause 3.5.2 request should neither delay nor disrupt work. Accordingly, there is no extension of time event or loss and expense matter to cover such a delay in response. In extreme cases it may amount to a breach of contract on the part of the employer.

NOTES TO CLAUSE 3.5

[1] 'Upon receipt...'
The contractor's request must be made on receipt of a purported instruction, i.e. straightaway.

Clauses 3.6 and 3.7

Variations

3.6   The Architect/the Contract Administrator may subject to clause 3.5.1 issue instructions[1] requiring a Variation and sanction[2] in writing any Variation made by the Contractor otherwise than pursuant to such an instruction. No such instruction or sanction shall vitiate this Contract.

The term Variation as used in the Conditions means:

3.6.1   the alteration or modification of the design or quality or quantity of the Works including
   − the addition, omission or substitution of any work,
   − the alteration of the kind or standard of any materials or goods to be used in the Works,
   − the removal from the site of any work executed or materials or goods brought thereon by the Contractor for the purposes of the Works other than work materials or goods which are not in accordance with this Contract;

3.6.2   the imposition by the Employer of any obligations or restrictions or the addition to or alteration or omission of any such obligations or restrictions so imposed or imposed by the Employer in the Specification/Schedules of Work/Contract Bills in regard to:
   − access to the site or use of any specific parts of the site;
   − limitations of working space;
   − limitations of working hours;
   − the execution or completion of the work in any specific order.

Valuation of Variations and provisional sum work – Approximate Quantity, measurement and valuation

3.7.1   .1   The amount to be added to or deducted from the Contract Sum in respect of
   − instructions requiring a Variation and
   − any Variation made by the Contractor and sanctioned in writing by the Architect/the Contract Administrator and
   − all work which under the Conditions is to be treated as if it were a Variation required by an instruction of the Architect/the Contract Administrator under clause 3.6 and
   − instructions on the expenditure of a provisional sum and
   − work executed by the Contractor for which an Approximate Quantity is included in the Contract Documents
   may be agreed between the Employer and the Contractor prior to[3] the Contractor complying with any such instruction or carrying out such work; but if not so agreed there shall be added to or deducted from the Contract Sum an amount determined
   − under Option A in clause 3.7.1.2 or
   − if Option A is not implemented by the Contractor, under Option B in clause 3.7.1.2.

3.7.1   .2   Option A: Contractor's Price Statement
   Paragraph:
   A1   Without prejudice to his obligation to comply with any instruction or to execute any work to which clause 3.7.1.1 refers, the Contractor may within 21 days from receipt of the instruction or from commencement of work for which an Approximate Quantity is included in the Contract documents or, if later, from receipt of sufficient information[4] to enable the Contractor to prepare his Price Statement, submit to the Quantity Surveyor his price ('Price Statement') for such compliance or for such work.
   The Price Statement shall state the Contractor's price for the work which shall be based on[5] the provisions of clauses 3.7.2 to 3.7.10 and may also separately attach[6] the Contractor's requirements for:
   .1   any amount to be paid in lieu of any ascertainment under clause 4.11 of direct loss and/or expense not included in any previous ascertainment[7] under clause 4.11;
   .2   any adjustment to the time for the completion of the Works to the extent that such adjustment is not included in any extension of time[8] that has been made by the Architect/the Contract Administrator under clause 2.3.
   (See paragraph A)

A2   Within 21 days of receipt of a Price Statement the Quantity Surveyor, after consultation with the Architect/the Contract Administrator, shall notify the Contractor in
writing
either
.1   that the Price Statement is accepted;
or
.2   that the Price Statement, or a part thereof, is not accepted.

A3   Where the Price Statement or a part thereof has been accepted the price in that
accepted Price Statement or in that part which has been accepted shall in
accordance with clause 3.7.1 be added to or deducted from the Contract Sum.

A4   Where the Price Statement or a part thereof has not been accepted:
.1   the Quantity Surveyor shall include in his notification to the Contractor the
reasons[9] for not having accepted the Price Statement or a part thereof and
set out those reasons in similar detail to that given by the Contractor in his
Price Statement and supply an amended Price Statement which is acceptable
to the Quantity Surveyor after consultation with the Architect/the Contract
Administrator;
.2   within 14 days from receipt of the amended Price Statement the Contractor
shall state whether or not he accepts the amended Price Statement or part
thereof and if accepted paragraph A3 shall apply to that amended Price
Statement or part thereof; if no statement within the 14 day period is made the
Contractor shall be deemed not to have accepted, in whole or in part, the
amended Price Statement;
.3   to the extent that the amended Price Statement is not accepted by the Contractor, the Contractor's Price Statement and the amended Price Statement
may be referred either by the Employer or by the Contractor as a dispute or
difference to the Adjudicator[10] in accordance with the provisions of clause 9A.

A5   Where no notification has been given pursuant to paragraph A2 the Price Statement is deemed not to have been accepted, and the Contractor may, on or after the
expiry of the 21 day period to which paragraph A2 refers, refer his Price Statement
as a dispute or difference to the Adjudicator in accordance with the provisions of
clause 9A.

A6   Where a Price Statement is not accepted by the Quantity Surveyor after consultation with the Architect/the Contract Administrator or an amended Price
Statement has not been accepted by the Contractor and no reference to the
Adjudicator[11] under paragraph A4.3 or paragraph A5 has been made, Option B
shall apply.

A7   .1   Where the Contractor pursuant to paragraph A1 has attached his requirements[12] to his Price Statement the Quantity Surveyor after consultation with
the Architect/the Contract Administrator shall within 21 days of receipt thereof
notify the Contractor
.1   either that the requirement in paragraph A1.1 in respect of the amount to
be paid in lieu of ascertainment under clause 4.11 is accepted or that the
requirement is not accepted and clause 4.11 shall apply in respect of the
ascertainment of any direct loss and/or expense;
and
.2   either that the requirement in paragraph A1.2 in respect of an adjustment
to the time for the completion of the Works is accepted or that the
requirement is not accepted and clause 2.3 shall apply in respect of any
such adjustment.

A7   .2   If the Quantity Surveyor has not notified the Contractor within the 21 days
specified in paragraph A7.1, clause 4.11 and clause 2.3 shall apply as if no
requirements had been attached to the Price Statement.

3.7.1   .2 Option B
The Valuation shall be made by the Quantity Surveyor in accordance with the provisions
of clauses 3.7.2 to 3.7.10.

Valuation rules
The valuation rules are:

3.7.2 'priced document' as referred to in clauses 3.7.3 to 3.7.9 means, where the Second recital, alternative A applies, the Specification or the Schedules of Work as priced by the Contractor or the Contract Bills; and, where the Second recital, alternative B applies, the Contract Sum Analysis or the Schedule of Rates;

3.7.3 omissions shall be valued in accordance with the relevant prices in the priced document;

3.7.4 (a) except for work for which an Approximate Quantity is included in the Contract Documents, for work of a similar[13] character to that set out in the priced document the valuation shall be consistent with the relevant values therein[14] making due allowance for any change in the conditions[15] under which the work is carried out and/or any significant[16] change in quantity of the work so set out;
(b) for work for which an Approximate Quantity is included in the Contract Documents:
− where the Approximate Quantity is a reasonably accurate forecast of the quantity of work required the valuation shall be consistent with the relevant values in the priced document;
− where the Approximate Quantity is not a reasonably accurate forecast of the quantity of work required the relevant values in the priced document shall be the basis for the valuation and that valuation shall include a fair allowance for such difference in quantity;
provided that clauses 3.7.4(a) and 3.7.4(b) shall only apply to the extent that the work has not been modified other than in quantity;

3.7.5 a fair valuation[17] shall be made
− where there is no work of a similar character set out in the priced document, or
− to the extent that the valuation does not relate to the execution of additional or substituted work or the execution of work for which an Approximate Quantity is included in the Contract Documents or the omission of work, or
− to the extent that the valuation of any work or liabilities directly associated with the instruction cannot reasonably be effected by a valuation by the application of clause 3.7.3 or clause 3.7.4;

3.7.6 where the appropriate basis of a fair valuation is Daywork, the valuation shall comprise
− the prime cost of such work (calculated in accordance with the 'Definition of Prime Cost of Daywork carried out under a Building Contract' issued by the Royal Institution of Chartered Surveyors and the Building Employers Confederation (now Construction Confederation) which was current at the Base Date) together with percentage additions to each section of the prime cost at the rates set out by the Contractor in the priced document, or
− where the work is within the province of any specialist trade and the said Institution and the appropriate [m] body representing the employers in that trade have agreed and issued a definition of prime cost of daywork, the prime cost of such work calculated in accordance with that definition which was current at the Base Date together with percentage additions on the prime cost at the rates set out by the Contractor in the priced document;

---
[m] There are three Definitions to which the second sub-paragraph of clause 3.7.6 refers namely those agreed between the Royal Institution and the Electrical Contractors Association, the Royal Institution and the Electrical Contractors Association of Scotland and the Royal Institution and the Heating and Ventilating Contractors Association.

3.7.7 the valuation shall include, where appropriate, any addition to or reduction of any relevant items of a preliminary nature; provided that where the Contract Documents include bills of quantities no such addition or reduction shall be made in respect of compliance with an instruction as to the expenditure of a provisional sum for defined work* included in such bills;

---
[*] See footnote [bb] to clause 8.3 (*Definitions*).

3.7.8   no allowance shall be made in the valuation for any effect upon the regular progress of the Works or for any other direct loss and/or expense[18] for which the Contractor would be reimbursed by payment under any other provision in the Conditions;

3.7.9   if compliance with any such instructions substantially changes the conditions under which any other work is executed, then such other work shall be treated as if it had been the subject of an instruction of the Architect/the Contract Administrator requiring a Variation under clause 3.6 to which clause 3.7 shall apply; clause 3.7.9 shall apply to the execution of work for which an Approximate Quantity is included in the Contract Documents to such extent as the quantity is more or less than the quantity ascribed to that work in the Contract Documents and, where the Contract Documents include bills of quantities, to compliance with an instruction as to the expenditure of a provisional sum for defined work* only to the extent that the instruction for that work differs from the description given for such work in such bills;

---

[*] See footnote [bb] to clause 8.3 (*Definitions*).

3.7.10 where the priced document is the Contract Sum Analysis or the Schedule of Rates and relevant rates and prices are not set out therein so that the whole or part of clauses 3.7.3 to 3.7.9 cannot apply, a fair valuation shall be made.

COMMENTARY ON CLAUSES 3.6 AND 3.7

For a summary of the general law see earlier page 155.

Clause 3.6 empowers the architect to issue instructions requiring a variation. In addition, it permits the architect to sanction in writing any variation made by the contractor but not authorised by an instruction. Variations therefore can be the result of a prior instruction or a sanction after the event.

The meaning of the term variation is given in clauses 3.6.1 and 3.6.2. Briefly, it means firstly (under clause 3.6.1):

(1) Alteration or modification of design
(2) Alteration or modification of quality
(3) Alteration or modification of quantity.

and expressly includes:

(a) Additions, omissions or substitutions of any work
(b) Alterations to kinds or standards of materials or goods
(c) The removal of completed work or of materials or goods from the site other than work, materials or goods which are not in accordance with the contract.

Secondly (under clause 3.6.2) it means the imposition by the employer of obligations or restrictions, if not previously imposed by the employer in the specification/schedule of work/contract bills, and the addition to, or alteration or omission of, such obligations or restrictions whether previously imposed or not in regard to:

(1) Access to the site
(2) Limitations of working space
(3) Limitations of working hours
(4) The execution or completion of work in any specific order.

Note that the architect may issue an instruction changing any of the 'particulars' in relation to named persons as sub-contractors which may be contained in the contract documents. This will be 'regarded' as a variation, presumably even where the nature of the change in the particulars is outside the meaning given to variation in this clause, e.g. a change in the named sub-contractor's price for the work between the main contract being let and the sub-contract being entered into (see clause 3.3.1). The position is similar in relation to instructions given under clause 3.3.3(b).

Although there is no express restriction on the extent to which an instruction can vary the works, there must, it is submitted, be some point at which a so-called variation would change the contract so fundamentally that it would not be regarded as a variation within the meaning of the contract: see *Sir Lindsay Parkinson and Co.* v. *Commissioners of Works* (1949). Depending on the nature of the contract, it will often be the case that it would require variations extreme in content or number before the courts would declare the changes as falling outside the scope of the variation clause – see for example *McAlpine Humberoak Ltd* v. *McDermott International Inc. (No. 1)* (1992), considered earlier in this book (see page 33).

In relation to the matters set out in clause 3.6.2, the contractor can make reasonable objection to any instruction imposing or varying such matters – see the commentary to clause 3.5.1 on page 157.

Although it is common practice to issue a certificate of practical completion when some relatively unimportant items still remain to be completed, it is generally thought that the architect has no power to order variations after the issue of this certificate. Certainly the overall structure of the contract provides ample logic for this view.

Firstly, a variation instruction after practical completion of the works, for example, could result in the works being rendered thereby not sufficiently complete to warrant a certificate of practical completion, i.e. a variation in such circumstances could appear to undo an existing practical completion certificate. Secondly, the extension of time and loss and expense provisions are geared either to delays in completion, i.e. delays to practical completion, or disturbance to regular progress respectively, whereas it is difficult to see how the progress of the works can be affected once they are complete. It clearly makes good sense that the works should not be capable of being varied by a variation after they have been completed.

However, the consequence of this is that where the architect issues a certificate of practical completion even though there are some outstanding items of work, he probably cannot issue variation instructions relating thereto. As the outstanding work, if any, will often relate to final finishes which are a frequent area of variations, it could in strict contractual terms, though perhaps not often in practice, give rise to problems. The use of clause 2.11 permitting partial possession and providing for deemed practical completion of part may be a solution in appropriate circumstances (see page 109).

Valuation of variations and provisional sum work – approximate quantity, measurement and valuation

Clause 3.7 deals with the valuation of the architect's instructions requiring a variation issued under clause 3.6, including work which is to be treated as if it were a variation, as well as sanctioned variations and the valuation of architect's instructions issued under clause 3.8 covering the expenditure of provisional sums included in the contract documents. It also deals with the measurement and valuation of work executed by the contractor for which an approximate quantity is included in the contract documents. The clause lays down a series of steps which have to be followed when valuing variations. They are as follows:

(1)   A valuation may be agreed between the employer and the contractor prior to the contractor complying with the instruction;

(2)   A valuation may be obtained by means of the procedure involving the contractor's price statement (known as option A);

(3)   If option A does not produce a result then a valuation is made by the quantity surveyor using as a basis the relevant prices in the 'priced document';

       or

(4)   Where there are no relevant prices then the valuation is by means of a fair rate or value for the work;

       or

(5)   By daywork.

Added emphasis has been placed on the possibility of agreement between employer and contractor as to the price for the work, presumably to stress the simple nature of the work for which IFC 98 is intended, as compared with the valuation provisions of JCT 98 (clause 13.5). Although that possibility is covered in clause 13.4.1.1 of JCT 98, it is not given the same prominence as in this clause where it appears as the first option open to the parties to the contract. It is important to note that it is the employer, and not the architect or the quantity surveyor on his behalf, who must be a party to the agreement, although the architect or quantity surveyor may well act as agent of the employer in reaching an agreement if so authorised for this purpose. If this is the case it is suggested that the contractor should obtain confirmation of this authority for his own protection.

   If an attempted agreement between employer and contractor in relation to the price to be paid for a variation does not materialise, the contractor must nevertheless 'forthwith comply' with the architect's instruction (clause 3.5.1) unless it is an instruction requiring a variation of the kind described in clause 3.6.2. In the absence of an agreement, the instruction will be valued either in accordance with option A (contractor's price statement) or valued by the quantity surveyor (option B) in accordance with the valuation rules contained in clauses 3.7.2 to 3.7.10. These two options are looked at here.

Option A

Clause 3.7.1.2 provides in option A for a contractor's price statement. It introduces the facility for the contractor, if he wishes, to produce the statement as a means of seeking to establish the value of complying with instructions for variations or as to the expenditure of provisional sums. The main benefit of this procedure so far as the contractor is concerned is that, even if it does not lead to an agreement in respect of the valuation, it is a means by which any disagreement can be identified within a fairly short time span, following which it can be the subject of an adjudicator's decision in accordance with the provisions of clause 9A.

Option A is available in respect of not only instructions requiring a variation (including work treated as if it were subject to such an instruction), and instructions on the expenditure of provisional sums, but also in connection with work for which an approximate quantity is included in the contract documents.

Even though the contractor has a right to proceed down the option A route, in effect it is a joint decision of the parties as, if the quantity surveyor chooses not to respond, or even if he responds if the employer does not accept the contractor's price statement in any event, the matter can only proceed to adjudication or, in the absence of adjudication, down the option B route. It might appear from looking at clauses 3.7.1.2A.4.3 and 3.7.1.2A5 that if agreement is reached between the quantity surveyor and the contractor, this cannot be challenged by the employer in adjudication. It is submitted however that this is not the correct interpretation of these provisions. They are targeted at failures to agree between the quantity surveyor and the contractor. If the quantity surveyor and contractor reach agreement, this contractually must be based on and be consistent with the valuation rules set out in clauses 3.7.2 to 3.7.10. If the employer is of the opinion that the agreement reached is not properly based on the operation of the valuation rules there is nothing, it is submitted, to prevent the employer taking that issue to an adjudicator, or indeed to an arbitrator or the courts as appropriate under the contract.

If the contractor operates option A, it is clear that this in no way relieves him of the obligation to comply with the instruction.

In summary the option A procedure is as follows:

(1) The contractor receives the instruction or commences approximate quantity work. Within 21 days of this date or, if at that time the contractor has insufficient information to be able to prepare his price statement then within 21 days from receipt of sufficient information, the contractor submits to the quantity surveyor his price statement.

(2) The price statement must state the contractor's price for the work and this price must be based on the provisions of clauses 3.7.2 to 3.7.10, which contain the rules for valuation purposes.

(3) The contractor also has an option to separately attach to the price statement any amount he would accept in lieu of any ascertainment of direct loss and/or expense and any adjustment he would agree to by way of extension of time to the existing contractual completion date.

(4) Within 21 days of the quantity surveyor's receipt of the price statement, he must, after consulting the architect, notify the contractor in writing whether his price statement is accepted or alternatively in whole or in part not accepted. Presumably if the contractor is informed that his price statement is not accepted in part, this would implicitly mean that the remainder was accepted and would be treated accordingly.

(5) If the quantity surveyor has not responded to the price statement within the 21 days it is deemed not to have been accepted and the contractor may refer the matter to adjudication.

(6) To the extent that the price statement is accepted the resulting agreed sum is to be added to or deducted from the contract sum. If a sum is to be added to the contract sum, when does it become payable? It is submitted that unless the employer and contractor have agreed to it being paid in advance of the instruction being implemented, payment will be made through interim payment certificates as the work proceeds. As the agreed sum should have been based on the operation of the valuation rules, there should be little difficulty in determining what the interim payments should be.

(7) To the extent that the contractor's price statement is not acceptable the quantity surveyor must do two things:
(a) provide the contractor with the reasons for not having accepted the price statement in whole or in part, providing similar details to that given by the contractor in his price statement; and
(b) supply the contractor with an amended price statement which is acceptable to the quantity surveyor after having consulted the architect.

(8) The contractor within 14 days from receipt of the amended price statement must state whether or not he accepts it or any part of it. To the extent that it is accepted, the agreed sum will be added to the contract sum.

(9) If the contractor does not give his answer within the 14 day period he is deemed not to have accepted the amended price statement.

(10) To the extent that the amended price statement is not accepted it may be referred by the employer or the contractor to adjudication in accordance with clause 9A.

(11) If and to the extent that option A does not produce agreement between the quantity surveyor and the contractor then, if there has been no reference to an adjudicator, option B applies.

Loss and expense and extensions of time

In relation to the contractor's attempt to agree a sum in respect of loss and expense in lieu of an ascertainment under clause 4.11 and an extension of time to the contractual completion date, the quantity surveyor must consult with the architect and again, within 21 days of receipt of the contractor's requirements in this regard must notify the contractor whether or not the sum required by the contractor in lieu of an ascertainment of loss and expense is agreed and whether the contractor's suggested adjustment to the completion date is or is not agreed.

It is surprising not to find an express requirement for the quantity surveyor and architect to obtain the consent and approval of the employer in considering whether to agree to the contractor's requirements. After all, this is a departure from the established contractual procedures and does not appear to be conditioned by the principles contained in clauses 2.3 and 4.11.

Clause A7.1 appears to give the quantity surveyor and architect a very wide and unwarranted discretion. Contrast it, for example, with the employer's consent required under clauses 2.10 and 3.9 and also the requirement for agreement between contractor and employer under clause 3.7.1.1.

If the sum in lieu of an ascertainment of loss and expense is agreed, when is it to be paid? Unlike the situation in relation to the price statement, there is no reference to this amount being added to the contract sum and this may be unfortunate in terms of the administration of the contract. If it was intended to be an adjustment to the contract sum it would have been better to expressly provide for this. Looking at what makes up the adjusted contract sum under clause 4.2, this agreed sum does not easily slot into any category. Clause 4.2.2 refers to an amount being added to the contract sum under clause 4.11 (which deals with reimbursement of direct loss and expense) to the extent that such amounts have been ascertained. The agreed sum is in lieu of any such ascertainment. If it was to be added to the contract sum one could therefore have expected an express reference in clause 4.2.2 making this clear. If despite all of this, it is regarded as an adjustment to the contract sum then presumably it will be paid in the next following interim payment certificate even though the loss and expense in respect of which the sum has been agreed may not yet have been suffered or incurred. If it is not to be paid in the next certificate, it is difficult to determine when it should be paid as it is not based on actual loss and expense being incurred, with the result that payments cannot be related to the actual incurring of loss and expense by the contractor. If the sum agreed in lieu of an ascertainment is not to be added to the contract sum then presumably it must be paid separately within a reasonable time of it having been agreed; all the more reason why the employer's consent should have been expressly required.

There should however be no problems in the architect immediately granting any agreed extension of time.

To the extent that there is no agreement in relation to a sum in lieu of an ascertainment or an adjustment to the contractual completion date, this will be dealt with in the usual way under clause 4.11 (for the ascertainment) and clause 2.3 (for the extension of time). It is important to note that following any such non-agreement, there is no provision for the quantity surveyor, architect or employer to respond with a suggested sum in lieu of ascertainment or adjustment to the contractual completion date. Furthermore, there is no provision for adjudication in respect of any such non-agreement. This is because it is not in truth a dispute or difference under the contract; it is simply a facility which may be used if both parties agree. Additionally, so far as the loss and expense element is concerned, it would be inappropriate for an adjudicator to be able to determine a sum in lieu of loss and expense which would then have to be paid, quite possibly before any delay or disruption costs were actually incurred by the contractor. This is not the basis upon which IFC 98 is drafted.

If the quantity surveyor fails to notify the contractor one way or the other within 21 days of receipt of the contractor's requirements in respect of loss and expense and extensions of time then clause 4.11 dealing with loss and expense and clause 2.3 dealing with extensions of time will operate in the normal way as if the contractor had not indicated any such requirements.

If either the contractor does not avail himself of option A or, alternatively, option A does not produce an agreement and adjudication is not sought, the instruction will be valued in accordance with clause 3.7.1.2 – option B.

Option B

Clauses 3.7.2 to 3.7.10 introduce the valuation rules and can be summarised as follows:

(1) Clause 3.7.3: Omissions shall be valued in accordance with the relevant prices in the priced document;

(2) Clause 3.7.4: Except for work for which an approximate quantity is included in the contract documents, where the work is of a similar character to that set out in the priced document, the valuation shall be consistent with the values therein but allowance must be made for:
(a)   Changes in the conditions under which the work is carried out
(b)   Any significant change in quantity.
In respect of work for which an approximate quantity is included in the contract documents, then
(a)   Where the approximate quantity is a reasonably accurate forecast the valuation should be consistent with the relevant values in the price document
(b)   Where the approximate quantity is not a reasonably accurate forecast then the relevant values in the price document shall be used but only as a basis for the valuation and the valuation shall include a fair allowance in respect of the difference in quantity.
Clause 3.7.4 only applies to the extent that the work has not been modified other than in quantity. In other words, it only applies to the extent that a quantity whether approximate under clause 3.7.4(b) or firm under clause 3.7.4(a) has been modified. Therefore, where work for which an approximate quantity is included is varied in the usual way (i.e. apart from the quantity differing because of an inaccurate forecast) the usual valuation rules set out above and below (excluding 3.7.4(b)) will apply. Of course, even where there has been no variation at all, the approximate quantity work still has to be measured and valued, this being an exception to the general lump sum nature of IFC 98.

(3) Clause 3.7.5: A fair valuation shall be made:
(a)   Where there is no work of a similar character in the priced document
(b)   To the extent that the instruction calling for valuation does not involve the addition, substitution or omission of work, e.g. variations under clause 3.6.2
(c)   To the extent that the valuation of the work or liabilities directly associated with the instruction cannot reasonably be achieved by a valuation under (1) or (2) above.

(4) Clause 3.7.6: Where the appropriate basis of a fair valuation is daywork, the valuation shall comprise:
(a)   The prime cost of the work calculated in accordance with the 'Definition of Prime Cost of Daywork carried out under a Building Contract' issued by the RICS and the BEC (now the Construction Confederation) current at the base date stated in the appendix to IFC 98 (see clause 8.3), plus any percentage addition to each section of the prime cost at the rates set out by the contractor in the priced document; *or*

(b) Where there is specialist work and the RICS and the appropriate body representing the employers in that trade have issued an agreed definition of prime cost daywork then calculated in accordance with that definition plus any percentage as in (a) above.

There is no express provision for the submission of daywork vouchers by the contractor, unlike JCT 98. The submission of vouchers will, however, almost certainly be necessary and the lack of a time limit for their submission could create difficulties unless it is covered elsewhere in the contract documents.

(5) Clause 3.7.7: The valuation shall include, where appropriate, any addition to or reduction of any relevant items of a preliminary nature.

If it can be established that the instruction has had an effect on the allowances included by the contractor in his preliminaries, then an element of the valuation should include for relevant items from those preliminaries. This could prove difficult if a breakdown of the preliminaries has not been established prior to entering into the contract.

(6) Clause 3.7.8: There shall be no allowance made in the valuation for any effect on the regular progress of the works or for any other loss and expense reimbursable under any other provision in the conditions.

(7) Clause 3.7.9: If compliance with the variation instruction substantially changes the conditions under which any other work is executed, such other work shall be treated as if it had been the subject of a variation instruction.

It is not difficult to envisage an instruction from the architect which affects the whole operation of the contract, e.g. an instruction under clause 3.6.2 affecting the 'execution or completion of the work in any specific order'. The architect would therefore be well advised to consider the effect that his instructions requiring a variation may have on the valuation of other work before issuing the instruction, particularly when the instruction may result in a complete revaluation of all the work remaining to be done if the conditions under which it is executed change substantially. The employer's protection is in the word 'substantially'. The fact that the conditions for the carrying out of the other work change is not enough. The change must be substantial which, it is submitted, in this context means more than just significant. It is only a little way short of requiring completely different conditions.

This clause is expressly stated to apply to approximate quantity work to the extent that the approximate quantity is greater or less than the approximate quantity stated in the contract documents. This will be so whether the approximate quantity work itself is the subject of a variation or not.

(8) Clause 3.7.10: If the priced document is a contract sum analysis or schedule of rates not containing any relevant rates or prices so that either the whole or part of the valuation rules summarised in (1) to (7) above cannot apply, then a fair valuation shall be made. This clause can also apply to approximate quantity work contained in the contract sum analysis or schedule of rates where there are no relevant rates or prices.

Somewhat oddly, the contractor is given no express right to be present whenever it is necessary to measure work for valuation purposes – compare clause 13.6 of JCT 98.

Provisional sums for defined work under SMM7

Where the valuation relates to an instruction for the expenditure of a provisional sum for defined work included in the bills of quantities, the valuation rules in clauses 3.7.2 to 3.7.10 make it clear that:

- By clause 3.7.7 the valuation shall not include an addition to or reduction of any relevant preliminary items as these are deemed to have been allowed for in the contractor's programming, planning and pricing of preliminaries – see General Rule 10.4 of SMM 7. Clearly, if the instruction differs from the description of the work in the defined provisional sum then this may well amount to a variation to which clause 3.7.7 and indeed the other sub-clauses in clause 3.7 dealing with the valuation of variations will apply as appropriate;
- By clause 3.7.9 compliance with the instruction shall not be treated as changing the conditions under which any other work is executed so as to lead to that other work being treated as if it had been the subject of an instruction requiring a variation, except to the extent that the instruction for the defined provisional sum work differs from the description given for such work in the contract bills. Quite apart from this express exception, it is difficult to see how compliance with an instruction to expend a provisional sum for defined work which is strictly in line with the description given for such work could change the conditions under which other work is executed, any more than carrying out any fully described and measured item could. Clearly, if the instruction changes the description this would in any event be a variation under clause 3.6 so that the valuation rules in clauses 3.7.2 to 3.7.10 (including therefore clause 3.7.9) would apply anyway.

Reference should also be made to the commentary on clauses 1.4 and 1.5 (see page 48).

NOTES TO CLAUSES 3.6 AND 3.7

[1] '...may subject to clause 3.5.1 issue instructions...'
These must be in writing and be authorised by the conditions of contract, as required in clause 3.5.1.

[2] '...may ... sanction...'
This vests in the architect a considerable discretion. It is a limited licence to accept the contractor's unauthorised departures from the specified requirements for the contract works. Clearly, however, as with the power to vary generally, the authority of the architect to give such sanctions is limited in two ways. Firstly the sanction must be issued in accordance with this clause. Secondly, as between the architect and his client, the employer, his authority will depend on his terms of engagement which will need to contain some form of express authority from his employer/client. This is not a matter with which the contractor need concern himself, unless perhaps he is put on notice that the architect does not have such authority. It also of course enables the architect to regularise a previous orally requested variation with which the contractor has complied. If the architect declines to exercise the discretion inherent in the word 'may', it is difficult to see any remedy for the contractor under the contract. The adjudication and arbitration

provisions, though giving the adjudicator and arbitrator wide powers – see for example clauses 9A.5.5.2 and 9B.2 – may not extend that far.

[3] '...prior to...'
There would be nothing to prevent the employer and contractor agreeing the amount after compliance with the instruction. Strictly speaking, such an agreement, unless reached by operating the option A procedure, would be outside the terms of the contract.

[4] '...or, if later, from receipt of sufficient information...'
The contractor cannot be expected to put in a price statement based on the valuation rules if, despite having received the instruction, or having commenced approximate quantity work, he has insufficient information on which to prepare his valuation and price. However, there will often be much room for differences of opinion as to how much information is sufficient. In the event, however, it is most unlikely that it will pose a practical problem; for the employer or his professional advisors to argue the point seems a fruitless exercise.

[5] '...based on...'
This suggests that the valuation rules in clauses 3.7.2 to 3.7.10 shall form the basis of the contractor's price, but that he need not slavishly follow them. This slight room for manoeuvre is appropriate in terms of achieving agreement between the parties. Complete discretion would be inappropriate as, if there were failure to agree, an adjudicator would have no contractual framework against which to make his decision.

[6] '...separately attach...'
Is the separate attachment regarding loss and expense and extensions of time part of the contractor's price statement or not? The fact that it is to be separately attached suggests that it does not form part of the contractor's price statement. This is confirmed by the different treatment given to these two matters, namely no provision for a response from the employer's side if they are not agreed and no provision for adjudication in relation to any failure to agree. In any event, even if this separate attachment were to be regarded as part of the contractor's price statement, any failure to agree could not, within the provisions of option A, be the subject of an employer counter-offer or be a matter which could be referred to an adjudicator – see the provisions of clause A7.

[7] '...included in any previous ascertainment...'
This is aimed at avoiding the possibility of double recovery in respect of delay and disruption costs.

[8] '...included in any extension of time...'
This is aimed at avoiding any double counting for extension of time purposes.

[9] '...the reasons...'
These must be related to the quantity surveyor's valuation applying the valuation rules in clauses 3.7.2 to 3.7.10.

[10] '... to the Adjudicator ...'
The adjudicator must use the valuation rules in clauses 3.7.2 to 3.7.10 as a basis for his decision. It is submitted that this express power for the parties to refer the issue to an adjudicator is to avoid any doubt as to whether adjudication is appropriate to any failure to agree under option A. Without such an express provision it could well be the position that if there is no agreement under option A then option B would apply so that there would be contractually no dispute or difference which could be referred either to an adjudicator or an arbitrator. If a party wished to arbitrate rather than adjudicate it would be as well to simply let option B apply in accordance with clause A6 and then to arbitrate if that did not produce a satisfactory result. Even then the opposing party could of course seek adjudication following the application of option B.

[11] '... no reference to the Adjudicator ...'
There is no time limit placed on the parties in relation to seeking adjudication. Under clause 9A and indeed under section 108 of the Housing Grants, Construction and Regeneration Act 1996, a party must be entitled to '... give notice at any time of his intention to refer a dispute to adjudication' – section 108(2)(a). Having said this, as noted above, it may well be that a failure to agree under option A is not something which inherently would be a dispute or difference which could be referred to an adjudicator and can be so referred only as a result of the express contractual right to do so. On this basis a time limit could have been inserted and would perhaps have been sensible. Without such time limit how is it possible to know when option B applies? Presumably, therefore, even after option B has been applied, a party could seek an adjudication under option A. In practice, however, the result would probably be the same as adjudicating in respect of the result given under option B.

[12] '... his requirements ...'
This is a reference to the requirements referred to in the second paragraph of clause A1. It might have been sensible to expressly cross-reference it to make this more obvious.

[13] '... similar ...'
For this provision to operate fairly between the employer and the contractor, this word must in practice be treated as meaning of the same character. This can be demonstrated by looking at any particular item of work. The item of work will be covered by a description. If the instruction requiring a variation alters in any way that description of the work, it must thereby become different and may well in fairness be deserving of a different rate. To interpret the words in any other way will be to prevent the quantity surveyor from applying a fair valuation where the varied description of the item, although arguably remaining of a similar character to the original, justifies a different rate. Furthermore, the last few lines of clause 3.7.4 make it clear that due allowance for any change in conditions is made only where the work has not been modified other than in quantity so that the character of the work itself must remain unchanged.

[14] '... consistent with the relevant values therein ...'
If the contractor has made a mistake in his rate he will nevertheless generally be

bound by it. This could be to his advantage or disadvantage: *Dudley Corporation* v. *Parsons and Morrin* (1967). This view is reinforced by the provisions of clause 4.1, which provides that any error or omission, whether of arithmetic or not, in the computation of the contract sum shall be deemed to have been accepted by the parties to the contract.

[15] '...change in the conditions...'
This can include a change in site conditions, e.g. more restricted access, or in the elements, i.e. carrying out work at winter rates rather than the summer rates contained in the priced document. The location of the work may also be relevant here.

[16] '...significant...'
This is a matter of degree. Clearly a substantial change in quantity can affect the unit rate in the priced document.

[17] '...a fair valuation...'
This will include, in appropriate circumstances, a reasonable sum for overheads and profit. The overall pricing structure of the contract will be a relevant factor in determining what that profit level might be: see for example *Sanjay Lachhani and Another* v. *Destination Canada (UK) Ltd* (1996). It could also include the payment of damages under necessarily cancelled supply contracts or sub-contracts: see *Tinghamgrange Ltd* v. *Dew Group Ltd and North West Water Ltd* (1995). The 'fair' valuation is not to be carried out on a general or global basis. Each item of work should be particularised and priced: see *Crittall Windows Ltd* v. *T.J. Evers Ltd* (1996).

[18] '...or for any other direct loss and/or expense...'
The word 'other', and indeed the whole phrase, is somewhat confusing. To begin with, it equates the words 'any effect upon the regular progress of the works' with 'direct loss and/or expense', whereas they are of course different things. Direct loss and/or expense may or may not result from the regular progress of the works being affected. Further, if there is a disturbance of regular progress due to a variation instruction which involves the contractor in direct loss and/or expense, this is reimbursable under clause 4.12.7 and it is difficult to see how any other direct loss and/or expense (note the same phrase is used) can be suffered, let alone reimbursed by payment under any other provision in the conditions. Perhaps the words used are a somewhat abstruse reference to clauses 4.2 to 4.5 which provide the actual mechanism for payment in respect of loss and expense.

Clause 3.8

Instructions to expend provisional sums
3.8 The Architect/The Contract Administrator shall[1] issue instructions as to the expenditure of any provisional sums.

COMMENTARY ON CLAUSE 3.8

This imposes an express duty on the architect to issue instructions with regard to the expenditure of provisional sums whether by way of naming a sub-contractor under clause 3.3.2 or otherwise.

Since the introduction into IFC 84 of SMM 7 where bills of quantities are used (Amendment 4 of July 1988) provisional sums may be defined or undefined (see clause 8.3 – Definitions and the footnote to that clause where SMM 7 General Rules 10.1 to 10.6 are set out – page 370 of this book). Defined provisional sums have been commented on earlier in relation to clause 1.4 (page 49); when considering clause 2.4.5 (see note [15] page 101); and when considering the valuation of provisional sums (see page 170).

A provisional sum for undefined work is the sum provided for work where the information required in accordance with General Rule 10.3 (defined work) cannot be given. The possibility of a quantity surveyor intending to provide for an item as a provisional sum for defined work and unwittingly creating a provisional sum for undefined work and the possible significance of this has been considered earlier when commenting on clause 1.4 (see page 49).

NOTES TO CLAUSE 3.8

[1] '... shall ...'
This word is peremptory, and it is submitted that the architect is bound to issue instructions. It may be, and often is, decided not to expend a provisional sum item at all, in which case an instruction must be issued omitting it.

Clause 3.9

Levels and setting out
3.9   The Architect/the Contract Administrator shall determine any levels which may be required for the execution of the Works, and shall provide the Contractor by way of accurately dimensioned drawings with such information as shall enable the Contractor to set out the Works[1]. The Contractor shall be responsible for, and shall, at no cost to the Employer, amend, any errors arising from his own inaccurate setting out. With the consent of the Employer[2] the Architect/the Contract Administrator may instruct that such errors shall not be amended and an appropriate deduction[3] for such errors not required to be amended shall be made from the Contract Sum.

COMMENTARY ON CLAUSE 3.9

While it is for the architect to determine any levels which may be required and to provide the contractor, by accurately dimensioned drawings, with sufficient information for him to be able to set out the works, it is for the contractor to actually set them out accurately. If he fails to do so, any errors must be corrected, unless the employer consents to adopting the error, at no cost to the employer. In reaching any such compromise the employer should weigh carefully the possible implications of accepting an error in setting out, e.g. if it contravenes any planning consent or condition or if it could give rise to a third party claim such as for trespass or infringement of a right of light. If the error is to be adopted by the employer then an appropriate deduction is to be made from the contract sum. Such an adjustment will often be agreed on, but if not, it will not necessarily be easy to assess, and can, it is submitted, be the subject of a reference to adjudication or arbitration. See also Note [3] which follows.

From the wording of the clause it would appear that the contractor has no right to insist on amending his error instead of suffering a reduction in the contract sum.

NOTES TO CLAUSE 3.9

[1] '...to set out the Works'
In JCT 98 (clause 7) these words are followed by the words 'at ground level'. The absence of these words is no doubt meant to make it clear that the contractor's responsibility for accurately setting out also covers the situation where the contract works are built below ground level or above it, e.g. an underground or rooftop car park.

[2] 'With the consent of the Employer...'
If the contractor fails to set out the works accurately he will be in breach of contract. In such a case the architect would not normally have any apparent or ostensible authority to relieve the contractor of his breach of contract. These words make it clear that if the contractor is to be relieved from his contractual obligation to correct his inaccuracy, then the instruction of the architect can only operate with the consent of the employer who is, unlike the architect, a party to the contract. As the consequences of errors in setting out can vary from the insignificant to the disastrous, it is submitted that the express requirement for the employer's consent is reasonable. Even so, it may be pointed out that perhaps it is not appropriate to introduce something which should be dealt with in the terms of engagement between the employer and his architect, into a contract between the employer and the contractor, which is where the extent of the architect's authority is to be found so far as the contractor is concerned.

[3] '...an appropriate deduction...'
If the architect instructs the contractor not to amend an error in his setting out, he must make an adjustment downwards of the contract sum. He cannot add to the contract sum. Under some earlier wording in JCT 80 (clause 7 before Amendment 4 of July 1987) it was just about conceivable that the architect had power, on ordering the contractor to amend his error, to adjust the contract sum upwards. The words used in JCT 80 could have been interpreted as giving the architect apparent or ostensible authority, particularly in an extreme case such as demolish and rebuild, to pay the contractor for his own error. Employers would have been rightly dismayed at such a possibility.

What is an appropriate deduction? The cost to the contractor of correcting a serious setting out error could be very large, e.g. the whole building being one metre too far to the west. To correct this could require total demolition and rebuilding. If the architect with the consent of the employer instructs the contractor not to correct the error can this enormous saving to the contractor be taken into account? It is submitted not as this introduces a ransom element into the valuation. It will nevertheless be very difficult on occasions for the architect or quantity surveyor to discern the correct principles to apply in every case in carrying out the valuation. The valuation itself takes the form of an adjustment to the contract sum (see clauses 4.2.2 and 4.5). It is therefore a matter for the architect or quantity surveyor and not the employer to determine what is an appropriate deduction.

Clause 3.10

Clerk of works
3.10   The Employer shall be entitled to appoint a clerk of works whose duty shall be to act solely as an inspector on behalf of the Employer under the directions of the Architect/the Contract Administrator.

COMMENTARY ON CLAUSE 3.10

A competent clerk of works is a great asset on a building project and he is often an essential member of the construction team. However, the contract pays him scant regard. He acts on behalf of the employer solely as an inspector under the directions of the architect. By the wording of this clause, he has no power at all to give directions to the contractor. He can only report on what he inspects.

Under JCT 98 (clause 12) it is contemplated that the clerk of works might give directions to the contractor but these are stated to be of no effect unless confirmed in writing by the architect within two working days of being issued, and in any event such directions must be in regard to a matter in respect of which the architect is empowered under those conditions to issue an instruction. This power under JCT 98 for the clerk of works to give such directions appears to be at odds with the reference in clause 12 of that contract to his acting solely as an inspector.

Under IFC 98 there is no reference at all to a power to give directions, even subject to confirmation. However, whether under JCT 98 or IFC 98, it is suggested that if directions are given and complied with which vary the contract works, then the architect may sanction this in writing as a variation.

In the case of *Kensington and Chelsea and Westminster Area Health Authority* v. *Wettern Composites and Others* (1984), the employer was held vicariously liable for the negligence of a clerk of works in his employ even though he was under the direction and control of the architect. As a result the employer's damages claim for negligent inspection on the part of the architect was reduced by 20% on the basis of the contributory negligence of the clerk of works in failing to detect inadequate wall fixings by sub-contractors.

Clause 3.11

Work not forming part of the Contract
3.11   Where the Contract Documents provide for work not forming part of this Contract to be carried out by the Employer or by persons employed or engaged by the Employer, the Contractor shall permit the execution of such work on the site of the Works concurrent with[1] his execution of the Contract Works. Where the Contract Documents do not so provide the Employer may nevertheless with the consent of the Contractor (which

consent shall not be unreasonably delayed or withheld) arrange for the execution of such work.

Every person so employed or engaged shall for the purposes of clauses 6.1 (*Injury and damage*) and 6.3 (*Insurance of the Works*) be deemed to be a person for whom the Employer is responsible and not a sub-contractor.

COMMENTARY ON CLAUSE 3.11

This clause deals with the position where the employer wants to carry out or have carried out on his behalf, certain work, concurrently with the contract works. It should be noted that work carried out under clause 3.11 can give rise to the making of an extension of time under clause 2.4.8 and reimbursement of loss and expense under clause 4.12.3. The clause provides for two different situations, as given here.

(1) Where the contract documents identify work to be executed by the employer or by other persons engaged by the employer

In such a case, the contractor must permit the execution of such work on the site. Depending on the information provided in the contract documents, the contractor will be expected to make allowance for any interface problems which could reasonably be foreseen so that he should not be granted an extension of time or loss and expense for disturbance to regular progress caused by the carrying out of such work where the delays or the disturbance could be anticipated. The contractor should take this into account in his programme. On the other hand, if the extent of the delay or disturbance caused is more than could reasonably be foreseen, there will be an entitlement to an extension of time and loss and expense in appropriate circumstances.

(2) Where the contract documents do not identify work to be carried out by the employer or persons engaged by the employer

Here, the employer may only have such work carried out if the contractor consents. However, this consent cannot be unreasonably delayed or withheld. As the contractor will clearly be entitled to claim an extension of time in relation to delays to completion and reimbursement of loss and expense in relation to disturbance of regular progress as appropriate, it will, it is submitted, be exceptional for the contractor to have a reasonable ground for delaying or withholding consent.

Work executed by a public authority employer's direct labour organisation could fall within this clause provided it did not form part of the contract works. If, however, the direct labour organisation was engaged by the contractor to carry out part of the works, this clause would be inapplicable. See the discussion of this point earlier in considering clause 2.4.8 (page 91).

NOTES TO CLAUSE 3.11

[1] '... concurrent with ...'
This makes clear that the contractor's possession of the site may well not be exclusive and can, by the contract documents, be limited. This is dealt with in greater detail in Chapter 4 dealing with possession and completion.

Clauses 3.12 to 3.14

Instructions as to inspection – tests

3.12   The Architect/the Contract Administrator may issue instructions requiring the Con-
        tractor to open up for inspection any work covered up or to arrange for or carry out any
        test[1] of any materials or goods (whether or not already incorporated in the Works) or
        of any executed work. The cost of such opening up or testing (together with the cost of
        making good in consequence thereof) shall be added to the Contract Sum unless
        provided for in the Specification/Schedules of Work/Contract Bills or the inspection or
        test shows that the materials, goods or work are not in accordance with this Contract.

Instructions following failure of work etc.

3.13.1 If a failure of work or of materials or goods to be in accordance with this Contract is
        discovered during the carrying-out of the Works[2], the Contractor upon such discovery
        shall state in writing to the Architect/the Contract Administrator the action which the
        Contractor will immediately take[3] at no cost to the Employer to establish that there is
        no similar failure in work already executed or materials or goods already supplied
        (whether or not incorporated in the Works). If the Architect/the Contract Administrator:
        – has not received such statement within 7 days of such discovery, or
        – if he is not satisfied with the action proposed by the Contractor, or
        – if because of considerations of safety or statutory obligations he is unable to wait
          for the written proposals for action from the Contractor,
        the Architect/the Contract Administrator may issue instructions requiring the Con-
        tractor at no cost to the Employer to open up for inspection any work covered up or to
        arrange for or carry out any test of any materials or goods (whether or not already
        incorporated in the Works) or any executed work to establish that there is no similar
        failure and to make good in consequence thereof.
        The Contractor shall forthwith[4] comply with any instruction under this clause 3.13.1.

3.13.2 If, within ten days of receipt of the instruction under clause 3.13.1, and without pre-
        judice to his obligation to comply therewith, the Contractor objects to compliance
        stating his reasons in writing, and if within 7 days of receipt thereof the Architect/the
        Contract Administrator does not in writing withdraw the instruction or modify the
        instruction[5] to remove the Contractor's objection, then any dispute or difference as to
        whether the nature or the extent[6] of the opening up for inspection or testing instructed
        by the Architect/the Contract Administrator was reasonable in all the circumstances
        shall be and is hereby referred[7] to a person appointed pursuant to the procedures
        under this Contract relevant to the resolution of disputes or differences.
        To the extent that the person appointed finds the same is not fair and reasonable as
        aforesaid[8] he shall decide the amount, if any, to be paid[9] by the Employer to the
        Contractor in respect of compliance (together with making good in consequence
        thereof) and the extensions of time[10], if any, for completion of the Works to be made in
        respect of such compliance. Any amount so awarded by such person shall be a debt
        due and payable to the Contractor by the Employer.

Instructions as to removal of work etc.

3.14.1 The Architect/the Contract Administrator may issue instructions in regard to the
        removal from the site of any work, materials or goods which are not in accordance with
        this Contract.

3.14.2 If any work is not carried out, as required by clause 1.1, in a proper and workmanlike
        manner the Architect/the Contract Administrator may issue such instructions to the
        Contractor as are reasonably necessary as a consequence thereof with which the
        Contractor shall comply at no cost to the Employer.

COMMENTARY ON CLAUSES 3.12 TO 3.14

Clauses 3.12, 3.13 and 3.14 deal with the uncovering of work for inspection, the
testing of materials or goods, and the removal of defective work, materials or
goods not in accordance with the contract, together with the giving of instructions

by the architect in such situations as well as where work is not carried out in a proper and workmanlike manner. There are equivalent clauses in JCT 98, namely clauses 8.3, 8.4 and 8.5. However, the treatment of instructions for opening up and testing following discovery of work or materials or goods which are not in accordance with the contract is significantly different – see clause 8.4.4 of JCT 98 and the code of practice referred to therein. The JCT 98 approach to this problem is both more succinct and more helpful than the approach adopted in clause 3.13 of IFC 98.

Both clauses 3.12 and 3.13 deal with opening up and testing etc. The difference is that following a failure of work or materials or goods to be in accordance with the contract, the architect can choose to issue an instruction under clause 3.12 or 3.13, whereas in the absence of such a failure the architect can only issue an instruction for inspection or testing under clause 3.12. In addition, the financial consequences can vary depending on which clause is used.

(i) Inspections and tests etc. where there has been no previous failure

By clause 3.12 the architect can issue instructions requiring the contractor to open up covered work for inspection and to test materials or goods, whether or not incorporated in the works. Such testing can therefore clearly cover materials and goods on site, the ownership in which has passed to the employer under the provisions of clause 1.10, and no doubt extends to off-site materials or goods in which the ownership has passed under clause 1.11.

It may be that there is some specialist plant or equipment being manufactured off-site, the ownership in which has not passed but which the architect requires to be tested before being transported to site for installation. Provided such materials or goods are intended for incorporation into the works, then, it is submitted, even if the ownership in them has not yet vested in the employer, an instruction for testing under this clause can validly be given provided it is within the power of the contractor to comply with it. However, in such a case, to avoid any practical problems arising, the requirement for such testing should be clearly set out in the contract documents so that the contractor can make the necessary arrangements when entering into the sub-contract for the supply of such materials or goods. Such a requirement may appear as an item in section I and/or section II of NAM/T.

Costs of opening up or testing and making good: extensions of time; loss and expense

The cost of opening up or testing and of making good in compliance with instructions given under clause 3.12 is to be added to the contract sum except in two situations:

(1) Where the contract documents make provision for such cost
(2) Where the inspection of the uncovered work or the results of the testing show that the work or materials or goods do not comply with the contract. In this case the contractor bears the costs.

In (1) above therefore, on a literal interpretation, it appears that the employer pays even if the opening up or testing discloses a failure to meet the contractual requirements, whereas in (2) above the contractor pays.

This clause does not apply where the contractor has been required to price for specific testing within the contract sum and which does not stem from an architect's instruction. It appears therefore to relate to the expenditure of a provisional sum, in which case, it is submitted, the wording attached to the description of the provisional sum needs to be carefully considered if the cost of the testing etc. (which shows the work or materials not to be in accordance with the contract) is not to be borne by the employer (always bearing in mind the provisions of clause 1.3 (priority of contract documents)).

Clause 2.4.6 provides for an extension of time for the contractor if the opening up or testing causes delay to the completion date unless the results disclose a failure to meet the contractual requirements. In making an extension of time the architect should take into account any requirement in the contract documents as to opening up or testing. A reasonably competent contractor should have foreseen that time would be required for testing, and the contractor should have incorporated such time in his programme, and it is only any excess time that should be allowed.

Clause 4.12.2 provides for reimbursement to the contractor of loss and expense due to compliance with an instruction under clause 3.12 unless the results show a failure to comply with the contractual requirements. Again, in determining the amount of loss and expense, account should be taken of the extent to which a reasonably competent contractor could foresee that any requirement in the contract documents would, if not adequately reflected in his programme, result in some disturbance to the regular progress of the works.

There is nothing in the contract to prevent the architect from issuing an instruction under clause 3.12 even where a failure of work or materials or goods to be in accordance with the contract has been discovered. Whether or not an instruction in such circumstances should be given under clause 3.12 or 3.13 depends on a number of factors which are discussed in more detail below. In any event, if there has been such a failure the architect would be well advised to seek express instructions from the client if it is intended to issue instructions under clause 3.12 rather than 3.13.

(ii) Inspections and tests etc. following a failure

Where it is discovered during the carrying out of the works that there is a failure of work, materials or goods to be in accordance with the contract, the architect may be able to issue instructions requiring the contractor, at no cost to the employer, to open up for inspection any other covered work or to arrange for, or carry out, tests on any other materials or goods to establish that there is no similar failure. However, unless owing to considerations of safety or because of statutory obligations it is necessary to issue instructions immediately, the architect must wait for up to seven days following the discovery for the contractor to state what action he proposes to take (immediately and at no cost to the employer) to establish that there is no similar failure. If the architect is not satisfied with the contractor's proposed action, or if the contractor fails to put forward a proposed course of action, the architect can proceed to issue instructions.

Whether the contractor wishes to object to the instruction or not, he is obliged to comply with it forthwith. However, within 10 days of receipt of the instruction he may object to compliance, and must state his reasons in writing. If within seven days of the architect's receipt of this objection he does not either withdraw the instruction or modify it so as to remove the contractor's objection, any resulting dispute as to whether the nature or extent of the opening up was reasonable in all the circumstances is automatically referred to '. . . a person appointed pursuant to the procedures under this Contract relevant to the resolution of disputes or differences' (see clause 3.13.2). These words pose problems.

Firstly, the dispute resolution procedures may involve adjudication under clause 9A and/or arbitration under clause 9B or litigation under clause 9C. In other words there is a choice to be made and yet the dispute or difference '. . . shall be and *is hereby referred* . . .'. It is automatically referred, but to where? Perhaps the party wishing to pursue the dispute or difference can start a procedure off without this being regarded as falling foul of this clause. It is worth noting that IFC 84 in article 5 (prior to Amendment 12 of April 1998) sought to *automatically* refer disputes or differences to arbitration; nevertheless in practice the arbitration procedure still had to be instigated by the service of a notice to concur; it did not start automatically.

Secondly, so far as adjudication is concerned, there is another potential problem; clause 3.13.2 provides that the contractor can only initiate the procedures for resolving disputes [including therefore adjudication] if he has objected to compliance with the instruction within 10 days of its receipt, and also provided the architect has not within seven days of the objection withdrawn the instruction or modified it to remove the contractor's objection. While such a time-table is sensible, if the dispute or difference is to be adjudicated, there is a problem with these time limits as the requirement under section 108 of the Act is for either party to be able to give notice of intention to refer a dispute or difference to adjudication *at any time*.

If therefore the contractor objects, say, on the fourteenth day following receipt of the instruction, the employer may argue that it is then too late to object and the contractor must comply with the instruction with no recourse to either adjudication or arbitration on the basis that there is no dispute or difference as described in the clause. This may work for arbitration, but in terms of adjudication it may be seen as an attempt to circumvent the statutory right to adjudicate. If this is the case the contractor, despite the time limits, could object at any time. This would prove most unsatisfactory in practice where time is likely to be very important.

This is one of the problems of the statutory provision which enables a party to serve notice of intention to adjudicate 'at any time' – section 108(2)(a).

If this is the position, it may render the contractual scheme for adjudication in clause 9A invalid as a whole, the contract not providing adjudication provisions which meet the requirements of section 108. The result would be that the statutory Scheme for Construction Contracts Regulations 1998 (Part I – Adjudication) would apply not only to this dispute but all others. This is such a drastic result that it is difficult to believe a court would reach such a conclusion unless it was inescapable.

It may be that on a true interpretation of this clause it will not be seen as an attempt to circumvent the parties' statutory rights to adjudicate at any time. They can adjudicate at any time once a dispute or difference has arisen. A dispute or difference does not arise unless and until there has been a timely objection by the contractor and a failure by the architect to withdraw or accommodate the objection. So construed, the adjudication provisions in the contract would remain valid.

The only problem is that if this is the correct way to construe the clause as to what amounts to a dispute or difference, and more particularly when it arises, it might similarly be possible in many instances throughout the contract to impose stringent time limits as a condition to the creation of a dispute or difference and this could be used as a means of circumventing the parties' statutory rights. As such it would be of dubious validity.

A possible way through this problem might have been to introduce some form of conclusiveness provision in the same manner as the 28 day time limit in relation to the final certificate (see clause 4.7.1).

If the dispute resolution tribunal finds that the nature or extent of the opening up for inspection or testing was not fair and reasonable in all the circumstances, it will decide the amount, if any, to be paid by the employer to the contractor. It also has power to deal with the extensions of time to be awarded.*

The position therefore is that the contractor's compliance with the instructions issued under this clause is to be at no cost to the employer except where the tribunal determines otherwise, unless of course the employer and contractor reach an agreement without operating the dispute resolution procedures. The policy behind the clause appears to be reasonable. A failure of work, materials or goods to be in accordance with the contract will generally amount to a breach of contract by the contractor (some may say only a temporary disconformity – see Lord Diplock's Speech in *Hosier and Dickinson Ltd* v. *Kay (P & M) Ltd* (1972)). The architect may reasonably enough wish to find out if there are any further similar failures especially where there are structural or safety considerations involved. Even if the opening up etc. showed that there had been no further failure, it might reasonably be argued that the employer should not have to meet the cost associated with it and pay the contractor's loss and expense as he would under clause 3.12 (even if it was the contractor's breach of contract which necessitated the opening up etc.). Under clause 3.13 in such circumstances the contractor meets the cost subject to his right to object to compliance and to have that objection considered by an independent tribunal if the architect is unwilling to accept it.

Unfortunately the drafting of the clause is not particularly clear. It appears to be too detailed. For instance, the whole of clause 3.13.2 could be removed by a requirement in clause 3.13.1 that the instruction must be reasonable in all the circumstances, taking into account considerations of health and safety, structural significance and cost. As such it compares unfavourably with JCT 98 clause 8.4.4. The rather involved wording may well stem from a suspicion on the contracting side of the industry, whether well founded or not, that this clause opens the door for the architect to exercise an unfettered licence to spend the contractor's money. Alternatively, it could be seen to involve a disproportionate degree of risk for the contractor, and consequently sub-contractors, particularly in specialist areas, where the architect's training and experience may not have equipped him to exercise such potentially drastic powers in the most sensible way. However, a requirement that the instruction should be reasonable in all the circumstances would normally ensure that the architect issued such instructions responsibly. Some of the difficulties raised by the wording of this clause are dealt with under notes [1] to [10] below.

---

*See the footnote to page 390.

Under which clause should architect issue instructions?

Should the architect issue instructions under clause 3.12 or 3.13? Of course, if there has been no failure of work or materials etc. to be in accordance with the contract the architect can only issue an instruction for opening up or testing etc. under clause 3.12. He only has a choice where a failure has been discovered and he wishes to ascertain whether or not there are further similar failures.

Depending on which clause is relied on, the financial consequences can vary considerably. For instance, if the instruction is given under clause 3.12 then, if the opening up or testing etc. shows that work or materials are not in accordance with the contract, the contractor must bear the cost of the opening up and testing, together with the costs of any disruption etc. caused. He may also be liable to pay liquidated damages if the opening up or testing etc. causes delay. This may be so even if the nature and extent of the opening up or testing required by the architect is over and above what might be strictly necessary. If the result of the opening up or testing etc. is that the inspection or test shows that the work, materials or goods are not in accordance with the contract, that fact in itself will mean that the contractor meets the cost.

In the same situation, if the instruction were given under clause 3.13.1, the contractor must have first been given the opportunity to put forward his own proposals, unless there were considerations of safety or statutory obligations involved. Furthermore, following the instruction, the contractor can object to compliance and if the architect does not accommodate the contractor's objection then the contractor can take the matter to an independent tribunal for a decision. The tribunal may decide that the nature and extent of the inspection or testing etc. required was unreasonable, even if it shows that the materials, goods or work were not in accordance with the contract, and may make a decision or an award requiring the employer to pay part of the cost of this, including, it is submitted, disruption costs of the contractor.

On the other hand, if compliance with an instruction under clause 3.13 reveals that the materials, goods or work are in accordance with the contract, provided the nature and extent of the testing required was reasonable in all the circumstances, compliance with the instruction will still be at the contractor's cost. Without the benefit of hindsight the architect may be in something of a dilemma. Generally however, provided the architect feels sure that the nature and extent of the testing is reasonable in all the circumstances, the employer is probably better served by the architect seeking the contractor's proposed course of action under clause 3.13.1 and, if dissatisfied with this, issuing an instruction under that clause accordingly.

In any event the architect should make it abundantly clear under which clause he is actually issuing the instruction.

Removal of work

Under clause 3.14.1 the architect may issue instructions requiring the removal of any work, materials or goods not in accordance with the contract. It is not sufficient for the architect to issue an instruction simply condemning the work, materials or goods even if that instruction also refers to clause 3.14.1. He must expressly require its removal. In *Holland Hannen and Cubitts (Northern) Ltd* v. *Welsh Health Technical Services Organisation and Others* (1981) Judge John Newey said:

'...In my opinion, an architect's power [under clause 6(4) JCT 63] is simply to instruct the removal of work or materials from the site on the ground that they are not in accordance with the contract. A notice which does not require the removal of anything at all is not a valid notice under clause 6(4).'

An instruction under clause 3.14.1 is not of course a variation – see clause 3.6.1.

Work not carried out in a proper and workmanlike manner

Work or materials provided by the contractor may be in accordance with the contract and yet the work may not have been carried out in a proper and work-manlike manner, e.g. use of a tower crane which constitutes a trespass over adjoining air space; or damage to the soil (see for example *Greater Nottingham Co-Operative Society* v. *Cementation Piling & Foundations Ltd* (1989). This issue has been considered earlier in the commentary on and note to clause 1.1 (see page 42).

By clause 3.14.2, if work is not carried out in a proper and workmanlike manner the architect may issue such instructions to the contractor as are reasonably necessary as a consequence thereof. The contractor must comply with such instructions at no cost to the employer.

While clients and their architects may well see this as a sensible and appropriate power to have, it is nevertheless potentially extreme in nature. The only qualification is that the instruction must be reasonably necessary as a consequence of the failure to carry out work in a proper and workmanlike manner. The scope of the instructions is not limited by the clause. There is bound to be room for argument. For example, supposing the contractor in excavating ready for foundations, adversely affects the surrounding sensitive soil conditions in a manner which breaches the obligation to carry out work in a proper and workmanlike manner. The result is that the existing foundation design is rendered inadequate and additional works, say piling, will be necessary if the building is to proceed. However it would also be possible at considerably less cost for the building to be slightly relocated on the site so that the foundations will not be affected by the damage to the soil. Can the architect instruct the contractor to instal piles at very great cost in such circumstances? Put another way, firstly, is cost relevant in determining what is reasonably necessary, and secondly, is the employer entitled to insist on the contractor complying with his contractual obligations when to modify them slightly would avoid considerable cost? It is tentatively submitted that both matters are relevant in determining whether the architect's instruction is reasonably necessary as a consequence of the contractor's failure.

NOTES TO CLAUSES 3.12 TO 3.14

[1] '...any test...'
Bearing in mind the important financial consequences which flow from the result of any inspection or test instructed under clause 3.12, it may be very important to determine whether or not an instruction refers to a single test or in fact to a series of different tests. For example, if the instruction calls for an identical test on six critical welded joints, two of which are discovered to be unsatisfactory and four of which are satisfactory, are there six tests, two of which are at the contractor's

expense and four at the employer's expense? Or is there just one single test which has disclosed defective work and which will therefore be wholly at the contractor's expense? While it must clearly be a question of degree in all the circumstances of any particular case, it is tentatively submitted that generally the determining factor will be the nature of the test, i.e. critical welds in the above example rather than the number of times the same test is repeated.

[2] '... during the carrying-out of the Works...'

An instruction cannot be issued after the date of practical completion. While this must surely be right in relation to inspections and tests under clause 3.12 where there has been no failure (even though incidentally clause 3.12 is not expressly so limited), under clause 3.13 which operates following the discovery of a failure it might be a useful power for the architect to compel the contractor to investigate the extent of a problem which is discovered during the defects liability period. In complying with his obligations under clause 2.10 to make good defects at no cost to the employer, the contractor may well investigate the extent of any failures. If he does not, and it is not known for certain whether further work is similarly defective, the architect, if he has suspicions about the work, must presumably advise the employer to have any necessary inspection or tests etc. carried out at the employer's cost, leaving the employer to claim (whatever the result of the tests etc.) the cost from the contractor as special damages for breach of contract, if the expenditure was reasonably and properly incurred as a result of the contractor's breach.

[3] '... the action which the Contractor will immediately take...'

If the contractor wishes to put forward his own proposals to demonstrate that there is no similar failure he must do so before the architect issues an instruction under clause 3.13.1. Of course in practice there will no doubt often be an attempt to agree an appropriate course of action. If the contractor does not put forward any proposals in accordance with clause 3.13.1, he can always submit his proposals as part of his objection in writing under clause 3.13.2 after receiving the architect's instruction, and no doubt the architect will have to consider them accordingly.

[4] '... forthwith...'

Whether or not an objection is made by the contractor, he must comply forthwith with the instruction. The exercise by the contractor of his right to object does not entitle him to suspend compliance with the instruction.

[5] '... modify the instruction...'

This will enable the architect to alter the nature or extent of the opening up or testing required following receipt of the contractor's objection. If the nature of the contractor's objection is that he thinks it unreasonable that he should bear all of the costs of compliance with such an instruction, it is submitted that the architect has no power to modify his instruction so as to place all or part of the costs of compliance on the employer. The architect only has power under clause 3.13.1 to issue an instruction requiring the contractor to open up for inspection or test etc. at no cost to the employer. If the cost is to be shared or is to fall on the employer, this will require the express agreement of the employer outside the provisions of clause 3.13.

[6] '...the nature or the extent...'
These are perhaps the most important words in clause 3.13. The objection by the contractor is almost certainly going to be based on the nature or extent of the opening up or testing required. Much will depend on the nature and extent of the failure discovered. In some circumstances the adoption of progressive sampling techniques may suffice.

[7] '...hereby referred...'
No doubt if the matter is referred to the independent tribunal the architect ceases to have any function in connection with the resulting dispute and there is an immediate confrontation between employer and contractor, all within as little as 17 days of the issue of the instruction.

[8] '...fair and reasonable as aforesaid...'
This is odd and could be a defect in the drafting. It refers to a finding by the independent tribunal that the nature and extent of the testing is not fair and reasonable 'as aforesaid' and yet there is no reference earlier in the clause to the question of fairness. The only reference is to whether or not the nature and extent of the opening up or testing was reasonable in all the circumstances. What then if the tribunal finds that the nature and extent was reasonable although not fair? Without the introduction of the word 'fair', the tribunal would doubtless have to look at all the circumstances objectively and not subjectively, so that any sub-jective peculiarities of the parties, e.g. their respective financial positions, should not be taken into account. However, if the tribunal must in fact consider whether or not the nature and extent of the opening up or testing is fair in all the cir-cumstances, perhaps this introduces a more subjective element. It seems inap-propriate for the tribunal to determine whether or not the nature and extent of the opening up or testing is fair rather than reasonable in all the circumstances. It is to be hoped that the tribunal's consideration will be limited to what is reasonable only in all the circumstances.

The actual results of the inspection or tests should have no bearing on the tri-bunal's determination of whether the nature and extent of the opening up for inspection or testing was reasonable in all the circumstances. Reasonableness must be determined at the time when the instruction is given, or at the latest at the end of seven days from the receipt by the architect of the contractor's objection.

[9] '...the amount, if any, to be paid...'
This relates to the costs of any tests etc. including making good, and presumably also covers any other costs, e.g. loss and expense suffered by the contractor. The independent tribunal can no doubt apportion between the employer and the contractor the costs incurred in testing etc. together with any disruption costs incurred by the contractor.

[10] '...extensions of time...'
By clause 2.4.6 the contractor is entitled to an extension of time for compliance with an instruction under clause 3.13.1 where the effect of compliance has been

that the completion date has been delayed thereby, except where the result of the opening up or testing reveals that the work, materials or goods were not in accordance with the contract. However, even where the results show that the contractual requirements have not been complied with, so that the architect has no power to make an extension of time, the independent tribunal appears to be given such power. This is novel indeed. It means that the tribunal can extend the completion date even though there is no event within clause 2.4 which permits this. Was this intended?

Clause 3.15

Instructions as to postponement
3.15   The Architect/the Contract Administrator may issue instructions in regard to the postponement of any work to be executed under the provisions of this Contract.

COMMENTARY ON CLAUSE 3.15

The architect can issue an instruction to postpone any or all of the work. If he does so, the contractor is entitled to an extension of time under clause 2.4.5 and to reimbursement of loss and expense under clause 4.12.5. Further, if the postponement results in the carrying out of the whole, or substantially the whole, of the uncompleted works being suspended for a continuous period of one month, then unless that postponement was caused by reason of some negligence or default of the contractor, the contractor will be entitled to give the employer a default notice which could lead to the contractor terminating his own employment under the contract and enabling him to make various financial claims against the employer – see clauses 7.9.2(b), 7.9.3, 7.9.4 and clause 7.11.

It is unlikely that clause 3.15 can be used where there is a delay in giving initial possession of the site to the contractor. This view is supported by the inclusion in this contract of an express power for the employer to defer, for a limited period, the giving of possession of the site to the contractor, where clause 2.2 is stated in the appendix to apply.

An instruction to postpone may be implicit in the architect's request for the contractor to reprogramme all or part of his work even though clause 3.15 is not expressly referred to. In the case of *M. Harrison & Co. (Leeds) Ltd* v. *Leeds City Council* (1980) the main contractor was required to reprogramme part of his work by delaying it so as to enable a nominated sub-contractor to come on to the site earlier than had originally been planned. It was held that this amounted in effect to a postponement of the work, though no mention was made of that word or of clause 21(2) of JCT 63 which was the relevant contract in that case. (Clause 21(2) of JCT 63 is similar in terms to clause 3.15 of IFC 98.)

The acknowledgement by the architect of a *de facto* suspension of work may also apparently amount to an instruction to postpone – see the first instance decision in *Jarvis* v. *Rockdale Housing Association* (1987) where the contractor was unable to proceed with the works following the installation of defective piles by a nominated sub-contractor involving the need for redesign.

Chapter 7
Payment

CONTENT

This chapter deals with section 4 which concerns payment and related matters. It deals with the following:

The contract sum
Interim payments
Retention
The final certificate
Fluctuations
Disturbance of regular progress (including reimbursement of direct loss and/or expense).

SUMMARY OF GENERAL LAW

(A) Entire and severable contracts

In many everyday contracts, payment by one party is due only on the complete fulfilment by the other party of its contractual obligations. It may involve a contract for the supply of materials or goods or the performance of a service. Such contracts are known as entire contracts. A classic example of this is to be found in the very old case of *Cutter* v. *Powell* (1795).

Facts:

A sailor agreed the following terms in his employment contract:

> 'Ten days after the ship *Governor Pary* ... arrives at Liverpool I promise to pay to Mr. T. Cutter the sum of 30 guineas, provided he proceeds, continues and does his duty as Second Mate in the said ship from here to the Port of Liverpool ...'

The sailor died before completion of the voyage and his personal representatives sought to recover a proportionate part of the agreed remuneration. It was held that they could not do so as this was an entire contract so that the sailor had to continue carrying out his duties until the ship arrived at Liverpool and failure to do this, even though as a result of his death, disentitled his personal representatives to any part of the remuneration.

On the other hand, some contracts, by their terms or by their nature, permit final payment against an interim valuation of work done or on the completion of a stage of the work having been reached. These are known as severable contracts.

Most standard forms of building contract will contain express provisions for payment by instalments, though this does not, of itself, prevent them from being entire contracts as the provision for payment by instalments will usually be treated as being for payment on account of a final sum. The fundamental principle in relation to entire contracts, i.e. payment in full on completion in full, can still apply to the last instalment or the release of the final balance. However, if the contract expressly envisages that the contractor may be entitled to some payment notwithstanding his failure to complete, then this will mean that the contract in question is not an entire contract but is severable. See *Tern Construction Group Ltd (In Administrative Receivership)* v. *R B S Garages Ltd* (1992), a case on JCT 80 (containing similar provisions to IFC 98) in which Judge John Newey QC held that the contract was not an entire contract on the basis that it provided for the possibility that the contractor would be paid or credited with the value of work done even where he failed to complete the works in total. He said:

> 'In my judgment the contract made between the parties using the JCT Standard Form, with its elaborate and detailed provisions dealing with many matters, but most importantly employers going into partial possession, determination of the contractors' employment without determination of the contract and payment by instalments, was not simply a contract for the contractors to perform all or nearly all their obligations before the employers performed any of theirs, which can usefully be described as "entire".'

Even however in a contract which is not an entire contract, this need not mean that instalment payments are treated as several in nature. In other words, in JCT and similar contracts, the principle that instalment payments are treated as payments on account of a final sum which is finally adjusted by, in effect, a valuation of the whole of the works at completion, remains intact. A consequence of this is that instalment payments for work done or materials supplied create no estoppel against the employer if the work or materials are discovered subsequently to be defective. A later valuation can take this decrease in value into account. The mechanism by which interim instalments are paid is generally through the issue of interim payment certificates by the architect named in the contract. Subject always to the express terms of the contract, especially any arbitration or adjudication clause, production of an interim certificate will generally be a condition precedent to the employer's obligation to make payment.

In contracts for work and materials which involve work being carried out over a significant period of time, the courts will readily imply a term, in the absence of an express term, that the presumed intention of the parties is for interim payments to be made – see for example the first instance decision in *Williams* v. *Roffey Bros and Nicholls (Contractors) Ltd* (1989). The need for the implication of such a term in construction contracts (as defined) has now been largely removed by section 109 of the Housing Grants, Construction and Regeneration Act 1996 which provides that a party to a construction contract is entitled to payment by instalments unless the contract duration is of less than 45 days. If the parties have not in their contract stated the amount of, or the intervals at which, or circumstances in which, interim payments become due, the Scheme for Construction Contracts (England and Wales) Regulations 1998 will apply and part II of the Regulations provides in paragraphs 2 to 4 inclusive for payments based on 28 day cycles reflecting the value of work carried out including the value of materials on site. All relevant JCT contracts, including IFC 98, make express provision for payment by instalments so that these provisions in the statutory scheme will not apply to such contracts.

On a strict application of the law relating to entire contracts, a contractor carrying out work on the employer's land, which he fails to complete, will not be entitled to payment. This could of course provide an unexpected and possibly unwarranted benefit to the employer. The rigours of the operation of this principle are qualified by the law in a number of ways. Firstly, if in such a case the employer sues the contractor for breach of contract to recover the increased cost of having the work completed, he will, in the assessment of damages, have to give some credit against what he would have had to pay had the contract been properly performed. Secondly, the doctrine of substantial performance may aid the contractor. The essence of this doctrine is that, provided the contractor has substantially performed his obligations, he will be permitted to sue for the price, giving credit for the outstanding work left incomplete.

Where the contract expressly provides for payment by instalments, provided any conditions required to be fulfilled before an instalment becomes due have been met, a debt is created. This is likely to be so even if the relevant payment clause does not expressly state that the obligation to pay amounts to a debt: see *Re: Clemence Plc* (1992). If the contract requires a certificate to be issued in respect of an instalment this may be a condition precedent to the debt coming into existence.

(B) The certification process

Most standard forms of building contract allow for interim payments by means of the certificate of the architect. This may relate to a stage which is reached in the work, in which case there is little or no act of valuation required. On the other hand, many contracts provide for interim certificates against valuations of work carried out and usually goods and materials delivered to the site and sometimes those delivered off site but intended for incorporation into the works. The certificate will be issued by the architect and may be dependent on the valuation carried out by the quantity surveyor if one is appointed under the contract concerned. Depending on the wording of the contract it may well be that the issue of the certificate is a condition precedent to the employer's obligation to pay the contractor, so that no debt becomes due in favour of the contractor unless and until a certificate is issued for the amount being claimed – see for example *Lubenham Fidelities and Investment Co. Ltd* v. *South Pembrokeshire District Council and Another* (1986), a case which has however received some criticism on this aspect of it (*Hudson's Building and Engineering Contracts*, 11th edition, para 6.193). Most building contracts which provide for interim certificates state the period at which such certificates will be issued, e.g. monthly.

Even if the issue of a certificate from the architect is a condition precedent to the employer's obligation to pay, on the basis that no debt becomes due in favour of the contractor unless and until a certificate is issued for the amount being claimed, it will generally be possible, through the adjudication or arbitration process, to enable an adjudicator or an arbitrator to ascertain any sum which ought to have been the subject of or included in any certificate. If the parties choose to litigate, the court will have the same powers and accordingly the contractor can ask the court to consider whether a certificate should have been issued and to order payment accordingly: see *Beaufort Developments (NI) Ltd* v. *Gilbert-Ash NI Ltd and Others* (1997) overruling *Northern Regional Health Authority* v. *Derek Crouch Construction Co. Ltd* (1984).

The court will substitute its own machinery for resolving a claim even in the absence of a certificate, where the contractual machinery has completely broken down and is incapable of operating. Certainly, if the failure of the architect to issue a certificate is due to some breach of contract on the part of the employer, the court is likely to provide a remedy: see *Croudace Ltd* v. *London Borough of Lambeth* (1986).

If the architect negligently undercertifies in an interim certificate, is this a breach of contract on the part of the employer? Further, even if the employer is fully aware that the sum certified is lower than it should be, possibly even if due to an arithmetical error, is it in order for the employer to pay the certified sum and not thereby be in breach? The Court of Appeal case of *Lubenham Fidelities and Investment Co. Ltd* v. *South Pembrokeshire District Council and Another* (1986) would suggest that in connection with both matters the employer would not be in breach of the contract, at any rate not if there exists an appropriate method within the contract by which the contractor can then challenge the certificate, for example, an adjudication or arbitration clause. However, it is at least arguable that following some more recent cases, this contention may have been undermined. A possible interpretation of the cases of *John Barker Construction Ltd* v. *London Portman Hotel Ltd* (1996) and *Balfour Beatty Civil Engineering Ltd* v. *Docklands Light Railway Ltd* (1996) is that if the third party certifier exercises his decision-making powers in relation to such matters as interim payments and extensions of time, in a partial or unfair or even perhaps unreasonable way, this will put the employer in breach of an implied term that the employer will ensure that the third party certifier acts lawfully, fairly and reasonably.

It is possible that this departure from what appeared to be the established legal position was part and parcel of the attempts by various courts to avoid the restrictive effects on jurisdiction brought about by the case of *Northern Regional Health Authority* v. *Derek Crouch Construction Co. Ltd and Others* (1984). This last mentioned case has now been overruled by the *Beaufort Developments* case referred to above and it may now be therefore that the position as confirmed in the *Lubenham Fidelities* case remains good law. The *Lubenham* case has, however, received some criticism (see earlier page 190). The situation is otherwise if the employer has sought to put pressure on the architect to undercertify payment or the quantity surveyor to undervalue work. In such a case the court will be prepared to determine what amount should properly be certified: see for example *A.G. Machin Design and Build Contractors Ltd* v. *Long* (1992).

Even if the inclusion of an amount in an interim certificate is a condition precedent to the contractor's right to claim it, subject to arbitration and adjudication as referred to above, it does not work in reverse. In other words, if a sum is included in an interim payment certificate, this does not amount to conclusive evidence that such a sum is due to the contractor. Therefore, if the contractor seeks payment of the certified sum, e.g. by way of summary judgment in the High Court, assuming that there is no arbitration clause, it is open for the employer to maintain that there is an arguable defence to the claim, so preventing the contractor from obtaining judgment for the certified sum. The employer needs to establish to the satisfaction of the court that he can, in good faith and on reasonable grounds, raise an arguable contention that the certificate is open to challenge: see *R. M. Douglas Construction Ltd* v. *Bass Leisure Ltd* (1990).

The most common basis for an employer refusing to pay an interim payment certificate in whole or in part is that the value of work has been included in it which ought not to have been because it has not been carried out properly. If the payment certificate is issued by a third party certifier, the employer will generally require some independent expert evidence in support of his contention if he wishes to satisfy a court that the contractor should not be entitled to summary judgment for the certified sum. If the dispute goes to arbitration and the arbitrator is satisfied that the employer is not putting forward his defence in good faith or on reasonable grounds, or where it is clear on the employer's own figures that a certain sum is undoubtedly due then, exceptionally, the arbitrator can make an award in favour of the contractor without having a full hearing: see *The Modern Trading Company Ltd* v. *Swale Building and Construction Ltd* (1990) and Arbitration Act 1996 especially sections 33 and 34.

Defective work

In relation to the issue of a payment certificate, it is undoubtedly the case that the certifier owes a duty of care to the employer. If there is included within the certificate payment for work which is defective and which the certifier ought to have known was defective and which thereby renders the valuation excessive, the certifier will be liable to the employer in damages. Further, there is no immunity from liability because of the apparent quasi-judicial role of the certifier, i.e. his need to act independently using his own professional judgment in the valuation and certificate process: see *Sutcliffe* v. *Thackrah* (1974). In that particular case the defendant acted both as architect and quantity surveyor.

If the certifier negligently undervalues the work done, thereby causing the contractor loss, the question of whether he may be liable to the contractor for a breach of duty of care under the tort of negligence is difficult to answer. On the one hand there is judicial support of considerable authority for the view that a duty of care is owed by the architect to the contractor in such circumstances to avoid negligently causing financial loss, e.g. Lord Salmon in the case of *Arenson* v. *Casson, Beckman* (1977) who said:

> '... The architect owed a duty to his client, the building owner, arising out of the contract between them to use reasonable care in issuing his certificate. He also, however, owed a similar duty of care to the contractor arising out of their proximity, see *Hedley-Byrne*. In *Sutcliffe* v. *Thackrah* the architect negligently certified that more money was due than was in fact due and he was successfully sued for the damage which this had caused his client. He might, however, have negligently certified that less money was payable than was in fact due, and thereby starved the contractor of money. In a trade in which cash flow is especially important, this might have caused the contractor serious damage for which the architect could have been successfully sued.'

This view was subsequently endorsed, firstly in the excellent judgment of Mr Justice Hunter in the Hong Kong case of *Shui On Construction Co. Ltd* v. *Shui Kay Co. Ltd* (1985), and secondly in the English case of *Michael Salliss & Co. Ltd* v. *E.C.A. Calil and F.B. Calil and William F. Newman & Associates* (1987), a decision in which the Official Referee said:

> 'But it is self evident that a contractor who is a party to a JCT contract looks to the architect ... to act fairly as between him and the building employer in matters such as certificates and extensions of time. Without a confident belief that that reliance will be justified, in an industry where cash flow is so important to the contractor, contracting could be a hazardous operation. If the architect unfairly promotes the building employer's interest by low certification or merely fails properly to exercise reasonable care and skill in his certification it is reasonable that the contractor should not only have the right as against the owner to have the certificate reviewed in arbitration but also should have the right to recover damages against the unfair architect.'

However, serious doubt has been cast over the *Sallis* case by the Court of Appeal in the case of *Pacific Associates, Inc. and Another* v. *Baxter and Others* (1988) in which it was held that an engineer under the FIDIC form of international civil engineering contract did not owe the contractor or sub-contractor any duty of care in relation to his duty to act impartially in dealing with contractual claims. A decision that there was no duty of care was quite understandable on the facts of that case. It is perhaps unfortunate, however, that judgments were not so qualified. Instead they extend to general principles so that at the present time it is unlikely that a duty will be owed except in the most exceptional circumstances. The prospect of an architect, in exercising functions under the building contract, in which he is clearly required to act impartially, owing a legal contractual and no doubt tortious duty to his employer client on the one hand but only at best some sort of moral duty to the contractor on the other, can only bring the law into contempt.

It is submitted that the existence of such a legal duty owed by an architect to the contractor would be sound in law and also accord with common sense.

It might also be possible to argue that a third party certifier who exercises his independent functions under the contract in a manner which is not lawful, or not fair, or perhaps not reasonable, may put his client in breach of the contract thereby entitling the contractor to sue the employer client for damages (see earlier page 191).

Although in the *Sutcliffe* case there was no quantity surveyor acting, nevertheless the role of the quantity surveyor in valuing, and the relationship between the quantity surveyor and the architect, were considered in detail in the first instance judgment of Judge Stabb QC in the *Sutcliffe* case under the name of *Sutcliffe* v. *Chippendale and Edmondson* (1971). The judge said as follows:

> 'I readily acknowledge and accept that any prolonged or detailed inspection or measurement at an interim stage is impracticable, and not to be expected. On the other hand ... the issuing of certificates is a continuing process, leaving each time a limited amount of work to be inspected and I should have thought that more than a glance round was to be expected .... The quality of the work was ... the responsibility of the architect and never that of the quantity surveyor and

since work properly executed is the work for which the progress payment is being recommended, I think that the architect is in duty bound to notify the quantity surveyor in advance of any work which he, the architect, classifies as not properly executed, so as to give the quantity surveyor the opportunity of excluding it.'

As a matter of practice it is advisable to have such notification in writing and also for the quantity surveyor, on his part, to notify the architect of any apparently defective work when carrying out his valuation.

(C) Retention

Standard forms of building contract which provide for payment by instalments will usually enable the employer to retain a certain percentage of the total value included in the interim certificate. This percentage is often fixed at three or five per cent, with the first half of it released at practical or substantial completion and the remaining half after the making good of any defects at the end of the defects liability period.

From the employer's point of view it is a useful system as it represents some protection against the inclusion of defective work in a valuation and which is therefore included in the amount of an interim certificate. It also provides security for the performance by the contractor of his obligations. Its main purpose, however, is to provide the employer with a fund during the defects liability period following practical or substantial completion, should the contractor fail to return and make good any defects of which he is notified.

Often, the express terms of the contract will give to the retention fund the status of trust money. In other words, the employer will hold the retention fund in a fiduciary capacity as trustee of it for the benefit of the contractor. In such a case, provided the retention money is kept as a separate identifiable fund (and it is the duty of a trustee to keep trust funds separate from other funds), a trustee in bankruptcy or a liquidator of the employer will not be able to include the retention fund as part of the employer's estate. It will have to be held for the benefit of the contractor and released to him at the appropriate time. A case in point is *Re: Tout and Finch* (1954).

Facts:

A sub-contract was entered into under the NFBTE/FASS ('green' form). Clause 11(h) provided that the main contractor's interest in retention money withheld under the sub-contract was held by him in a fiduciary capacity as trustee for the sub-contractor. The main contractor went into voluntary liquidation after completion but before the defects liability period had expired and before, therefore, the release of the final balance of the retention money.

Held:

The provisions of clause 11(h) created an equitable assignment of the appropriate part of the retention money in favour of the sub-contractor. In other words, the beneficial interest in the retention fund belonged in equity to the sub-contractor, and even though at the time that the contractor went into liquidation the retention money had not been released to the sub-contractor, the court regarded it, in equity, as having already been so transferred. Accordingly, the sub-contractor was entitled to it.

The protection of the trust money from the creditors of the employer (or the main contractor in the case of a sub-contract) is a very significant benefit to the party concerned. However, this benefit can be lost should the employer (or main contractor as the case may be) not place the retention money in a separate identifiable fund. If this is not done and the fund cannot be identified, then the other contracting party will rank only as an ordinary creditor without any priority in respect of the retention money due to him.

If the contract does not contain any express provisions dealing with the status of retention money, then it is submitted that it will not be treated as trust money in the hands of the employer.

More is said about the trust status of retention in the commentary on clause 4.4 (see page 210).

(D) Deductions from payments: set-off and counterclaim

Authorised deductions

If a building contract provides for payments against interim certificates it will also usually entitle the employer to make certain deductions, generally in relation to specific ascertained amounts. The most important such entitlement will be the employer's right to deduct liquidated and ascertained damages. However, other deductions may also be authorised, e.g. recovery of insurance premiums paid by the employer where the contractor has failed to take out any necessary insurance required by the contract; and the cost of employing another contractor to carry out work which the contractor has failed to do despite having received a valid instruction.

Even where the contract expressly authorises a deduction or withholding, if it is achieved by being set against sums otherwise due under the contract, prior notice will be required pursuant to section 111 of the Housing Grants, Construction and Regeneration Act 1996. The notice must be given not later than the 'prescribed period' before the final date for payment. The contract may provide for such a period. If it does not then by virtue of the Scheme for Construction Contracts (England and Wales) Regulations 1998 (Part II) paragraph 10, the prescribed period is not later than seven days before the final date for payment under the contract. The notice must also set out the amount proposed to be withheld and the grounds for withholding it.

Set-off and counterclaim

An equitable set-off is a cross claim which is in the nature of a defence. Strictly so-called, it is to be used only as a shield to an action and not as a cause of action in its own right. It is used only as a defence and not as a counterclaim as such, so that if the claim were abandoned, the set-off would not stand on its own in the same way as a counterclaim. It is not always easy to determine if a cross claim is a set-off or a counterclaim or both. This subject received detailed consideration in the House of Lords case of *Hanak* v. *Green* (1958).

Lord Morris said that a cross claim was to be treated as a set-off if it arose:

> '... directly under and affected the contract on which the plaintiff herself relies .... Some counterclaims might be quite incompatible with the plaintiff's claim, in no way connected with it and wholly unsuitable to be used as a set-off, but the present class of action involving building or repairs, extras and incidental work so often leads to cross claims for bad or unfinished work, delay or other breaches of contract that a set-off would normally prove just and convenient. ...'

A cross claim may not amount to a set-off but may nevertheless be the subject of a counterclaim, e.g. where it arises out of a different and separate contract between the same parties, not being part of a series of similar contracts.

A vital question concerning cross claims is the extent to which one party can withhold payment to the other by reliance on a cross claim, even to the extent of a refusal to pay certified sums. The terms of the contract in question may expressly deal with such rights. Any such terms will need careful drafting so as not to fall foul of the requirements of section 111 of the Housing Grants, Construction and Regeneration Act 1996, referred to above. Most JCT forms of contract and sub-contract and also those linked to them, e g. NAM/SC, the JCT Standard Form of Sub-Contract Conditions for Sub-Contractors named under IFC 98, and DOM/1, the domestic form of sub-contract for use between domestic sub-contractors and main contractors using the JCT 98 Standard Form of Building Contract, have express terms which mirror the provisions of section 111 of the Act, though the prescribed period is shorter than that which would otherwise apply, namely seven days, under the Scheme for Construction Contracts (England and Wales) Regulations 1998.

(E) Contractor's claims arising under the contract and for breach of contract

Most standard forms of building contract expressly entitle the contractor to recover sums in addition to the value of the work done when he is involved in extra cost, expenditure or losses due to certain matters, some of which may also amount to a breach of contract by the employer. Frequently, words are used such as 'direct loss and/or expense', 'direct loss and/or damage', 'increased cost' etc. The extent of recovery by the contractor will depend on the precise words used. However, most standard forms of contract also preserve, either expressly or by their silence on the point, the contractor's common law rights to claim damages for breach of contract.

The general legal principles on which damages are awarded for breach of contract are considered later in Chapter 10, dealing with determination of the employment of the contractor. The application of these general principles is likely to overlap to a considerable degree with any discussion of the particular principles on which recovery can be claimed under the express wording of the contract itself. Accordingly, these particular principles are discussed in detail in relation to the specific relevant clauses in IFC 98 and any possible difference between recovery under the express terms of the contract and at common law will, where appropriate, be highlighted.

CONSIDERATION OF THE RELEVANT CLAUSES OF IFC 98

Clause 4.1

Contract Sum
4.1 The Contract Sum shall not be adjusted or altered in any way otherwise than in accordance with the express provisions of the Conditions and subject to clause 1.4 (*Inconsistencies*) any error or omission, whether of arithmetic or not, in the computation of the Contract Sum shall be deemed[1] to have been accepted by the parties hereto.

COMMENTARY ON CLAUSE 4.1

The contract sum is referred to in article 2 of the agreement attached to the conditions of contract. It is expressed as being exclusive of value added tax and it can only be varied in the manner specified in the conditions (see clause 4.5). Clause 4.1 expressly provides that the contract sum is not to be adjusted or altered in any way other than in accordance with express provisions of the conditions of contract so that there can be no adjustment of the contract sum by implication.

Subject to the exception contained in clause 1.4 (see page 47) the parties are bound by any errors or omissions incorporated in the contract sum unless there exists sufficient ground to persuade a court to grant the equitable remedy of rectification. This remedy is discretionary and the courts will have to be satisfied that the written contract fails in some way to express what was the clear intention of both parties. It does this firstly in order that the written document properly reflects the agreement actually made between the parties, and secondly to prevent a party unfairly holding the other to a written contract which he knows does not accurately reflect the agreement reached. Cases on the rectification of building contracts are rare but one such case is *A. Roberts and Co. Ltd* v. *Leicestershire County Council* (1961).

Facts:

The contractor's revised tender contained a completion period of 18 months. The county council decided that the period for completion should be 30 months, that is, the same date but one year later than the date put forward by the contractor. The county council did not refer to any date in its letter of acceptance. Instead, the formal contract, when drawn up, contained a completion date one year later than that put forward by the contractor. The contractor did not notice the change of year and sealed and returned the contract. Before the county council itself sealed the formal contract, it held a meeting with the contractor during which the contractor referred to his plans to complete in 18 months. The county council's officers did not mention the later date inserted in the formal contract. The county council subsequently sealed the formal contract.

Held:

Rectification would be ordered. A contracting party is entitled to rectification of a contract if he can prove that he believed a particular term to be included and the other party concluded the contract without that term being included in the knowledge that the first party believed that it was included.

The exception in clause 1.4 referred to in this clause deals with inconsistencies, errors or omissions in and between contract documents and other documents issued under clause 1.7 and clause 3.9, and departures from the Standard Method of Measurement referred to in clause 1.5. This is a significant exception and the reader is referred to those clauses and the commentary on them.

It is important to distinguish the contract sum from the contractor's tender sum. The sum put forward by the contractor in his tender does not become the contract sum until the tender is accepted. Depending on the tendering procedures adopted, the tender sum may be adjusted by agreement between the parties, e.g. on the discovery of an error in the make-up of the tender sum. Alternatively, the contractor may be asked to elect to maintain the tender sum notwithstanding the error or to withdraw his tender.

NOTES TO CLAUSE 4.1

[1] '. . . shall be deemed . . .'
In other words, the parties shall be regarded as having been aware of the errors or omissions and, subject of course to clause 1.4, to have accepted them.

Interim and final payments and the final certificate

Clauses 4.2 to 4.8

Interim payments

4.2   (a) Subject to any agreement between the parties as to stage payments, the Architect/ the Contract Administrator shall, at intervals of one month, unless a different interval is stated in the Appendix, calculated from the Date of Possession stated in the Appendix,[1] certify the amount of interim payments to be made by the Employer to the Contractor specifying to what the amount relates and the basis on which that amount was calculated and the final date for payment pursuant to a certificate shall be 14 days from the date of issue of each certificate[2].

If the Employer fails properly to pay[3] the amount, or any part thereof, due to the Contractor under the Conditions by the final date for its payment the Employer shall pay to the Contractor[4] in addition to the amount not properly paid simple interest thereon for the period until such payment is made. Payment of such simple interest shall be treated as a debt due to the Contractor by the Employer. The rate of interest payable shall be five per cent (5%) over the Base Rate of the Bank of England which is current at the date the payment by the Employer became overdue. Any payment of simple interest under this clause 4.2(a) shall not in any circumstances be construed as a waiver by the Contractor of his right to proper payment of the principal amount due from the Employer to the Contractor in accordance with, and within the time stated in, the Conditions or of the rights of the Contractor in regard to suspension of performance of his obligations under this Contract to the Employer pursuant to clause 4.4A or to determination of his employment pursuant to the default referred to in clause 7.9.1(a).

4.2   (b) Clause 4.2(b) does not apply where the Employer is a local authority. Where it is stated in the Appendix that clause 4.2(b) applies, the advance payment identified in the

Appendix shall be paid to the Contractor on the date stated in the Appendix[5] and such advance payment shall be reimbursed to the Employer by the Contractor on the terms stated in the Appendix[6]. Provided that where the Appendix states that an advance payment bond is required such payment shall only be made if the Contractor has provided to the Employer such bond from a surety approved by the Employer on the terms agreed between the British Bankers' Association and the JCT and annexed to the Appendix unless pursuant to the Fifth recital a bond on other terms is required by the Employer.

4.2 (c) Interim valuations shall be made by the Quantity Surveyor whenever the Architect/ the Contract Administrator considers them to be necessary for the purpose of ascertaining the amount to be stated as due in an interim payment.

Without prejudice to the obligation of the Architect/the Contract Administrator to certify the amounts due in interim payments, the Contractor, not later than 7 days before the date of a certificate of interim payment, may submit to the Quantity Surveyor an application which sets out what the Contractor considers to be the amount of the valuation pursuant to clauses 4.2.1 and 4.2.2. If the Contractor submits such an application the Quantity Surveyor shall make an interim valuation. To the extent that the Quantity Surveyor disagrees with the valuation in the Contractor's application the Quantity Surveyor at the same time as making the valuation shall submit to the Contractor a statement, which shall be in similar detail[7] to that given in the application, which identifies such disagreement.

4.2 (d) The amount of the interim payment to be certified shall be the total of the amounts in clauses 4.2.1 and 4.2.2, at a date not more than 7 days before the date of the certificate, less the amount of any advance payment or part thereof due for reimbursement to the Employer in accordance with the terms for such reimbursement stated in the Appendix pursuant to clause 4.2(b) and less any sums previously certified for payment[8]. These amounts are:

4.2.1 95% of

Work properly executed
(a) the total value of the work properly executed[9] by the Contractor including any work so executed to which Option B in clause 3.7.1.2 applies or to which a Price Statement or any part thereof accepted pursuant to clause 3.7.1.2 paragraph A2 or amended Price Statement or any part thereof accepted pursuant to clause 3.7.1.2 paragraph A4.2 applies but excluding any restoration, replacement or repair of loss or damage and removal and disposal of debris which in clauses 6.3B.3.5 and 6.3C.4.4 are treated as if they were a Variation, together with, where applicable, any adjustment of that value under clause 4.9(b) (*Formulae adjustment*). Where it is stated in the Appendix that a priced Activity Schedule is attached thereto the value of the work to which the Activity Schedule relates shall be the total of the various sums which result from the application of the proportion of the work[10] in an activity listed in the Activity Schedule properly executed to the price for that work as stated in the Activity Schedule;

Materials and goods delivered to Works
(b) the total value of the materials and goods which have been reasonably and properly[11] and not prematurely delivered to or adjacent to the Works for incorporation therein and which are adequately protected against weather and other casualties[12];

Off-site materials or goods - 'the listed items' [n]
(c) the value of any materials or goods or items prefabricated for inclusion in the Works which have been listed by the Employer in a list supplied to the Contractor and annexed to the Contract Bills/the Specification/Schedules of Work ('the listed items'), before delivery thereof[13] to or adjacent to the Works, provided that the following conditions have been fulfilled:

---

[n] See also clause 1.11

.1 the Contractor has provided the Architect/the Contract Administrator with reasonable proof[14] that the property in uniquely identified[15] listed items is vested in the

Contractor so that, pursuant to clause 1.11, after the amount in respect thereof included in a certificate for interim payment has been paid by the Employer[16], the uniquely identified listed items shall become the property of the Employer; and, if so stated in the Appendix, has also provided from a surety approved by the Employer a bond in favour of the Employer on the terms agreed between the JCT and the British Bankers' Association and annexed to the Appendix unless pursuant to the Fifth recital a bond in other terms is required by the Employer;

.2   the Contractor in respect of listed items which are not uniquely identified[17] has provided the Architect/the Contractor Administrator
with reasonable proof that the property in such listed items is vested in the Contractor so that, pursuant to clause 1.11, after the amount in respect thereof included in a certificate for interim payment has been paid by the Employer, such listed items shall become the property of the Employer; and
the Contractor has provided from a surety approved by the Employer a bond in favour of the Employer on the terms agreed between the JCT and the British Bankers' Association and annexed to the Appendix unless pursuant to the Fifth recital a bond in other terms is required by the Employer;

.3   the listed items are in accordance with this Contract;

.4   the listed items at the premises where they have been manufactured or assembled or stored
either
are set apart
or
have been clearly and visibly marked individually or in sets by letters or figures or by reference to a pre-determined code
and identify
.1   the Employer and to whose order they are held; and
.2   their destination as the Works;

.5   the Contractor has provided the Employer with reasonable proof that the listed items are insured against loss or damage for their full value under a policy of insurance protecting the interests of the Employer and the Contractor in respect of the Specified Perils[18], during the period commencing with the transfer of property in the items to the Contractor until they are delivered to, or adjacent to, the Works.

4.2.2   100% of
any amounts payable to the Contractor or to be added to the Contract Sum under clauses
2.1   (*Insurance premium*)
3.12   (*Inspection*)
4.9(a)   (*Tax etc. fluctuations*)
4.10   (*Fluctuations: named persons*)
4.11   (*Disturbance of progress*)
5.1   (*Statutory obligations*)
6.2.4   (*Insurance – liability etc. of Employer*)
6.3A.4   (*Insurance monies*)
6.3B.2   (*Insurance premiums*)
6.3B.3.5   (*Restoration etc. of loss or damage*)
6.3C.3   (*Insurance premiums*)
6.3C.4.4   (*Restoration etc. of loss or damage*),
to the extent that such amounts have been ascertained, together with any deduction made under clause 3.9 (*Levels*) or 4.9(a) or 4.10.

4.2.3   (a) Not later than 5 days after the date of issue of a certificate of interim payment, the Employer shall give a written notice to the Contractor which shall, in respect of the amount stated as due in that certificate of interim payment, specify the amount of the payment proposed to be made, to what the amount of the payment relates and the basis on which the amount is calculated.
(b) Not later than 5 days before the final date for payment of the amount due pursuant to

clauses 4.2.1 and 4.2.2 the Employer may give a written notice to the Contractor which shall specify any amount proposed to be withheld and/or deducted from that due amount, the ground or grounds for such withholding and/or deduction and the amount of the withholding and/or deduction attributable to each ground.

(c) Where the Employer does not give a written notice pursuant to clause 4.2.3(a) and/ or to clause 4.2.3(b) the Employer shall pay[19] the amount due pursuant to clauses 4.2.1 and 4.2.2.

Interim payment on Practical Completion

4.3   (a) The Architect/The Contract Administrator shall, within 14 days after[20] the date of Practical Completion, certify payment to be made by the Employer to the Contractor of $97\frac{1}{2}$% of the total value referred to in clause 4.2.1(a) and 100% of any amounts payable pursuant to clause 4.2.2 together with any deduction under clauses 3.9 (*Levels*) or 4.9(a) (*Tax etc. fluctuations*) or 4.10 (*Fluctuations: named persons*) less the amount of any advance payment made pursuant to clause 4.2(b) and less any sums previously certified for payment. The final date for payment of the amount pursuant to the certificate shall be 14 days from the date of issue of the certificate.

(b) Not later than 5 days after[21] the date of issue of the certificate the Employer shall give a written notice to the Contractor which shall, in respect of the amount stated as due in that certificate, specify the amount of the payment proposed to be made, to what the amount of the payment relates and the basis on which that amount is calculated.

(c) Not later than 5 days before the final date for payment of the amount due pursuant to clause 4.3(a) the Employer may give a written notice to the Contractor which shall specify any amount proposed to be withheld and/or deducted from that due amount, the ground or grounds for such withholding and/or deduction and the amount of the withholding and/or deduction attributable to each ground.

(d) Where the Employer does not give a written notice pursuant to clause 4.3(b) and/or to clause 4.3(c) the Employer shall pay[19] the amount due pursuant to clause 4.3(a).

(e) If the Employer fails properly to pay the amount, or any part thereof, due to the Contractor under the Conditions by the final date for its payment the Employer shall pay to the Contractor in addition to the amount not properly paid simple interest thereon on the terms set out in clause 4.2(a).

Interest in percentage withheld

4.4   Where the Employer is not a local authority[22] the Employer's interest in the percentage of the total value not included in the amounts of the interim payments to be certified under clauses 4.2 and 4.3 shall be fiduciary as trustee for the Contractor (but without obligation to invest) and the Contractor's beneficial interest therein shall be subject only[23] to the right of the Employer to have recourse thereto[24] from time to time for payment of any amount which he is entitled under the provisions of this Contract to withhold and/or deduct from any sum due or to become due to the Contractor.

Right of suspension by Contractor

4.4A   Without prejudice to any other rights and remedies which the Contractor may possess if the Employer shall, subject to any notice issued pursuant to clause 4.2.3(b) or 4.3(c)[25], fail to pay the Contractor in full by the final date for payment as required by the Conditions and such failure shall continue for 7 days after the Contractor has given to the Employer, with a copy to the Architect[26]/the Contract Administrator, written notice of his intention to suspend the performance of his obligations under this Contract to the Employer and the ground or grounds on which it is intended to suspend performance, then the Contractor may suspend such performance of his obligations under this Contract to the Employer until payment in full occurs. Such suspension shall not be treated as a suspension to which clause 7.2.1(a) refers or a failure to proceed regularly and diligently with the Works to which clause 7.2.1(b) refers.

Computation of adjusted Contract Sum

4.5   Not later than 6 months after Practical Completion[27] of the Works the Contractor shall provide the Architect/the Contract Administrator, or, if so instructed by the Architect/the Contract Administrator, the Quantity Surveyor, with all documents reasonably required for the purposes of the adjustment of the Contract Sum. Not later than 3 months after receipt by the Architect/the Contract Administrator or the Quantity Surveyor as the case may be of the aforesaid documents a statement of all the final Valuations under clause 3.7 (*Valuation of Variations*) shall be prepared by the Quantity Surveyor and a copy of such statement and a copy of the computations of the adjusted Contract Sum shall be sent forthwith to the Contractor.

The adjustment of the Contract Sum shall be in accordance with clause 3.7 and, where applicable, clause 4.9(b), and with the amounts, together with any withholdings[28] and/or deductions, referred to in clause 4.2.2 as finally ascertained less all provisional sums and the value of any work for which an Approximate Quantity is included in the Contract Documents and any amount to be deducted under clause 2.10 or under clause 3.9.

Issue of final certificate

4.6.1   .1 The Architect/The Contract Administrator shall, within 28 days of the sending of such computations of the adjusted Contract Sum to the Contractor or of the certificate issued by the Architect/the Contract Administrator under clause 2.10 (*Defects liability*), whichever is the later, issue a final certificate certifying the amount due to the Contractor or to the Employer as the case may be and stating to what the amount relates and the basis on which the amount due under the final certificate has been calculated. The amount to be certified shall be the Contract Sum adjusted as stated in clause 4.5 less the amount of any advance payment made pursuant to clause 4.2(b) and less any sums previously certified for payment.

4.6.1   .2 Not later than 5 days[29] after the date of issue of the final certificate the Employer shall give a written notice to the Contractor which shall, in respect of any amount stated as due to the Contractor from the Employer in the final certificate, specify the amount of the payment proposed to be made, to what the amount of the payment relates and the basis on which that amount is calculated.

4.6.1   .3 The final date for payment of the said amount by the Employer to the Contractor or by the Contractor to the Employer as the case may be shall be 28 days from the date of issue of the said certificate. Not later than 5 days before the final date for payment of the said amount the Employer may give a written notice to the Contractor which shall specify any amount proposed to be withheld and/or deducted from any amount due to the Contractor, the ground or grounds for such withholding and/or deduction and the amount of withholding and/or deduction attributable to each ground.

4.6.1   .4 Where the Employer does not give a written notice pursuant to clause 4.6.1.2 and/or to clause 4.6.1.3 the Employer shall pay[19] the Contractor the amount stated as due to the Contractor in the final certificate.

4.6.2   If the Employer or the Contractor fails properly to pay the amount due, or any part thereof, by the final date for its payment the Employer or the Contractor as the case may be shall pay to the other in addition to the amount not properly paid simple interest thereon for the period until such payment is made. The rate of interest payable shall be five per cent (5%) over the Base Rate of the Bank of England which is current at the date the payment by the Employer or by the Contractor as the case may be became overdue. Any payment of simple interest under this clause 4.6.2 shall not in any circumstances be construed as a waiver by the Contractor or by the Employer as the case may be of his right to proper payment of the aforesaid amount due from the Employer to the Contractor or from the Contractor to the Employer in accordance with this clause 4.6.

4.6.3   Liability for payment of the amount pursuant to clause 4.6.1 and of any interest pursuant to clause 4.6.2 shall be treated as a debt due to the Contractor by the Employer or to the Employer by the Contractor[30] as the case may be.

Effect of final certificate

4.7.1   The final certificate for payment shall be conclusive, except for any matter which is the subject of proceedings including adjudication commenced before or within 28 days[31] after the date of the final certificate for payment,

- that, where and to the extent that any of the particular qualities of any materials or goods or any particular standard of an item of workmanship was described expressly in the Contract Drawings or in either the Specification or the Schedules of Work or the Contract Bills, or in any of the Numbered Documents, or in any instruction issued by the Architect/the Contract Administrator under the Conditions, or in any drawings or details issued by the Architect/the Contract Administrator under clause 1.7 or 3.9, to be for the approval of the Architect/the Contract Administrator, the particular quality or standard was to the reasonable satisfaction of the Architect/the Contract Administrator, but such certificate shall not be conclusive that such or any other materials or goods or workmanship comply or complies with any other requirement or term of this Contract, and

- that any necessary effect has been given to all the terms of this Contract that require additions to adjustments of or deductions from the Contract Sum, save in regard to any accidental inclusion or exclusion of any item[32] or any arithmetical error in any computation, and
- that all and only such extensions of time, if any, as are due under clause 2.3 have been given, and
- that the reimbursement of direct loss and/or expense, if any, to the Contractor pursuant to clause 4.11 is in final settlement of all and any claims which the Contractor has or may have arising out of the occurrence of any of the matters referred to in clause 4.12 whether such claim be for breach of contract, duty of care, statutory duty or otherwise.

4.7.2 Where pursuant to clause 9A.7.1 either Party wishes to have a dispute or difference on which an Adjudicator has given his decision on a date which is after the date of issue of the final certificate finally determined by arbitration or legal proceedings, either Party may commence arbitration or legal proceedings within 28 days of the date on which the Adjudicator gave his decision[33].

Effect of certificates other than final

4.8 Save as provided in clause 4.7, no certificate of the Architect/the Contract Administrator shall of itself be conclusive evidence that any work, materials or goods to which it relates are in accordance with this Contract.

COMMENTARY ON CLAUSES 4.2 TO 4.8

Clause 4.2 commences with reference to the possibility of stage payments, although no detailed machinery is provided, as it is in the case of interim certificates dependent on valuations. For the simple type of contract for which IFC 98 is envisaged, it might have been anticipated that machinery for stage payments as a basis for interim payment would have been given equal weight to detailed valuations.

Advance or mobilisation payments

Provision is made in clause 4.2(b) for the employer to make an advance or mobilisation payment. This is dealt with by means of an appendix item which should state the date for the payment, the amount and also the terms on which the advance payment is to be reimbursed. The contract goes on to provide that in calculating the amount to be included in interim payment certificates, there should be netted off the amount of any advance payment in accordance with the agreed reimbursement terms.

The appendix may also require the contractor to put in place an advance payment bond. If this requirement is included in the contract, the contractor is not entitled to the advance payment unless and until such bond has been provided. The bondsman has to be approved by the employer and the terms of the bond will generally be those agreed between the Tribunal and the British Bankers' Association. This form of bond is annexed to the appendix to the conditions (a copy of the Advance Payment Bond is set out in Appendix 1 to this book – see page 407). Should the employer and contractor agree on a different form of bond, this is covered by the fifth recital to the articles of agreement which requires the contractor to have been given a copy before entering into the contract. For all intents and purposes the bond is of the 'on demand' rather than conditional type. In other words, provided the formalities for calling the bond are strictly complied with, the bondsman will pay under the bond even if there is a dispute between the employer and the contractor as to whether the employer is entitled to call the bond. However, even under 'on demand' bonds, a call can be rejected if there is evidence of bad faith, e.g. fraud or deceit.

It is important to note that the optional advance payment provisions are expressed not to apply where the employer is a local authority.

The machinery for interim payments

Except where stage payments have been agreed, interim certificates regulate the interim payments to be made to the contractor by the employer. Unless otherwise stated in the appendix, certificates of the amount of interim payments are to be issued by the architect at intervals of one month starting from the date of possession stated in the appendix. The architect's certificate must specify as to what the amount relates and the basis on which that amount was calculated (clause 4.2(a)). This requirement is stated in clause 4.2.3(a) (interim payment certificates prior to practical completion); clause 4.3(b) (interim payment on practical completion); and clause 4.6.1.2 (payment on the issue of final certificate). It relates to the employer's obligation not later than five days after the issue of a certificate to state what sum is proposed to be paid, to what the amount of the payment relates and the basis on which the amount is calculated, in order that the contract complies with the requirements of section 110 of the Housing Grants, Construction and Regeneration Act 1996 requiring construction contracts to provide an adequate mechanism for determining what payments become due and when.

The employer must pay the contractor the due amount within 14 days of the date of the certificate. This is expressly stated to be the final date for payment in respect of interim certificates. For the final certificate the final date for payment is 28 days from the issue of the certificate. Section 110 of the Act requires construction contracts to provide for a final date for payment in respect of any sums which become due.

If the employer fails to pay in full the amount due under any payment certificate by the final date for payment, the contractor is entitled, by clause 4.2(a), to simple interest thereon for the period of non-payment. The rate of interest is 5% over the base rate of the Bank of England current at the date when the payment became overdue. It is expressly stated that this entitlement to interest in no way acts as a waiver by the contractor of his right to the proper payment at the proper time; nor does it affect the contractor's right, in respect of a failure to pay interim certificates, including that payable on practical completion, to suspend performance of his obligations in connection with the non-payment pursuant to clause 4.4A or to determine his employment pursuant to the determination provisions in clause 7.9.1(a). So far as failure to pay in respect of the final certificate is concerned, as the contractor will by then have fulfilled his contractual obligations, there is no necessity to include a right to suspend or to determine employment (see clause 4.6.2).

The amount to be included in interim payment certificates

It is for the architect to certify, on a monthly basis, the amount of interim payments. The architect will generally instruct the quantity surveyor to make the interim valuations and also to ascertain any reimbursement of loss and expense to which the contractor may be entitled (see clauses 4.11 and 4.12). While the contract states that it is the quantity surveyor who shall carry out the valuation, clause 4.2(c) goes on to state that he shall only do so whenever the architect considers such valuations to be necessary for the purpose of ascertaining the amount to be stated as due in an interim payment. In other words, the architect need not use the quantity surveyor for carrying out interim valuations if such valuations are not necessary in determining the amount of the interim payment certificate. Only rarely will an interim payment not involve an interim valuation, e.g. where the interim payment deals only with reimbursement to the contractor of loss and expense. However, this principle that valuations will only be made when the architect considers them necessary, is subject to the contractor's right to submit to the quantity surveyor an application setting out what the contractor considers should be included pursuant to clauses 4.2.1 and 4.2.2.

If the contractor makes such an application not later than seven days before the date of a certificate for payment, the quantity surveyor must make an interim valuation and, to the extent that there is any disagreement with the valuation in the contractor's application, the quantity surveyor must submit to the contractor a statement identifying the disagreement. The quantity surveyor must supply an explanation for such disagreement in similar detail to that provided by the contractor in respect of the valuation. If the contractor chooses not to make the application this in no way relieves the architect of his obligation to certify amounts due in respect of interim payments at the appropriate time.

In practice it is normal for contractors to make applications for interim payments even where this is not contractually necessary.

By virtue of clause 4.2(d), a valuation, if any, must be carried out not more than seven days before the date of the certificate. The amount included in the interim payment certificate is to be the total of the following less sums previously certified for payment and less the amount of any advance payment due for reimbursement in accordance with the terms set out in the appendix:

(1) *95 per cent of:*
    (a) the value of the work (including varied work) properly executed by the contractor, together with, where applicable, any adjustment of the value under the formulae adjustment for fluctuations (see 4.9(b) and supplemental condition D) (*per clause 4.2.1(a)*)
    (b) the value of materials and goods reasonably and properly (and not prematurely) delivered to or adjacent to the works for incorporation therein, provided they are adequately protected against weather and other casualties (*per clause 4.2.1(b)*)

(c) the value of off-site goods and materials in respect of items listed by the employer in a list supplied to the contractor and annexed to the contract bills, the specification or schedules of work (*per clause 4.2.1(c)*).

In respect of clause 4.2.1(c) the inclusion of the value of any listed items before their delivery to or adjacent to the works depends on certain conditions having been fulfilled:

(i) the contractor providing reasonable proof of ownership so that following payment by the employer of the sum properly due in respect thereof, the property in the listed items will become the property of the employer under clause 1.11. This question of title and the passing of it between main contractor and employer has been dealt with earlier in considering clause 1.11 (see page 58).

(ii) the contractor providing a bond in favour of the employer from a surety approved by the employer either on terms agreed between the Tribunal and the British Bankers' Association (which is annexed to the appendix) (a copy of this form of bond is set out in Appendix 2 to this book – see page 410) or in other terms as required by the employer. If such other terms are required then the details need to be recited by reference to the fifth recital to the articles of agreement. IFC 98 in its terms requires such a bond to be provided as a standard contractual obligation in respect of items in the list which are uniquely identified. Where the items are not uniquely identified then the bond is only supplied if the requirement for it is stated in the appendix to the contract. The bond is in the nature of an 'on demand' bond as is the case also in respect of the advance payment bond (for which see earlier page 203). Reference should be made to the commentary on clause 1.11 where the differing bond requirements in respect of uniquely specified items and items which are not uniquely specified is discussed (see page 58).

(iii) the listed items for which payment is sought meeting the contractual requirements as to quality etc.

(iv) the listed items being at the premises where they have been manufactured, assembled or stored and either set apart or clearly and visibly marked individually or in sets by letters or figures or by reference to a predetermined code. They must also identify the employer and to whose order they are held as well as their destination as the works.

(v) finally, the contractor providing reasonable proof of insurance against loss or damage for the full value of the listed items under an insurance policy protecting the employer and the contractor in respect of 'Specified Perils'. This cover must operate from the time when the listed items become the contractor's property until they are delivered to, or adjacent to, the works.

(2) *100 per cent of any amounts payable to the contractor to be added to the contract sum under the following clauses:*

2.1	(insurance premium)
3.12	(inspection)
4.9(a)	(tax etc. fluctuations)

4.10   (fluctuations: named persons)
4.11   (disturbance of progress)
5.1   (statutory obligations)
6.2.4   (insurance: liability etc. of employer)
6.3A.4   (insurance monies)
6.3B.2   (insurance premiums)
6.3B.3.5 (restoration etc. of loss or damage)
6.3C.3   (insurance premiums)
6.3C.4.4 (restoration etc. of loss or damage)
to the extent ascertained, together with any deduction made under clauses 3.9 (levels) or 4.9(a) (tax etc. fluctuations) or 4.10 (fluctuations: named persons) (per clause 4.2.2).

Notice of proposed payment and of proposed withholding or deduction from payments due

Clause 4.2.3 deals with the requirements of sections 110 and 111 of the Housing Grants, Construction and Regeneration Act 1996.

Clause 4.2.3(a) provides that not later than five days after the issue of a certificate the employer must give a written notice to the contractor stating, in respect of the amount stated as due in the certificate, the amount which the employer proposes to pay together with the basis on which payment is being made. This is intended to meet the requirements of section 110(2) of the Act. However, the Act itself states that the notice must be given not later than five days after the date on which a payment becomes due, i.e. the due date (the date of the certificate) or the date on which a payment would have become due if:

(1) The other party had carried out his obligations under the contract; *and*
(2) No set-off or abatement was permitted by reference to any sum claimed to be due under one or more other contracts.

These requirements are implicitly covered by IFC 98. As for the first requirement, this requires the notice to be given not later than five days after a sum would have become due if the other party had carried out his obligations under the contract. So, if no sum is actually due, for instance where the value of the works has remained unchanged or has decreased (e.g. following the discovery of defective work) but it would have become due had the contractor carried out his obligations under the contract, then this will still involve the issue of a certificate albeit showing nothing due or even negative in amount. The notice is still required stating that nothing is proposed to be paid and stating the basis of the calculation. As for the second requirement, the fact that some other contract may enable a set-off or abatement to be utilised against payments due under the IFC contract will not prevent a certificate being issued in the normal manner without regard to the set-off or abatement. The employer's notice would then have to state the basis for paying less than the certified sum.

If the employer fails to comply with clause 4.2.3(a), no doubt this will be a breach of the contract. However, it is difficult to see what damage the contractor will suffer, except perhaps that as the notice ought to indicate the basis on which the proposed amount to be paid is calculated, the contractor may be engaged in wasted time and effort in trying to establish why there is a difference between the certified (due) sum and the amount paid by the employer. Even here, however, there is no indication either in section 110 of the Act or in clause 4.2.3(a) as to what is meant by '…the basis on which the amount is calculated'. Is a detailed breakdown required or is it adequate simply to refer to the fact that the proposed payment has been calculated in accordance with, for example, clause 4 of the conditions?

If what the employer is proposing to pay is less than the certified sum and will involve a withholding or deduction from sums otherwise due, then the requirements for notice of withholding or deduction to reflect the requirements of section 111 of the Act (see for example clause 4.2.3(b)) will in any event require details to be given. Indeed, section 111 of the Act expressly provides that it is possible for the section 111 and section 110(2) notices to be combined. However they are not combined in IFC 98 because of the different time limits applying to them. The notice of proposed payment has to be given not later than five days after the date of the certificate; whereas the notice of intended withholding or deduction has to be given not later than five days before the final date for payment. In the case of an interim certificate therefore the latter notice has to be given on or before the ninth day following the date of the issue of the certificate.

Clause 4.2.3(a) deals with notice of proposed payment in respect of interim certificates prior to practical completion. A similar provision in respect of the payment on practical completion is to be found in clause 4.3(b) and in relation to the final certificate in clause 4.6.1.2.

Clause 4.2.3(b) contains the requirement for the employer to give notice of intended withholding or deduction from sums otherwise due not later than five days before the final payment date in respect of such sum. For interim payment certificates the final payment date is 14 days from the certificate (due) date. The notice must specify the amount proposed to be withheld or deducted and the grounds for doing so. This is intended to meet the statutory requirements of section 111 of the Act and probably does so. A similar provision is included in respect of payment following practical completion (clause 4.3(c)) and in respect of the final certificate (clause 4.6.1.3).

In the case of all payment certificates, there is express provision that if the employer fails to give the required notice he shall pay the amount due pursuant to the payment clause (see clauses 4.2.3(c), 4.3(d) and 4.6.1.4).

What is the consequence of the employer not giving a written notice of proposed payment under clauses 4.2.3(a) or 4.3(b) or 4.6.1.2? The contract states that in such a case he shall pay the 'amount due' (see clauses 4.2.3(c) and 4.3(d) relating to interim certificates); or the 'amount stated as due' (per clause 4.6.1.4 relating to the final certificate). It is submitted that in relation to interim certificates, while the failure to serve the notice may be a breach of contract, it will not have the effect of rendering due any sum which would not otherwise be properly due, e.g. a sum wrongly included in an interim certificate. The sum due is whatever sum is legally due and that does not automatically equal the certified sum (see earlier in this Chapter under '(B) The certification process' – page 190). The position is more difficult in relation to the final certificate where the wording in clause 4.6.1.4 requires the employer to pay the 'amount stated as due'. It could be argued here that the failure to serve a notice of proposed payment (or deduction) will render the certified sum payable whether properly due or not. This is an extreme construction and would still face the hurdle posed by the words in clauses 4.6.2, 4.2(a) and 4.3(e): '...if the Employer fails properly to pay the amount ... due...'.

The employer will be in difficulty if, at the point five days before the final date for payment, he is unaware of any basis on which he may be entitled to withhold or deduct, but subsequently, when it is no longer possible to fulfil the notice requirements, a matter comes to his attention which would have entitled him to withhold or deduct. Presumably, in such a case he is nevertheless bound to pay having lost his right to set-off. It can be seen therefore that while the statutory requirements in section 111 are intended to regulate the exercise of common law or express contractual rights to withhold or deduct, and are not intended to affect the substantive law in relation to set-off, nevertheless its effect is to do so in a situation such as this.

If, however, the employer intends not to pay the certified sum in full on the basis that the part not to be paid was never 'due' then, it is submitted, there is no requirement either under the Act or the contract for a notice of intended withholding or deduction. Such a notice is only in respect of sums which 'become' due under a payment certificate. If, for example, the basis for not paying the certificate in full is because the employer contends that there has been an overvaluation with defective work being valued as if it were complying work, then no notice would be required. The employer would simply be abating rather than setting off. The position would be the same where the employer is contending that there should have been a downward adjustment of the contract sum which has not been carried out, e.g. in not requiring the contractor to correct inaccurate setting out – clause 3.9. It is tentatively submitted that the position could be otherwise under JCT 98.

For a discussion of sections 110 and 111 of the 1996 Act, see earlier page 27.

Interim payment certificate on practical completion

Within 14 days of practical completion the architect must certify payments of $97\frac{1}{2}$ per cent of the total value referred to in (1) above: in other words, half of the five per cent withheld under clause 4.2.1 is released at this time. Further, 100 per cent of the amounts referred to in (2) above (adjusted in the same way as stated there) must be certified, less of course any sums previously certified and less the amount of any advance payment made under clause 4.2(b). This is in effect an updating exercise by the architect or the quantity surveyor including adjustments of the contract sum in respect of reimbursement of loss and expense under clause 4.11. As with interim payment certificates prior to practical completion, the date of the certificate is the due date and the employer must pay by the final date for payment, namely 14 days thereafter (per clause 4.3(a)). As in the case of interim payments prior to practical completion, there is a requirement for the employer to indicate how much he proposes to pay (clause 4.3(b)) and also a requirement for the employer to give advance notice of any intended withholding or deduction (clause 4.3(c)). See above under the previous heading for a commentary in respect of these notices.

Retention money

For some general comments on this subject see under the heading 'Summary of General Law', page 194.

The status of retention money is dealt with in clause 4.4. The first point to note is that the contract treats the status of retention money differently depending on whether the employer is, or is not, a local authority.

If the employer is not a local authority, then the employer's interest in the retention money is fiduciary as trustee for the contractor. There is, however, an express provision to the effect that the employer has no obligation to invest the retention fund. It has been strongly argued (see John Parris: *The Standard Form of Building Contract JCT 80*, 2nd edition, pages 143 to 144) that despite this express provision (which is also to be found in JCT 98 clause 30.5.1) the employer must account to the contractor for any interest earned by the retention money on the basis that it is contrary to the general law that a trustee should benefit from the trust fund. However, it is submitted that, although the legal position is not beyond doubt, the terms of the contract will suffice to prevent the application of the ordinary rules relating to the duties of a trustee to account to the beneficiary for profits received out of the trust fund. It has been said that the rule:

> 'is not that reward ... is repugnant to the fiduciary duty, but that he who has the duty shall not take any secret remuneration or financial benefit not authorised by the law, or by his contract, or by the trust deed under which he acts, as the case may be' – per Lord Norman: *Dale* v. *I.R. Commissioners* (1953).

A trustee may therefore be able to establish his right to remuneration by virtue of the provisions of the trust instrument, in this case the conditions of contract.

Where the employer is a local authority, there are no express provisions dealing with the employer's interest in the retention fund. It is submitted that in such a case there is no obligation on the employer either to keep the retention fund in a separate account or to account for interest earned thereon.

A number of cases in recent years have confirmed that an express contractual provision stating that retention is to be held by the employer as a trustee is effective and that whether or not the contract states that the trust money can be required to be put into a separate identifiable fund, the employer's duty as a trustee is to take that step in any event and the courts will grant an injunction to the contractor compelling the employer to do so: see for example *Wates Construction (London) Ltd* v. *Franthom Property Ltd* (1991); *Finnegan Ltd* v. *Ford Sellar Morris Developments Ltd* (1991). However, if a clause in a standard form of contract expressly requiring the retention to be placed into a separate fund has been visibly deleted, i.e. it can be physically read even though deleted from the contract, this could well be enough to show that the parties intend to modify the usual trustee duties so that the employer will not have to place the retention into a separate fund: see *Herbert Construction (UK) Ltd* v. *Atlantic Estates Plc* (1993).

The 5 per cent withheld is released as to one half following the certificate of practical completion and the balance in the final certificate issued under clause 4.6.

Deductions by employer

The contract permits the employer to deduct certain sums from the amount shown in interim payment certificates. The circumstances where he may do so are as follows:

- Liquidated damages (clause 2.7)
- The cost of employing others following the contractor's failure to comply with an instruction (clause 3.5.1)
- The contractor's default in insuring against injury to persons and property (clause 6.2.3)
- The failure of the contractor to insure the works (clause 6.3A.2)
- Following determination of the contractor's employment under clauses 7.2 to 7.4 (clause 7.6(f)).

The contract makes no specific reference to any other rights of cross claim, whether of ascertained sums or otherwise. Whether the employer can set-off any other amounts against payment certificates will depend therefore on a construction of the contract as a whole in the light of the general law on this topic (see 'Summary of General Law' page 195). It is submitted that on balance it is likely that the general common law right of set-off will be available to the employer in appropriate cases. Whether the contract expressly authorises the deduction or the employer is exercising his general common law right of set-off, a notice of intention to set-off complying with the requirements of section 111 of the Housing Grants, Construction and Regeneration Act 1996, is required. See page 29 for a discussion of section 111 generally and see above (page 207) for a commentary in relation to the contractual requirements in IFC 98 in respect of such notices.

Contractor's right to suspend performance of contractual obligations for non-payment

Clause 4.4A provides that without prejudice to any other rights or remedies which the contractor may possess, if the employer fails to pay the contractor in full by the final date for payment and no proper notice of intention to set-off has been given (for which see above), then the contractor may give a written notice to the employer of his intention to suspend performance of his obligations under the contract. This notice must identify the grounds on which it is intended to suspend performance and if payment in full is not made for seven days after the notice has been given, then the contractor can suspend performance until such payment in full occurs. As would be expected, any such suspension is not to be treated as a ground for determining the contractor's employment under clauses 7.2.1(a) or 7.2.1(b) of IFC 98.

This contractual right to suspend is clearly intended to mirror the statutory right in section 112 of the Housing Grants, Construction and Regeneration Act 1996 under which a party to a construction contract who is not paid in full by the final date for payment can suspend performance of his obligations until payment has been made. Even without such an express contractual right therefore, the statutory right would of course exist to enable the contractor to suspend in such circumstances.

There is a consequential extension of time available for delays attributable to the exercise by the contractor of this right to suspend (clause 2.4.18), and a right for the contractor to reimbursement of loss and expense should the suspension materially affect the progress of the works (clause 4.12.10 – which, however, only applies provided the suspension is not frivolous or vexatious).

It should be noted that suspension is not just of construction work but of contractual obligations and could therefore extend to matters such as providing site security, advance ordering of goods and materials, or renewing insurance premiums. The effects of such a suspension on the employer could therefore be very serious and the employer should be vigilant to ensure that payments which are due are made on time.

Before exercising such a right to suspend, the contractor should be very sure of his grounds as, if he is wrong in suspending, it is likely to amount to a very serious breach of contract, quite possibly entitling the employer to determine the contractor's employment or to treat the breach as a repudiation of the contract by the contractor, enabling the employer to treat the contract as at an end. Section 112 of the Act is discussed generally earlier in this book (see page 29).

The adjusted contract sum and the final payment certificate

The contractor must send to the architect or, if so instructed by him, to the quantity surveyor, all documents reasonably required for the purposes of the adjustment of the contract sum. The contractor can do this before practical completion of the works or within six months thereafter. The quantity surveyor must prepare a statement of all the final valuations under clause 3.7 (valuation of variations), and a copy of the computations of the adjusted contract sum must be sent to the contractor not later than three months after the contractor has provided the required documents (clause 4.5).

The final adjustment of the contract sum is generally in line with that for interim payments except that there is no retention percentage withheld and any valuation and ascertainment is final. The same deductions as are permitted from interim valuations, namely under clause 3.9 (levels) or 4.9(a) (fluctuations) or clause 4.10 (fluctuations: named persons) are to be made, but in addition, as this is the final adjustment to the contract sum, there will also be a deduction of all provisional sums as well as the value of any approximate quantity work included in the contract documents, and finally any amount to be deducted under clause 2.10 (defects liability) where the architect has instructed that a defect shall not be made good by the contractor, with an appropriate deduction being made from the contract sum in respect thereof.

The final adjustment of the contract sum should not under clause 4.5 take longer than nine months and may of course be much earlier.

Having sent to the contractor a copy of the statement and computation of the adjusted contract sum, the architect must, within 28 days of such date or of the certificate under clause 2.10 (certificate of making good defects), whichever is the later, issue a final certificate certifying the amount due to the contractor or the employer, as the case may be.

The amount certified must include 100 per cent of the total value referred to in (1) above (see page 205); in other words, the release of the second half of the retention, together with the amounts referred to in (2) above (see page 206) as finally ascertained and adjusted as therein required, less any sums previously certified and less the amount of any advance payment made under clause 4.2(b). The amount certified is payable either to the contractor or to the employer, as the case may be, within 28 days from the date of the final certificate. If the balancing payment is due from the employer to the contractor, the employer may deduct from this such sums as the contract permits, provided of course the appropriate notice of intention to withhold or deduct is given pursuant to clause 4.6.1.3.

There is no specific provision relating to agreement on any final account between employer and contractor. The only stipulation is that a certificate be issued within 28 days of the computations of the adjusted contract sum having been sent to the contractor and that the certificate states in respect of what the amount relates and the basis on which the amount due has been calculated.

As in the case of interim payments, there is a requirement for the employer not later than five days after the issue of the final certificate to give written notice to the contractor of the amount of the proposed payment and the basis on which it is calculated (clause 4.6.1.2), and for the employer, if he intends to withhold or deduct sums from the final certificate, to give written notice in advance not later than five days before the final date for payment of the final certificate (clause 4.6.1.3). There is no reciprocal notice requirement where the balancing payment is due from the contractor to the employer. The purpose and effect of these notices has already been discussed in dealing with interim payment certificates earlier (see page 207). Again, as in the case of interim payment certificates, if the amount due is not paid in full by the final date for payment, the contractor can claim interest (clause 4.6.2) (see page 204). However, in the case of the final certificate, it could be that a sum is due from the contractor to the employer and accordingly there is also a contractual provision entitling the employer to claim interest from the contractor in this situation.

Clearly, in practice, there will be an attempt by the contractor and the architect or the quantity surveyor to reach agreement. As the amount to be included in the final certificate is of course governed by what is authorised under the conditions of contract, it will be unlikely that the architect will issue a final payment certificate containing a sum other than that reflected in the quantity surveyor's final adjustment of the contract sum, unless there is an obvious error in it, even if the contractor disagrees with the quantity surveyor's computations. The contractor's remedy will be to await the certificate and then attack this as provided in clause 4.7 (see below).

Effect of final certificate (clause 4.7)

To a limited extent only, the issue of the final certificate for payment is conclusive evidence in any proceedings, whether by adjudication, arbitration or litigation, that:

(1)   The quality of materials or the standard of workmanship is to the reasonable satisfaction of the architect

(2)   All terms of the contract that require additions to or adjustments of, or
      deductions from, the contract sum, with the exception of any accidental
      inclusion or exclusion of any items or any arithmetical error in any compu-
      tation, have been complied with
(3)   All appropriate extensions of time have been given
(4)   Any reimbursement to the contractor of direct loss and/or expense is in final
      settlement of all claims the contractor has arising out of such matters as gave
      rise to the reimbursement, even if such claims could be put in terms of breach
      of contract, duty of care, statutory duty or otherwise.

The limits to its conclusiveness result firstly from the matters to which its application is expressly restricted. Secondly, the employer or the contractor may take
steps to restrict its effect.

Matters covered by conclusiveness of final certificate if no steps taken by employer or contractor to avoid its effect

(a) Quality of materials and standards of workmanship

The final certificate is conclusive as to the particular qualities of materials or goods
or particular standard of an item of workmanship but only where the contract
documents or an instruction of the architect, or further information issued by the
architect under clause 1.7.2 (provision of further drawings or details) or clause 3.9
(levels and setting out), expressly describe it as being for the approval of the
architect. Once the final certificate has been issued, it is conclusively presumed
that it is to the reasonable satisfaction of the architect. However, the final certificate is not conclusive that those or any other materials or goods or workmanship
comply with any other requirement or term of the contract.

 This clause was amended (Amendment 9 to IFC 84 issued July 1995) in order to
give the conclusiveness more restricted effect following two cases in which the
court rather surprisingly gave previous wording a very wide interpretation.
Admittedly, the wording of the previous clause 4.7 was less specific than it is now,
but even so the court's interpretation came as a surprise to many. In the case of
Colbart Ltd v. *H. Kumar* (1992) it was held that the previous clause 4.7 made conclusive all materials and workmanship where the approval of the architect was
not just expressly required by the contract documentation but where his approval
was inherent, e.g. in issuing interim certificates, a certificate of practical completion or a certificate of making good defects. In all of these instances the architect
was expressing a kind of approval and this could be enough to satisfy the provisions of the previous clause 4.7, so making it impossible thereafter for the
employer to introduce evidence in any proceedings to the effect that any work or
materials did not comply with the contract specification.

 This view was reinforced by the Court of Appeal in the case of *Crown Estate
Commissioners* v. *John Mowlem & Co. Ltd* (1994), a decision which logically made the
final certificate conclusive as to all contract requirements in respect of goods or
materials or workmanship being met including objective criteria set out in the
contract documents and implied terms as to quality and fitness. The effect was to
relieve the contractor of liability for failing actually to comply with the contractual

requirements as to materials and workmanship unless proceedings were commenced by the employer either before or within 28 days of the issue of the final certificate. This is to be compared with the period under the limitation acts of six years (for contracts under hand) and 12 years (for contracts executed as a deed) from the date on which the breach took place, which is likely to be the date of practical completion, or possibly later in respect of defects made good. It is not surprising therefore that the Tribunal introduced amended wording.

This issue has been treated differently in Scotland. The Court of Session in the case of *Belcher Food Products Ltd* v. *Miller and Black & Others* (1998) decided that the effect of the final certificate in relation to quality of materials and workmanship was much more limited. The court considered wording (in a JCT 63 contract) similar to that considered by the court in the above two cases. In the Opinion of Lord Gill, the conclusiveness applied only in connection with an express provision reserving matters for the opinion of the architect. The conclusiveness did not extend to whether or not the contractor had complied either with an objective and defined standard, e.g. a British Standard specified in the contract, or to any implied terms as to quality where the contract was silent. In addition he held that in any event even to the extent that the final certificate was conclusive, that conclusiveness related only to the opinion of the architect and that it did not follow that the final certificate had the further effect of being conclusive evidence for all purposes that the standard or quality in question had been met. In other words, it apparently did not bind the employer.

This latter point is a surprising finding and is clearly at odds with the *Crown Estate Commissioners* v. *John Mowlem* case. As for the first point, this again is at odds with the *Crown Estate Commissioners* case but at least has the virtue of being far more in line with what was originally intended by the Tribunal, and accords far more with what would be the reasonable expectation of the parties. Following Amendment 9, it is now clear that the conclusiveness relates only to matters which are expressly reserved for the architect's approval and even that approval will not relieve the contractor of liability if the goods, materials or workmanship in question fail to satisfy some other requirement of the contract which may be applicable, e.g. a British Standard. It is now therefore restricted to the situation to which, it is submitted, it was always intended to apply, namely where compliance by the contractor with the contract can only be determined by the subjective approval of the architect; for example, where in an extension to an existing building, the contractor is asked to colour match new tile work with existing tile work. In such a situation a perfect match can never be guaranteed and it would therefore be appropriate for the contract documentation to require the contractor to colour match the tiles to the approval of the architect. Once that approval is obtained and the final certificate is issued and the employer has taken no steps to commence proceedings, including adjudication, within 28 days thereafter, it is reasonable for the architect's approval to have conclusive effect so that the contractor knows that he has complied with the obligation to colour match the tiles. Even here, some employers may point out that if the architect's decision is wrong in that the tiles clearly do not match, but in the circumstances the decision is not a negligent one (e.g. the architect was unaware that he was suffering from a temporary partial colour blindness), the employer is left without a remedy. It is submitted, nevertheless, that this provision strikes a reasonable balance.

Even in relation to those matters of quality and standards where the final certificate is conclusive evidence, it may not extend to consequential losses, e.g. loss of use suffered by the employer owing to defects appearing after practical completion but before the issue of the final certificate. On this basis, the final certificate is conclusive that, as at its date, the particular quality and standards referred to are satisfactory and this can include previously defective work which has been made good; but the certificate will not, it is argued, prevent the employer from recovering his reasonably foreseeable losses resulting from the defect. This view of the effect of the final certificate on consequential losses is given some support by the judgment of Lord Diplock in the House of Lords' case of *Hosier and Dickinson Ltd* v. *Kaye (P and M) Ltd* (1972).

(b) That any necessary effect has been given to all the terms of the contract that require adjustments to the contract sum

It should be noted firstly that it is only those terms of the contract which affect the contract sum to which the conclusiveness relates under this inset to clause 4.7. Secondly, it does not apply to the accidental inclusion or exclusion of any item or arithmetical error in any computation.

Under this head, therefore, as the conclusiveness of the final certificate refers to any necessary effect having been given to all the terms of the contract requiring additions to, or adjustments of, or deductions from, the contract sum, its issue would not prevent the contractor from challenging any decision of the architect in relation to extensions of time or any failure to make extensions of time, by way of proceedings at any time after the issue of the final certificate, subject always to the general limitation period under the Limitation Act 1980. Such proceedings, concerned as they would be with extensions of time and therefore the employer's entitlement to liquidated damages, could not affect the contract sum. It is for this reason that the third inset to clause 4.7 was introduced by Amendment 3 to IFC 84 dated July 1988 (see (c) on page 217).

On the wording of this inset of clause 4.7, and indeed on similar wording in JCT 98 (clause 30.9.1.2), it appears that as the provisions of clause 4.11, dealing with disturbance of regular progress, are without prejudice to other rights and remedies which the contractor may possess, he could where appropriate claim damages for breach of contract after the issue of the final certificate as this would not require any adjustment to the contract sum. Accordingly, in order to ensure that this could not be done, the fourth inset to clause 4.7 was introduced in Amendment 3 to IFC 84 dated July 1988 (see (d) on page 217).

The final certificate is not conclusive so as to prevent the employer claiming any liquidated damages to which he was entitled prior to the issue of the final certificate, provided of course that the architect has issued the appropriate certificate of non-completion under clause 2.6, and the employer has complied with the written notices required under clause 2.7 and clauses 4.2.3(b), 4.3(c) or 4.6.1.3 as appropriate.

However, if the architect purports to issue a certificate or further certificate under clause 2.6 after the issue of the final certificate, the clause 2.6 certificate is likely to be invalid. The reason for this is that it is arguable that after the issue of the final certificate, the architect becomes *functus officio* (without office) so that the clause 2.6 certificate may be of no effect and, being a condition precedent to the employer's right to deduct liquidated damages under clause 2.7, the employer could not then deduct these even if the employer had sent the appropriate written notices. The employer may still be able, however, to claim damages at common law for any delay on the part of the contractor in completion where this amounts to a breach of contract. The issue of the final certificate does not appear to prevent a claim for unliquidated damages for delay.

In the case of *Fairweather* v. *Asden Securities Ltd* (1978) Judge Stabb held that, once the final certificate was issued, the architect was *functus officio* so that he could not then issue a certificate under clause 22 of the 1963 JCT contract (for this purpose similar to clause 2.6 of IFC 98), which certificate was a condition precedent to the employer's right to claim liquidated damages. Although the final certificate under that contract (clause 30(7) JCT 63 1972 revision) was significantly different to that in later revisions of JCT 63 and in JCT 98 and IFC 98, the same point could be argued. The decision in this case has been criticised (e.g. by the learned editors of *Building Law Reports* – see 12 BLR, page 41 *et seq.*). However, in the case of *A. Bell & Son (Paddington) Ltd* v. *CBF Residential Care and Housing Association* (1989) it was assumed without argument that the architect could not issue such a certificate of non-completion after the final certificate had been issued.

(c) That all and only such extensions of time, if any, as are due under clause 2.3 have been given

This is intended to ensure that the final certificate is conclusive as to any claim by the contractor for further extensions of time or any claim by the employer that extensions of time given should be reduced.

(d) That the reimbursement of direct loss and/or expense, if any, to the Contractor pursuant to clause 4.11 is in final settlement of all and any claims which the contractor has or may have arising out of the occurrence of any of the matters referred to in clause 4.12 whether such claim be for breach of contract, duty of care, statutory duty or otherwise

The effect of the second inset to clause 4.7 prevents loss and expense claims being reopened after the issue of the final certificate as to do so would be to further adjust the contract sum and this the final certificate prevents. However, something more is needed to prevent a claim by the contractor for damages for breach of contract where a clause 4.12 matter entitling the contractor to reimbursement of loss and expense also amounts to a breach of contract. This could apply in a number of instances, e.g. late instructions (clause 4.12.1), or failure to give ingress to or egress from the site over adjoining or connected land (clause 4.12.6). The fourth inset to clause 4.7 prevents such claims. It might just be possible that this exclusion of liability could be attacked as failing to meet the requirements of reasonableness (see section 3 of the Unfair Contract Terms Act 1977 and the discussion earlier under note [3] under the notes to clause 1.1 (page 42). This is, however, a remote possibility.

Further, the possibility of framing a claim for loss and expense in the form of an action in negligence or for breach of statutory duty is also prevented by this inset.

The overall effect of the amendments to this clause introduced in Amendment 3 to IFC 84 dated July 1988 is to render the final certificate much more conclusive in terms of extensions of time and actions by the contractor for breach of contract where the factual background would have justified a loss and expense claim under clauses 4.11 and 4.12.

Steps which may be taken by the employer or the contractor to avoid the conclusiveness of the final certificate

Even the limited matters on which the final certificate can be conclusive can be saved from such a fate if they are the subject of proceedings (whether by way of adjudication, arbitration or litigation) commenced either before the date of the final certificate or within 28 days thereafter. If any matter is to be the subject of proceedings it is vital that the issues in the proceedings, i.e. in the notice of intention to adjudicate, or in the notice requesting the concurrence of the other party in the appointment of an arbitrator or the endorsement on the claim form, as the case may be, are drafted sufficiently widely to include all the possible areas of dispute which it is sought to resolve by the proceedings. If the issues are too narrowly defined it could result in an important and relevant matter being inadmissible.

Adjudication might not finally dispose of the dispute or difference between the parties. Either party can take the dispute or difference which was the subject of the adjudicator's decision to arbitration or litigation (depending on which option is chosen in the appendix to IFC 98). It is quite likely that the adjudicator's decision on a dispute or difference will be made later than 28 days from the date of the issue of the final certificate. It is therefore necessary to provide for this situation so that a party's wish to arbitrate or litigate the dispute or difference following the adjudicator's decision is not evidentially predjudiced by the opening words of clause 4.7.1. Accordingly, clause 4.7.2 provides that where an adjudicator's decision is given on a date after the issue of the final certificate, either party may commence arbitration or legal proceedings within 28 days of the date on which the adjudicator gave his decision. Presumably this would be interpreted as restricting the issues which can be arbitrated or litigated to those which were the subject of adjudication although this is not expressly stated to be the case.

Finally, the provisions of clause 4.8 should be noted – that except for the final certificate, no certificate of the architect of itself is to be conclusive evidence that any work materials or goods to which it relates are in accordance with this contract.

NOTES TO CLAUSES 4.2 TO 4.8

[1] '... the Date of Possession stated in the Appendix...'
If the date of possession stated in the appendix is deferred under clause 2.2 for, say, four weeks, then the first certificate is likely to be of nil amount unless any loss and expense attributable to the deferment has already been ascertained.

[2] '...14 days from the date of issue of each certificate...'
In calculating the 14 days, public holidays are excluded (see clause 1.14 and commentary thereon – page 63).

[3] 'If the Employer fails properly to pay...'
Somewhat surprisingly, there is no express obligation on the employer to pay. Rather there are express consequences if he fails properly to pay, e.g. the right for the contractor to claim interest; the right for the contractor to suspend performance of his contractual obligations; and the right to determine his own employment. An obligation to pay is therefore undoubtedly implied and it is submitted that, even if implied, the 14 day payment period is a time, after the date of the certificate, during which or within which the employer is required to pay. Accordingly, the 14 day period excludes public holidays (see clause 1.14 and the commentary thereon – page 63).

[4] '...the Employer shall pay to the Contractor...'
The payment of interest, it is submitted, is a direct obligation between employer and contractor and is not required or permitted to be added to the contract sum.

[5] '...shall be paid to the Contractor on the date stated in the Appendix...'
The advance payment, even though made only after the contract is entered into, is not the subject of a certificate of the architect. It may have been better to have included it in a certificate. Not only would this make it more straightforward in terms of taking it into account in subsequent certificates, it would also have meant that if the employer was late in making the advance payment, the interest provisions on late payment in favour of the contractor would have applied, whereas it seems quite likely that they do not apply to the advance payment provisions. The interest provisions are contained in clauses dealing specifically with payments under certificates of the architect. In addition, non payment of the advance payment does not fall within clause 7.9.1(a) (contractor entitled to determine own employment on employer's failure to discharge the amount properly due in respect of a certificate). However, depending on the seriousness of the employer's breach of contract, it may amount to a repudiation by the employer, entitling the contractor to treat the contract as a whole as at an end.

[6] '...on the terms stated in the Appendix...'
In practice, unless the repayment provisions are particularly straightforward, the appendix is likely to refer to a separate agreement containing a repayment schedule and dealing with related matters.

[7] '...in similar detail...'
If the quantity surveyor disagrees with the contractor's application he must respond with a statement in similar detail to the contractor's application. This may require a significant amount of work on the part of the quantity surveyor, which is likely to put pressure on professional fees.

[8] '...less any sums previously certified for payment'
Interim payment certificates are issued on account of the final certificate. This means that interim payment certificates should contain the total value of work and materials executed, together with other sums required to be included in interim payment certificates, with a deduction from this running total of any amounts previously certified. In this way, adjustments can be made in the current

valuation to take into account errors in previous certificates, e.g. fluctuations wrongly included or the value of work or materials included in earlier certificates and then found to be defective.

The inclusion in an interim payment certificate of the value of work is not of course conclusive evidence that the work concerned has been carried out in accordance with the contract – see clause 4.8.

[9] '...properly executed...'
Clearly, the architect's duty to the employer will include proper routine visits between interim payment certificates to ensure that no obviously defective work is mistakenly included in the quantity surveyor's valuations (discussed in more detail under the heading 'Defective work', page 192). For further consideration of the architect's duties in relation to inspection of work, the reader is referred to standard works covering this subject, e.g. *Hudson's Building and Engineering Contracts; Keating on Building Contracts,* and to the numerous relevant cases, such as:

Sutcliffe v. *Chippendale and Edmondson* (1971)
East Ham Corporation v. *Bernard Sunley & Sons Ltd* (1965)
Moresk Cleaners v. *Hicks* (1966)
London Borough of Merton v. *Lowe* (1981)
Brickfield Properties Ltd v. *Newton* (1971)
Holland Hannen and Cubbitts (Northern) Ltd v. *Welsh Health Technical Services Organisation* (1981)
Kensington and Chelsea and Westminster Area Health Authority v. *Wettern Composites and Others* (1984)
EDAC v. *William Moss Group Ltd and Others* (1984)
Tesco Stores Ltd v. *The Norman Hitchcox Partnership Ltd and Others* (1997)

[10] '...the application of the proportion of the work...'
This makes it clear that where an activity schedule is used, payment is against the portion of the activity which has been completed at the appropriate valuation date. The contractor does not have to fully complete an activity before being paid something in respect of it.

[11] '...reasonably and properly...'
What is reasonable and proper in relation to the delivery of materials and goods to the site will depend on a number of factors and all the relevant circumstances need to be taken into account, e.g. it may be reasonable for a contractor to stack up materials on site at a time of shortage. Also, where there is a particularly tight building programme it may well be reasonable for the contractor to hold greater stocks of materials on site than he would otherwise do if there were greater leeway in the overall construction period. As a matter of practice, if the architect intends to disallow any of the materials or goods on site, he should notify not only the contractor but also the quantity surveyor so that they can be excluded from his valuation.

[12] '...adequately protected against weather and other casualties'
The architect must decide whether such reasonable protection has been afforded. However, if the quantity surveyor, in carrying out his valuation, is of the view that

materials or goods have not been adequately protected, then as a matter of practice he should notify the architect.

If the materials or goods are delivered to, placed on or adjacent to the works, and are intended for incorporation therein, they are site materials within the meaning of those words in IFC 98 – see clause 8.3. Accordingly, if they are lost or damaged due to an event within the all risks insurance cover for the works required under clauses 6.3A, 6.3B or 6.3C, the policy should pick up the cost of replacement or rectification with the contractor getting an extension of time if appropriate and, where 6.3B or 6.3C applies, the replacement will be treated as an instruction requiring a variation and will attract therefore reimbursement of direct loss or expense where suffered or incurred by the contractor, though only in relation to the period following that time at which the contractor has the obligation to replace lost or damaged site materials (see page 307). This will be the case whether or not the site materials are adequately protected against weather or other casualties, even where this is instrumental to the loss or damage occurring, unless the absence of protection causes the site materials to suffer from deterioration, rust or mildew which is outside the all risks cover required under the contract. If the loss or damage is outside the scope of the policy then the risk will, it is submitted, fall on the contractor who will have to replace or rectify at his own cost – see clause 1.10.

[13] '...before delivery thereof...'
Nothing is included within this clause to prevent the value of listed items being included in interim certificates even though they are not due to be delivered to the site for a considerable period. It seems unsatisfactory to have a situation where a listed item, e.g. an expensive piece of air conditioning equipment which is not to be installed until, say, week 35 of a 50 week contract, must nevertheless be paid for in, say, week 10 of the contract because the contractor has the equipment stored off-site.

[14] '...reasonable proof...'
There is always debate about what amounts to 'reasonable proof' of ownership. Does it mean that the architect has to be reasonably satisfied that the contractor has title to the goods? If this is the governing criterion then no matter what amount of proof the contractor can reasonably produce, it might be inadequate to reasonably satisfy the architect. The emphasis however is on providing reasonable proof, rather than satisfying the architect. This seems to suggest that much will depend on how much proof is available. If all the available proof is provided and it indicates, though not conclusively, that title is vested in the contractor, it is difficult to see how the architect can refuse to operate this provision, even if there are still some nagging doubts. There is at least the consolation of a bond which can be required for uniquely identified listed items and a bond which is a standard requirement for listed items not uniquely identified. If the contractor does provide reasonable proof and it subsequently transpires that title was not vested in him, this will be a breach of contract by the contractor. In dealing with the question of reasonable proof, the Tribunal's Guidance Note to Amendment no. 12 to IFC 84 issued April 1998 states at page 34: 'In practice, however Employers will only pay pursuant to clause 4.2.1(c) if the proof provided of the Contractor's ownership is

as water-tight as possible'. The guidance note goes on to indicate the type of evidence which the contractor should supply, e.g. a copy of the supply contract between supplier and contractor; or perhaps a written statement from the supplier that any conditions requiring to be fulfilled before property passes to the contractor have been fulfilled and that the relevant items are not subject to any charge or incumbrance.

[15] '...uniquely identified...'
This for example would include a central heating boiler from a specified supplier.

[16] '...has been paid by the Employer...'
It is submitted that this phrase is to be interpreted in the sense that the Employer must satisfy his obligation to pay. This may be done either by making a physical payment to the contractor, or alternatively, by discharging the obligation in the course of properly withholding or deducting sums otherwise due under payment certificates, e.g. a proper deduction of liquidated damages. However the matter is not straightforward – see note [2] in the notes to clause 1.11 (page 59).

[17] '...not uniquely identified...'
This would include for example a quantity of bricks.

[18] '...Specified Perils...'
Note that the insurance cover is limited to specified perils, e.g. fire, flood, explosion etc. and does not extend to other matters which would be covered by a policy giving an all risks type of cover, e.g. theft, impact damage or damage due to vandalism. For expensive items of property, particularly if they are easily transportable, the employer may well wish to seek his own insurance cover on an all risks basis or, where reasonably practicable to do so, amend the contract to require the contractor to obtain the wider cover.

[19] '...the Employer shall pay...'
It is submitted that if the employer fails to pay and the reason for non-payment is because the sum was never due, e.g. a proper application of the law of abatement, then the employer would not be liable to pay notwithstanding clause 4.2.3(c), clause 4.3(d) or clause 4.6.1.4.

[20] '...within 14 days after...'
In calculating this period, public holidays will be excluded. See clause 1.14 and commentary thereon (page 63).

[21] 'Not later than 5 days after...'
See under previous note.

[22] 'Where the Employer is not a local authority...'
The restriction to local authorities is too narrow. Bearing in mind that the imposition of a trust status on retention money is primarily to protect the contractor's interest should the employer become insolvent, it would have been reasonable to extend the scope of the non-trust retention money to public bodies generally.

[23] '... subject only ...'
These words may be sufficient to prevent the non-local authority employer from claiming a common law right of set-off against the retention money, unless the words '... provisions of this Contract ...' extend to implied as well as express provisions. In any event this will not prevent the employer from claiming a set-off in the event that the contractor goes into liquidation or becomes bankrupt. This is because Rule 4.90 of the Insolvency Rules 1986 (in respect of winding-up) and section 323 of the Insolvency Act 1986 (in the case of bankruptcy) provides that where there have been mutual credits, mutual debts or other mutual dealings between a debtor and any person proving a debt in the winding-up or bankruptcy, an account must be taken of what is due from the one party to the other in respect of such mutual dealings and the balance of the account, and no more can be claimed or paid on either side respectively (note also clause 7.6(f)).

[24] '... the right of the Employer to have recourse thereto ...'
In the case of *Wates Construction (London) Ltd v. Franthom Property Ltd* (1991) the Court of Appeal held on similar though not identical wording, that the employer's right to have recourse to the retention fund did not in itself make the employer a beneficiary as well as a trustee of the fund.

[25] '... pursuant to clause 4.2.3(b) or 4.3(c) ...'
This confirms that the right to suspend does not apply if a notice under clause 4.2.3(b) or 4.3(c) has been given by the employer of his intention to withhold or deduct. The right to suspend set out in clause 4.4A is generally intended to mirror the statutory right to suspend performance for non-payment contained in section 112 of the Housing Grants, Construction and Regeneration Act 1996. Section 112 itself refers to the power being exercisable where '... no effective notice to withhold payment has been given ...' and no doubt this reference to clause 4.2.3(b) and 4.3(c) is meant to mirror this situation.

[26] '... with a copy to the Architect ...'
Before the contractual right to suspend can be utilised, the contractor must not only give notice to the employer but must also give a copy to the architect. Presumably if he fails to give a copy to the architect he cannot utilise this contractual right to suspend. However, it is submitted that if the contractor did nevertheless suspend, it would not be a breach of contract as the contractor could exercise his statutory right under section 112 of the Housing Grants, Construction and Regeneration Act 1996 provided he meets the requirements of that section.

[27] 'Not later than 6 months after Practical Completion ...'
While it is unlikely that the contractor can supply all the documents reasonably required for a final adjustment of the contract sum before practical completion, he can nevertheless submit what relevant documentation he has at any particular time. Indeed, it is often desirable that the preparation of the final account be a continuous operation from the outset of the contract, especially where agreement can be reached with the contractor at each stage. It makes the issue of the final payment certificate very much easier for the architect.

[28] '...together with any withholdings...'
Prior to Amendment 12 issued April 1998, IFC 84 referred only to 'deduction' rather than to 'withholdings and/or deductions'. It is understood that the purpose of expressly referring to 'withholdings' was to make the contract wording consistent with the provisions of section 111 of the Housing Grants, Construction and Regeneration Act 1996 which requires advance notice to be given where a party to a construction contract intends to 'withhold payment...'. In other clauses of the contract which regulate the employer's ability to set-off against sums stated as due in certificates, e.g. clauses 4.2.3(b), 4.3(c), 4.4 and 4.6.1.3, this is perfectly understandable. However, to use it as it is used in clause 4.5 is unnecessary as it deals with the calculation to be made in order to compute the adjusted contract sum. It is only once it is included in the certificate that it becomes a sum due and it is only once it is a sum due that section 111 of the Act has any relevance. Note the wording in section 111(1) ('...a sum due...'). It seems likely that this is in fact a drafting slip which will be corrected in due course.

[29] 'Not later than 5 days...'
In calculating these days, public holidays are excluded – see clause 1.14 and the commentary thereon (page 63).

[30] '...to the Employer by the Contractor...'
Interim payments are on account of the final amount payable in accordance with the conditions of contract. Accordingly, it is possible that an over-valuation or other mistake could result in an over-payment during the course of the contract. This would be corrected in the final account by a payment being required from the contractor to the employer. This adjustment in respect of an over-payment due to an over-valuation can be made even if the contractor has thereby over-paid a named or domestic sub-contractor and is unable, as a result of the sub-contract works having been completed, to deduct it from further sums due to the sub-contractor: see *John Laing Construction Ltd* v. *County and District Properties Ltd* (1982).

[31] '...within 28 days...'
Section 27 of the Arbitration Act 1950 (now repealed) gave the court certain discretion to extend any time limit in commencing arbitration, where a failure to meet it barred claims. It was held in *Crown Estate Commissioners* v. *John Mowlem & Co. Ltd* (1994) based on similar wording in JCT 80 that section 27 did not apply as the effect of the conclusiveness wording did not bar the claim as such but only prevented evidence being adduced as to conclusive matters, though the effect is the same. Accordingly, the court had no such discretion. The position is now governed by section 12 of the Arbitration Act 1996 which gives discretion not only in relation to the barring of a claim but also where the claimant's right is extinguished. Does the expiry of the time limit in clause 4.7 extinguish a right? The claimant still retains his right to bring a claim though his right to adduce evidence in support of it is extinguished. It depends what 'right' refers to. It is tentatively submitted that it refers to the right to arbitrate and accordingly section 12 has no application and the court has no discretion. However, a further point to note is that in clause 4.7.2, a cross reference to clause 4.7.1 seems to assume that the 28 day

period refers to the ability to commence proceedings; in other words, that it acts as a bar to the right itself. As there was no equivalent provision in the contract before the court in the *Crown Estate* case, this could affect the way in which a court construes the meaning of the words used in clause 4.7.1. If despite the clause 4.7.2 wording, 4.7.1 were nevertheless to be construed as only an evidential bar, this may have the odd result that a party wishing to take a dispute or difference which has been the subject of an adjudicator's decision after the final certificate to an arbitrator or the courts pursuant to clause 4.7.2 and who misses the 28 day deadline, will face a bar on the right to commence proceedings rather than an evidential bar. If this is the position and the venue for the dispute or difference is arbitration it may be possible to obtain relief under section 12 of the Arbitration Act 1996.

Where the option chosen for dispute resolution is litigation, there is of course no equivalent to section 12 of the Arbitration Act 1996. Whether the claim be put forward in arbitration or litigation, the 28 day time limit, even if an evidential bar rather than a bar to bringing the claim, nevertheless has real impact and can effectively deprive one party or the other from bringing the merits of the claim before a tribunal for consideration. A question may be raised whether such a provision could be subject to the test of reasonableness under the Unfair Contract Terms Act 1997, or subject to the requirements of good faith under the Unfair Terms in Consumer Contracts Regulations 1994.

As for the former, section 3 of the 1977 Act applies as between contracting parties where one of them deals as consumer or on the other's written standard terms of business. It provides that the non-consumer or the party relying on its own written standard terms of business, as the case may be, cannot when in breach of contract, exclude or restrict any liability in respect of the breach except in so far as the contract term relied on satisfies the requirements of reasonableness (as to which see section 11 and schedule 2 to which a court or an arbitrator in practice will have regard). It might be argued in favour of this term's reasonableness that it applies equally to both parties, although the practical effect of its application is that the employer is more often likely to be seriously prejudiced than the contractor by the operation of the 28 day limit. As to whether or not IFC 98 could be regarded as being either party's own standard terms of business, see the earlier reference to *Overland Shoes Ltd* v. *Schenkers Ltd* (1998) (see page 43) indicating that in some circumstances a standard form of contract such as IFC 98 could be regarded as the employer's written standard terms of business.

As to the Unfair Terms in Consumer Contracts Regulations 1994, provided the employer is acting as a 'consumer' within the Regulations, the term would have to pass the requirements of good faith (see regulation 4 and schedules 2 and 3). Here again a relevant consideration may be that the term applies equally to both sides, but again its effect in practice as being likely to prejudice the consumer (employer) more than the contractor would be a relevant factor. It has been suggested earlier (see page 43), that the Regulations may apply to an independently drafted standard form of contract, even where it is put forward for use by or on behalf of the consumer.

[32] *'... accidental inclusion or exclusion of any item ...'*
Does this mean an item which is, or perhaps should be, included in the final

account or does it relate to items of work or materials? It is submitted that it covers both matters.

[33] '*... may commence arbitration or legal proceedings within 28 days of the date on which the Adjudicator gave his decision*'
This has the effect of extending the period during which a challenge can be made without it being subject to the conclusiveness rule where an adjudicator's decision is made after the date of the final certificate provided the adjudication was commenced before or within 28 days after the date of the issue of the final certificate. This is necessary to preserve a 28 day unfettered right for the employer or contractor to refer an adjudicated dispute to arbitration or litigation. Note, however, that unlike the opening reference to 28 days at the beginning of clause 4.7.1, here it refers to commencement and not conclusiveness. It assumes that the effect of clause 4.7.1 is to bar proceedings rather than provide an evidential barrier. See under note [31] above as to the possible significance of this.

Clause 4.9

Fluctuations
4.9   The Contract Sum, less any amount included therein for work to be executed by a named person as a sub-contractor under clause 3.3, shall be adjusted in accordance with
(a) Supplemental Condition C (*Tax etc. fluctuations*), unless
(b) Supplemental Condition D (*Formula fluctuations*) is stated in the Appendix to apply and Contract Bills are included in the Contract Documents.

C   **Contributions, levy and tax fluctuations**
Clause 4.9(a)
Supplemental Conditions C1 to C7 (together with Supplemental Conditions D1 to D13) are printed in a booklet*: 'Intermediate Form of Building Contract IFC 1998 Edition: Supplemental Conditions C and D: Fluctuation Clauses'.

D   **Use of price adjustment formula**
Clause 4.9(b)
Supplemental Conditions D1 to D13 (together with Supplemental Conditions C1 to C7) are printed in a separate booklet*: 'Intermediate Form of Building Contract IFC 1998 Edition: Supplemental Conditions C and D: Fluctuation Clauses'.
The Formula Rules referred to in clause D1 are printed in a separate booklet*: 'Consolidated Main Contract Formula Rules'.

---
* (Note neither of these separate booklets is reproduced in this book)

Clause 4.10

Fluctuations: named persons
4.10   In respect of any amount included in the Contract Sum for work to be executed by a named person as mentioned in clause 4.9, the Contract Sum shall be adjusted by the net amount which is payable to or allowable by the named person under clause 33 or 34 (as applicable) of the Sub-Contract Conditions NAM/SC.
Provided that there shall be excluded from the net amount referred to above any sum which would have been excluded under clause 33.4.7 or 34.7.1 of the Sub-Contract Conditions NAM/SC had not the period or periods for completion of the Sub-Contract Works been extended by reason of an act, omission or default of the Contractor as referred to in clause 12.2.1 of the said Sub-Contract Conditions.

COMMENTARY ON CLAUSES 4.9 AND 4.10

It is not proposed in this book to deal in detail with the application of the fluctuations provisions set out in supplemental conditions C and D of IFC 98. This commentary is limited therefore to a brief outline only.

A discussion of supplemental conditions C and D must commence with a discussion of the fluctuations clauses in JCT 98 on which the supplemental conditions are closely based.

The fluctuations clauses in JCT 98 (and in its predecessors JCT 63 and JCT 80) are now very familiar to those who deal with building contracts. Where, under JCT 98, it is decided that full fluctuations should apply, it is possible to choose one of two possible methods of price adjustment. The first method in JCT 98 is covered by clause 39 and allows adjustment of the contract sum based on actual changes in rates of wages and in the prices of materials payable by the contractor. The second method is covered by clause 40 and provides for adjustment of the contract sum by reference to published indices, this method being known generally as the formula method. Where neither of these two methods of price adjustment is required and the contract sum is based on what is generally called (not altogether accurately) a 'fixed' or 'firm' price basis, JCT 98 in clause 38 permits an adjustment of the contract sum by reference to changes in contributions, levies and taxes imposed under or by virtue of any Act of Parliament.

Turning to IFC 98 the first point to appreciate is that there is no labour and materials fluctuations clause for this contract, the only options being either the fixed or firm price basis in supplemental condition C (the equivalent of clause 38 of JCT 98) or the formula method in supplemental condition D (equivalent to clause 40 of JCT 98). It is somewhat surprising that the materials and wages fluctuations option has been omitted. Although rapidly losing popularity in favour of the formula method, it would have been a useful inclusion in the contract, particularly for refurbishment schemes and for those contracts where the employer particularly wishes to limit the list of materials which are to be subject to adjustment.

Clause 4.9(b) is so drafted that if bills of quantities are not a contract document then the contract must be on the fixed or firm price basis with adjustments to the contract sum being restricted to changes in taxes, levies etc. and this must place a serious limitation on the use of IFC 98. However, it would be possible to incorporate formula fluctuations even where there are no contract bills, provided a contract sum analysis is used and provided careful drafting is employed to adapt Amendment 3 to JCT 80 Without Quantities Form – March 1987 (now consolidated into JCT 98), which enables formula fluctuations to be used in that contract. (See also page 15).

Where the formula method is used, the contract sum must be adjusted in accordance with the provisions of supplemental condition D and the formula rules referred to earlier. Briefly, this method of adjustment is based on the 49 'work categories' – see Rule 3 and Appendix A to the formula rules – the work included in the bills of quantities being broken down as far as possible into the separate categories. The items in the bills of quantities are annotated so that all items are as far as possible divided into a series of work categories or, alternatively, into broader work groups. Indices which broadly reflect the changes in the cost of materials and the rates of wages within each category are published monthly by the Property Services Agency of the Department of the Environment. Valuations are then prepared at monthly intervals and the valuation divided into the various work categories. The differences between the indices for the various work categories for that month and those applicable during the month on which the tender was based are then applied to the value of the work completed within each work category during that month in order to arrive at the amount due in respect of fluctuations. The figure thus produced is then adjusted to take into account those items introduced in the bills which, although subject to formula adjustment, cannot be allocated to a particular work category, e.g. preliminaries or lump sum adjustments on the tender summary. These items, grouped under the heading of 'balance of adjustable work', are adjusted in accordance with Rule 38 or 26 of the formula rules. Where the contract involves a local authority, a further adjustment is made by deducting a percentage for what is termed the 'non-adjustable element'. This abatement, which is stated as a percentage, must be indicated in the appendix to IFC 98 and must not exceed 10 per cent.

The formula rules require the contract bills to indicate whether Part I or Part II of section 2 of the rules shall apply. The contract bills must also specify the base month. This information must be included in the appendix to IFC 98. The base month identifies the calendar month, the index numbers of which are deemed to equate to the price levels represented by the rates and prices contained in the contract bills. They are index numbers on which the contractor will have based his tender. If tenderers have not been informed of the base month, it will normally be the month prior to that in which the tender is due to be returned.

Part 1 of section 2 of the rules applies where it is intended to use the work category method of adjustment. Part II of section 2 applies where what is known as the work groups method is to apply. This method in effect condenses the work categories by combining them into groups. The contract bills must state the method of grouping to be adopted.

The need to measure work in detail for each valuation is paramount where the formula method is used. Work which has been completed between valuations must be accurately measured in each separate category in order that the appropriate indices can be applied. However, should the amount of any adjustment included in a previous certificate require correction, then the correction shall be given effect in the next certificate – see supplemental condition D5 and Rule 5 of the formula rules. Should there be a delay in the publication of the monthly bulletin containing the indices, adjustment of the contract sum must be made in each interim certificate during such period of delay on a fair and reasonable basis and can be adjusted later on the recommencement of the publication – see supplemental condition D10.

Interim valuations must be made before the issue of each interim certificate, and accordingly the words 'whenever the Architect/the Contract Administrator considers them to be necessary' in clause 4.2 shall be deemed to have been deleted – see supplemental condition D6. Both supplemental conditions C and D provide (see C4.7 and C4.8, and D12 respectively) that adjustment of the contract sum will not continue after the date for completion of any extended time for completion of the works. In other words, the operation of the fluctuations provision is frozen at that date. However, it is important to appreciate that the freezing of these provisions is subject to clause 2.4 remaining unamended and to the architect having, in respect of every written notification by the contractor under clause 2.3, fixed or confirmed in writing such date for completion as he considers to be in accordance with that clause. Accordingly, if the employer wishes to amend the printed text of clause 2.4 without this having the effect of unfreezing the application of the fluctuations provisions, he must also ensure that the relevant supplemental condition is suitably amended. Further, it is of course of the utmost importance that the architect properly fulfils his duties under clause 2.3. Clearly, a failure to properly amend the supplemental condition or a failure by the architect to properly exercise his duties could involve the employer in substantial additional costs in paying fluctuations during a period of over-run by the contractor.

Even where the application of the fluctuations provisions is frozen, it is prudent for the quantity surveyor to continue to separate work completed after that time so that adjustment can readily be made should the architect subsequently make further extensions of time for completion of the works under clause 2.3.

It should be appreciated that both types of fluctuations contained in IFC 98 are 'rise and fall' provisions. In other words the contract sum can be adjusted downwards as well as upwards in appropriate circumstances.

It should be noted that, in relation to supplemental condition C (fluctuations for tax etc.), the contractor is required to give written notice to the architect of the occurrence of any of the events which trigger off an adjustment of the contract sum – see C4.1. The notice is required to be given within a reasonable time after the occurrence of the event to which the notice relates, and it is expressly stated that the giving of a written notice in that time is a condition precedent to any payment being made to the contractor in respect of the event in question: see C4.2. Supplemental condition C4.3 states that the quantity surveyor and the contractor may agree what shall be deemed for all purposes of the contract to be the net amount payable or allowable by the contractor in respect of the occurrence of any of the listed events. This does not give the quantity surveyor authority to so agree where the contractor has failed to serve the written notice within a reasonable time after the occurrence of the event in question: see *John Laing Construction Ltd* v. *County and District Properties Ltd* (1982).

By Amendment 12 to IFC 84 issued April 1998 (now consolidated into IFC 98), an additional clause was inserted in Supplemental Condition C (namely C2.3) providing for the contract sum to be adjusted in respect of changes in the incidence and rate of land fill tax under the Finance Act 1996. There is a proviso to the effect that no upwards adjustment will be made if the contractor could reasonably be expected to have disposed of the waste other than by land fill. This takes account of the social policy behind the application of land fill tax, i.e. to discourage this form of waste disposal. To simply allow increases in land fill tax to be automatically fed through the contractual chain to the client would not achieve the social purpose behind the incidence of this form of taxation.

Clause 4.10 deals with fluctuations for named persons as sub-contractors. It states that the contract sum shall be adjusted by the net amount payable to or allowable by the named person under clause 33 or 34 (as applicable) of the sub-contract conditions NAM/SC. Named persons are treated separately as the sub-contract fluctuations may be on a different basis to the main contract fluctuations. The contractor will be reimbursed for the actual amount of fluctuations payable by him to the named sub-contractor. However, there is a proviso in clause 4.10 to the effect that the employer is not required to reimburse the contractor the fluctuations which he (the contractor) has paid to the named sub-contractor to the extent that those fluctuations are payable by reason of the period or periods for completion of the sub-contract works having been extended by any act, omission or default of the contractor as referred to in clause 12.2.1 of NAM/SC.

Clauses 4.11 and 4.12

Disturbance of regular progress

4.11   If, upon written application being made to him by the Contractor within a reasonable time of it becoming apparent, the Architect/the Contract Administrator is of the opinion[1] that the Contractor has incurred or is likely to incur direct loss and/or expense, for which he would not be reimbursed by a payment under any other provision of this Contract, due to
(a) the deferment of the Employer giving possession of the site under clause 2.2 where that clause is stated in the Appendix to be applicable; or
(b) the regular progress[2] of the Works or part of the Works being materially affected by any one or more of the matters referred to in clause 4.12,
then the Architect/the Contract Administrator shall ascertain[3], or shall instruct the Quantity Surveyor to ascertain, such loss and expense incurred and the amount thereof shall be added to the Contract Sum provided that the Contractor shall in support of his application submit such information required by the Architect/the Contract Administrator or the Quantity Surveyor as is reasonably necessary for the purposes of this clause 4.11.
The provisions of this clause 4.11 are without prejudice to any other rights or remedies which the Contractor may possess.

Matters referred to in clause 4.11

4.12   The following are the matters referred to in clause 4.11:

4.12.1   .1 where an Information Release Schedule has been provided, failure of the Architect/ the Contract Administrator to comply with clause 1.7.1;

4.12.1   .2 failure of the Architect/the Contract Administrator to comply with clause 1.7.2;

4.12.2   the opening up for inspection of any work covered up or the testing of any of the work, materials or goods in accordance with clause 3.12 (including making good in consequence of such opening up or testing), unless the inspection or test showed that the work, materials or goods were not in accordance with this Contract;

4.12.3   the execution of work not forming part of this Contract by the Employer himself or by persons employed or otherwise engaged by the Employer as referred to in clause 3.11 or the failure to execute such work;

4.12.4   the supply by the Employer of materials and goods which the Employer has agreed to supply for the Works or the failure so to supply;

4.12.5   the Architect's/the Contract Administrator's instructions under clause 3.15 issued in regard to the postponement of any work to be executed under the provisions of this Contract;

4.12.6   failure of the Employer to give in due time ingress to or egress from the site of the works, or any part thereof through or over any land, buildings, way or passage adjoining or connected with the site and in the possession and control of the Employer, in accordance with the Contract Documents after receipt by the Architect/the Contract Administrator of such notice, if any, as the Contractor is required to give, or failure of the Employer to give such ingress or egress as otherwise agreed between the Architect/the Contract Administrator and the Contractor;

4.12.7　compliance with the Architect's/the Contract Administrator's instructions issued under clauses

　　1.4　(*Inconsistencies*) or
　　3.6　(*Variations*) or
　　3.8　(*Provisional sums*)
　　　　except, where the Contract Documents include bills of quantities, for the expenditure of a provisional sum for defined work* included in such bills,
　　　　or, to the extent provided therein, under clause 3.3 (*Named sub-contractors*);

4.12.8　the execution of work for which an Approximate Quantity is included in the Contract Documents which is not a reasonably accurate forecast of the quantity of work required;

4.12.9　compliance or non-compliance by the Employer with clause 5.7.1;

4.12.10　suspension by the Contractor of the performance of his obligations under this Contract to the Employer pursuant to clause 4.4A provided the suspension was not frivolous or vexatious.

---

[*] See footnote [bb] to clause 8.3 (*Definitions*).

COMMENTARY ON CLAUSES 4.11 AND 4.12

Clause 4.11 gives the contractor an entitlement to recover from the employer direct loss and/or expense arising from the employer deferring giving possession or where the regular progress of the works is materially affected by any of the matters itemised in clauses 4.12.1 to 4.12.10. These are all matters which, if they cause disturbance to the contractor, will directly or indirectly be the responsibility of the employer.

This commentary begins with a general discussion of the scope of the contractor's ability to recover any increased expenditure or losses suffered by him as a result of the deferment of possession or of one of the matters in clause 4.12 occurring, and it will then deal with the machinery and operation of the two clauses.

Scope of recovery

The governing words limiting the extent of recovery are 'direct loss and/or expense'. These words are the same as appear in JCT 98 and they are unquestionably intended to have the same meaning. The first point to appreciate is that the right of recovery does not extend to indirect, or what may be called consequential, loss or expense. The case of *Croudace Construction Ltd* v. *Cawoods Concrete Products Ltd* (1978) gives some assistance in drawing this distinction.

Facts:

The plaintiffs were main contractors for the erection of a school. They entered into a sub-contract with the defendants for the supply and delivery of masonry blocks. The sub-contract contained a clause which included the following words:

'...if any materials or goods supplied to us should be defective or not of the correct quality or specification ordered our liability shall be limited to free replacement of any materials or goods shown to be unsatisfactory. We are not under any circumstances to be liable for any consequential loss or damage caused or arising by reason of late supply or any fault, failure or defect in any materials or goods supplied by us or by reason of the same not being of the quality or specification ordered or by reason of any other matters whatsoever.'

The main contractor sued the sub-contractor for losses alleged to have arisen because of late delivery and defects in the materials and goods supplied, seeking as part of the claim to recover loss of productivity and additional costs of delay in executing the main contract works and also the cost to them of meeting other sub-contractors' claims which were brought about by the delays of the defendant sub-contractor. A preliminary issue was ordered to be tried as to:

'Whether on a proper construction of such Contract or Contracts, including the Defendants' Standard Conditions of Trading, the Plaintiffs are entitled to recover damages under any, and if so, which, of the Heads of Damage which they had pleaded.'

Held:

Consequential loss or damage meant that loss or damage which did not result directly and naturally from the breach of contract complained of. The Court of Appeal held that the meaning of the words 'consequential loss or damage' had already been decided in the case of *Millar's Machinery Co. Ltd* v. *David Way & Son* (1935) and that they were bound by that decision. They also agreed with its reasoning. The plaintiffs could therefore recover the losses which they had pleaded.

It is clear therefore that 'consequential' will be treated as meaning indirect and will be interpreted quite restrictively as including only such heads of loss or expense which do not flow naturally from the breach without other intervening cause and independently of special circumstances. The *Croudace* case has been approved and followed by the Court of Appeal in the cases of *British Sugar Plc* v. *N.E.I. Power Projects Ltd and Another* (1997) and *Deepak Fertilisers and Petrochemical Corporation* v. *Davy McKee (London) Ltd and Another* (1998).

This interpretation of the words 'direct loss and/or expense' brings them very close to the general common law position as to the measure of damages for breach of contract. Some of the matters listed in clause 4.12 will arise owing to a breach of contract by the employer. At common law it is well settled that, while the governing purpose of damages is to put the party whose rights have been violated in the same position, so far as money can do so, as if his rights had been observed, this overall position is qualified in that the aggrieved party is only entitled to recover such part of the loss actually resulting from the breach as was at the time of the contract reasonably foreseeable as liable to result from it. This will depend firstly on imputed knowledge and secondly on actual knowledge.

So far as imputed knowledge is concerned, a reasonable person is taken to know that in the ordinary course of things certain losses are liable to result from a breach of the contract. This is known as the first limb of the rule in the leading case of *Hadley* v. *Baxendale* (1854) (dealt with in more detail in Chapter 10, page 323). Secondly, the contracting parties may have particular knowledge which would lead them to the conclusion that a breach of contract would result in losses being suffered over and above those which might be thought to flow naturally from the breach in the ordinary course of things (the second limb): *Victoria Laundry Ltd* v. *Newman Industries Ltd* (1949). In a claim for damages for breach of contract both types of damages are recoverable although in relation to the second, i.e. that depending on the actual knowledge of the parties, specific evidence must be adduced to demonstrate this. The measure of damages for breach of contract at common law is dealt with briefly later in this book (see page 325). However, as the second type of damage is arguably indirect in nature it could be said to be outside the meaning of 'direct loss and/or expense' and so irrecoverable under clause 4.11.

Further, as has been stated above, the words 'direct loss and/or expense' are interpreted in line with the first limb of the rule in *Hadley* v. *Baxendale* (1854). This position was also confirmed by the Court of Appeal in the case of *F. G. Minter* v. *W.H.T.S.O.* (1980).

There is further reason for giving these words a restricted scope. In the case of *Wraight Ltd* v. *P.H. and T. (Holdings) Ltd* (1968) Mr Justice Megaw said:

> 'In my judgment, there are no grounds for giving the words 'direct loss and/or damage caused to the contractor by the determination' any other meaning than that which they have, for example, in a case of breach of contract.'

It is therefore apparent that the words 'direct loss and/or damage' mean the same as damages at common law for breach of contract. IFC 98 includes, as does JCT 98, the phrase 'direct loss and/or damage' as being recoverable by the contractor in the event of his employment being determined for certain stated reasons (see IFC clause 7.11.3(d) and JCT 98 clause 28.4.3.4). It is a general principle in construing the meaning of words in a contract to presume that the same words have the same meaning and that different words are intended to have a different meaning: in other words, that the words 'direct loss and/or expense' mean something different to the words 'direct loss and/or damage'. If this is so, then the former words appear to be more restrictive in their scope than do the latter.

Despite the present trend therefore of allowing under these words such items as the loss of profit which could have been earned on another contract had the disruption and delay not occurred, it may not yet be beyond doubt that such claims, whilst recoverable where the word 'damage' is used (as in the case of *Wraight* v. *P.H. and T. (Holdings) Ltd* above), will not be recoverable where the word 'expense' is used instead. However, it is equally arguable that it is the common interpretation of the word 'loss' which appears in both phrases, and which therefore can be interpreted in both places as being consistent with common law damages for breach of contract.

Heads of typical loss and expense claims

(1) Loss of productivity

Delay and disruption can lead to additional expenditure on labour and plant. While clearly recoverable, it is not always easy to establish the correct figures. This head of claim may refer to the contractor's own plant as well as that which is on hire. As is the case under every head of damage, the contractor will be in a much stronger position to substantiate his claim if he has adequate records. It cannot be too often emphasised just how important the keeping of detailed and efficient records can be to substantiate a valid claim whether to the satisfaction of the quantity surveyor, who is attempting to ascertain its value, or, should it be necessary, in adjudication, arbitration or other proceedings.

(2) Site-overheads

This head of claim will cover such matters as extra involvement of supervisory and administrative staff engaged on the site, the cost of accommodation, services etc.

(3) Head office overheads

Delay or disruption will almost certainly mean increased involvement of staff from head office dealing with problems caused by disruption. Further, the pricing of the contract in question will have included a contribution from the earnings on that contract to cover general head office overheads such as rent, rates, light, office equipment etc. This contribution will usually be related to the contract period. In other words, over a fixed period of time, it is anticipated that a certain level of contribution will be achieved. If there is delay and disruption, the same period of time will produce a lower contribution and therefore a loss will be incurred.

The measure of that loss is not always easy to determine and has led to the use of formulae. The best known of these is that set out in *Hudson's Building and Engineering Contracts*, 11th edition, paragraph 8–182, known as the 'Hudson Formula'. It is not possible in a book of this kind to debate in detail the validity or usefulness of such formulae. Suffice it to say that while their use, and in particular that of the Hudson Formula, has been criticised from time to time, they are simply an attempt at an approximation of the losses where the actual losses are extremely difficult, if not impossible, to quantify precisely. The key to their use is that they can only ever be regarded as the next best thing to actual proof, but so regarded, they may each in their own right have some merit. In any event, they should be used only in conjunction with whatever actual proof is available in the circumstances. In no sense are they to be treated as a substitute for actual proof where this can be obtained.

In the English cases of *Finnegan Ltd* v. *Sheffield City Council* (1988) and *St Modwen Developments Ltd* v. *Bowmer & Kirkland Ltd* (1996) and the Canadian case of *Ellis-Don* v. *The Parking Authority of Toronto* (1978) the use of a formula received some judicial support. Because such a claim is based on a loss of contribution from a particular contract towards the head office running costs, it is necessary to establish that had there been no delay on the particular contract, not only would resources thereby have become freed in order that they could be used to achieve a contribution from other contracts, but also that the state of the construction market was such that other work would have been available at the right price on which to deploy them. See *Whittal Builders Co. Ltd* v. *Chester-le-Street District Council* (1985), a case in which the contractor, having satisfied the court in relation to these matters, was awarded a sum based on a formula. Another relevant factor to keep in mind is that where prolongation costs are attributable to compliance by the contractor with variation instructions, an additional contribution to head office overheads may well be recouped in the applicable rates forming the basis of the valuation of the variation. Care must be taken therefore that there is no double recovery.

It can seen therefore that the use of any formula method must be treated with some caution; see the useful, interesting and critical analysis of the use of formula methods by Judge Humphrey Lloyd QC in *Alfred McAlpine Homes North Ltd* v. *Property and Land Contractors Ltd* (1995). A somewhat different approach more favourable to the use of a formula is to be found in the judgment of Judge Fox-Andrews QC in *St. Modwen Developments Ltd* v. *Bowmer & Kirkland Ltd* (1996).

(4) *Interest and financing charges*

When the contractor incurs loss and expense, this has to be financed by him either from his own capital resources or alternatively by increased borrowing. In either case, it is self-evident that the use of money costs money. Depending on the level of inflation, the cost of being stood out of money which would otherwise be available for other uses, or the cost of borrowing which would not otherwise have occurred can at best be a source of irritation to the contractor and at worst can be damaging to the point of bankruptcy or liquidation. It is therefore a matter of great importance as to how such interest or financing charges should be recovered.

Applying basic legal principles as to the measure of damages and ignoring existing case law and the tradition of the common law in relation to interest, there would appear to be little doubt that it would in the ordinary course of things be obvious that a breach of contract by the employer involving the contractor in the financing of losses and the incurring of additional expenditure would involve him in direct costs of financing and interest charges in which he would not otherwise have been involved. Unfortunately, the hostility of the traditional common law approach to interest, culminating in the House of Lords' decision in *London Chatham and Dover Railway Co.* v. *S. E. Railway Co.* (1893), has meant that this head of damages at common law has been irrecoverable, unless perhaps it can be demonstrated as being within the actual and particular knowledge of the con-tracting parties, enabling it to be claimed as special damages: see for example *Wadsworth* v. *Lydall* (1981).

In the case of the *President of India* v. *Lips Maritime Corp* (1987) in the Court of Appeal, Lord Justice Neill stated the current principles as follows:

'What then is the present law as to the recovery of damages at common law for a breach of contract which consists of the late payment of money? I would ven-ture to state the position as follows:

(1) A payee cannot recover damages by way of interest merely because the money has been paid late. The basis for this principle appears to be that the court will decline to impute to the parties the knowledge that in the ordinary course of things the late payment of money will result in loss. I would express my respectful agreement with the way in which Hobhouse J explained the surviving principle in *International Minerals and Chemical Corp.* v. *Karl O Helm AG* ...

In my judgment, the surviving principle of legal policy is that it is a legal presumption that in the ordinary course of things a person does not suffer any loss by reason of the late payment of money. This is an artificial presumption, but is justified by the fact that the usual loss is an interest loss and that compensation for this has been provided for and limited by statute.

(2) In order to recover damages for late payment it is therefore necessary for the payee to establish facts which bring the case within the second part of the rule in *Hedley* v. *Baxendale*. In *Knibb* v. *National Coal Board* (1986), Sir John Donaldson MR stated the effect of the decision in *President of India* v. *La Pintada Compania Navigacion SA* in these terms:

"From this (the decision) it emerges that there is no general common law power which entitles courts to award interest ... but that if a claimant could bring himself within the second part of the rule in *Hadley* v. *Baxendale* (1854) ... he could claim special damages, notwithstanding that the breach of contract alleged consisted in the nonpayment of a debt."

It may be said that the line between the two parts of the rule in *Hedley* v. *Baxendale* has become blurred so that the division has lost much of its utility ... But it is clear, as Staughton J recognised in the instant case, that the court must find the dividing line because it is only if the claim falls within the second part of the rule that the loss can be recovered.

(3) It is important to keep in mind that the question in each case is to determine what loss was reasonably within the contemplation of the parties at the time when the contract was made. As I understand the matter, the principle in *London and Dover Railway Co.* v. *South Eastern Railway Co.* in its modern and restricted form, goes no further than to bar the recovery of claims for interest by way of general damages. Thus, in the case of a claim for damages for the late payment of money the court will not determine in favour of the plaintiff that damages flow from such delay "naturally, that is, according to the usual course of things" (*Hadley* v. *Baxendale*). But a plaintiff will be able to recover damages in respect of a special loss if it is proved that the parties had knowledge of facts or circumstances from which it was reasonable to infer that delay in payment would lead to that loss.

Moreover, I do not understand that, provided that knowledge of the facts and circumstances from which such an inference can be drawn can be proved, it is necessary further to prove that the facts or circumstances were unusual, let alone unique to the particular contract.

As I have stated earlier, the question in each case is to determine what loss was reasonably within the contemplation of the parties at the time when the contract was made. In dealing with this question the court will not impute to the parties the knowledge that damages flow "naturally" from a delay in payment. But where there is evidence of what the parties knew or ought to have known the court is in a position to determine what was in their reasonable contemplation. For this purpose, the court is entitled to take account of the terms of the contract between the parties and of the surrounding circumstances, and to draw inferences. In drawing inferences as to the parties' actual or imputed knowledge, the court is not obliged to ignore facts or circumstances of which other people doing similar business might have been aware.'*

---

*Since coming into force on 1 November 1998, the Late Payment of Commercial Debts (Interest) Act 1996 makes it possible for companies having 50 or less employees to claim against large enterprises (having more than 50 employees) a statutory rate of interest (at the time of writing 8% above Bank of England base rate) on outstanding debts.

In the case of *Holbeach Plant Hire Ltd* v. *Anglian Water Authority* (1988) it was held that financing costs (for example, in running an overdraft) associated with the hiring of plant and equipment and the purchase of materials and the payment of wages, were capable, if they satisfied the criteria set out in the *Lips Maritime* case referred to above, of being special damages and therefore recoverable provided there is evidence that the parties had such losses in contemplation or would have done so had they addressed the point at the time of entering into the contract.

Surprisingly, in the Scottish Court of Session case of *Ogilvie Buildings Ltd* v. *Glasgow City District Council* (1994) the Scottish Court held that a claim for bank charges and loss of interest came within the first limb of the rule in *Hadley* v. *Baxendale*. Sensible though this decision may be, it is clearly contrary to House of Lords' authority and, at any rate so far as England and Wales is concerned, it is unlikely to be followed. The position as stated in the *President of India* case was confirmed by the Court of Appeal in *I.M. Properties Plc* v. *Cape and Dalgleish (a Firm)* (1998).

In any event a claim for financing charges in respect of loss and expense is different in nature to a claim for interest for late payment of a debt, which is the situation in which the common law displays such hostility. Until the architect has issued his certificate including the ascertainment of the loss and expense, there is no debt due and accordingly the strict common law approach has no obvious application. However, as regards any delay by the employer in paying an interim certificate containing an ascertainment of loss and expense, the common law does apply so that, provided payment is in fact made before proceedings are instituted by the contractor, he cannot claim interest in respect of late payment of the ascertained sum unless it can be squeezed into the category of special damage under the second limb of the rule in *Hadley* v. *Baxendale* (1854). Similarly, there can be no claim at common law for interest where the employer (in circumstances where he is vicariously liable for the actions of his architect) is in breach of contract following late certification by the architect. However, in this latter situation it may be possible, depending on the circumstances, to include such interest or financing charges as part of a claim for loss and expense: *Rees and Kirby Ltd* v. *Swansea City Council* (1985).

So far as the actual claim for loss and expense is concerned, it is now clear that financing charges, or the cost of being stood out of one's money, are a recoverable head of claim and rightly so. This is established beyond doubt in the Court of Appeal in the case of *F.G. Minter Ltd* v. *Welsh Health Technical Services Organisation* (1980). The Court of Appeal held that the historic common law hostility towards awarding interest was no bar to the recovery of financing charges because there should be no presumption in favour of an anomaly and anachronism. Subject therefore to the particular contractual machinery for recovering loss and expense, in principle interest and financing charges are properly recoverable. The only question therefore is the period for which they are recoverable and this will depend on the wording of the contractual provisions concerned.

So far as IFC 98 is concerned, the important words in clause 4.11 are '... the Contractor has incurred or is likely to incur direct loss and/or expense'. Interest and financing charges will therefore be recoverable, subject to the machinery of the clause, up until at least the date of ascertainment by the quantity surveyor. The contractor would therefore be well advised to state a daily interest rate or some other method by which the cost can be calculated.

A consideration of the judgments of the Court of Appeal in the *Minter* case and in *Rees and Kirby Ltd* v. *Swansea City Council* (1985) prompts the following suggested guidelines for the recovery of interest or financing charges as part of a claim for reimbursement of direct loss or expense.

(i) Timing of application

(a) The application should be made 'within a reasonable time' of it becoming apparent that direct loss or expense has been or is likely to be incurred.

(b) If the direct loss or expense is of a continuing nature it was unwise and possibly fatal, under JCT 63, for the contractor to wait until the loss or expense had stopped before making the application. Under the provisions of IFC 98 and JCT 98 successive applications are probably not required as the initial application is likely to cover continuing losses, including financing charges.

(c) However, even if the contractor has failed to make his application within a reasonable time, unless the employer takes the point at an early stage, he may find himself estopped from challenging the contractor's application on this ground. This will be particularly so if the employer has treated the application as being valid and the contractor has relied on this to his detriment.

(ii) Contents of application

Generally, a liberal approach is to be adopted in relation to the contents of the written application. However, the cost of being stood out of money, i.e. the financing or interest charges, should be referred to in the application even if not strictly necessary under IFC 98 or JCT 98.

(iii) Period for which financing charges can be recovered

Financing charges can be recovered from the date of their being incurred up to the date of the primary loss being reimbursed with the exception of any period during that time where the financing charge is being incurred for a reason not directly attributable to the incurring of the primary loss to which the finance charge initially related. This could for example relate to the situation where the contractor takes an unreasonable time to provide necessary details of his loss or expense (but note that in the *Rees* case itself a period of almost 12 months to produce a claim was held not to be unreasonable).

There is no automatic cut off point for recovery of financing charges at practical completion.

(iv) Compound and simple interest

During the period for which the financing charges are recoverable (see above) compound interest is payable based on periodic rests. It is suggested that usually the rate of interest and frequency of rests will be either those which applied to the contractor or those which accord with normal banking practice, whichever produces the lower figure. If the contractor pays particularly high rates of interest in circumstances where this was not reasonably foreseeable by the employer then it is unlikely that the employer will be responsible for other than normal commercial rates.

If this compounded figure having been ascertained and included in an interim payment certificate is not paid by the final date for payment by the employer and proceedings are issued before payment has been made, the contractor can make a claim in those proceedings for simple interest on the compounded financing charge, although it has to be remembered that the payment of this interest and the rate is ultimately within the discretion of the court. In practice, so far as IFC 98 and JCT 98 are concerned, the contractor is likely to claim simple interest pursuant to express contractual provisions enabling the contractor to claim at 5% over the base rate of the Bank of England. This is almost certainly more advantageous than the rate awarded by the court or an arbitrator.*

In conclusion, there appears to be a paradox. On the one hand, at common law, interest or financing charges cannot be recovered as general damages for breach of contract under the first limb of the rule in *Hadley* v. *Baxendale* (1854). If recoverable at all it must be as special damages under the second limb of the rule as being in the actual contemplation of the parties, even if not flowing naturally in the ordinary course of things from the breach. On the other hand, interest or financing charges are clearly now recoverable as part of direct loss and expense and yet, as we have seen, direct loss and/or expense has been construed as falling within only the first limb of the rule in *Hadley* v. *Baxendale* (1854) under which interest or financing charges are not recoverable. Perhaps the most rational explanation to this paradox is to emphasise that a claim for financing charges forming part of loss and expense incurred is different in nature to a claim for interest in respect of late payment of a debt in breach of contract.

(5) Loss of profit

If the words 'direct loss and/or expense' are to be interpreted in line with the measure of damages at common law for breach of contract, then there is no doubt at all that the contractor can claim for a loss of profit which he would have earned on other contracts had there been no delay and disruption to the current contract, provided such loss is foreseeable. The contractor's entitlement to claim loss of profit for breach of contract at common law is perfectly reasonable and just, and provided there is clear evidence the contractor can properly include this head of claim in his claim for loss and expense under the contract. Again, the use of formulae is commonplace. The comments relating to formulae under heading (3), *Head office overheads* (see page 234), are also relevant here.

(6) The costs of the claim

Often contractors will include a sum in their claims for the cost of preparing the claim itself. Where the preparation of the claim is undertaken by in-house staff of the contractor, this cost will no doubt be reflected somewhere in the claim for increased head office managerial time. If the preparation of the claim is undertaken by independent consultants a specific sum in respect of their fees for the

---

* But see also footnote to page 236.

preparation of the claim may be included. However, in either case, the architect or quantity surveyor may well object to such a claim. As the loss and expense must be due either to a deferment of possession or to the regular progress of the works being affected by one of the matters referred to in clause 4.12, it is certainly arguable that the cost of preparation of a claim is due not to any of the matters referred to, but to the contractor's decision to claim reimbursement of loss and expense. It can perhaps be mentioned in passing that there is authority for the proposition that professional fees paid to a claims consultant for work done as an expert witness in the preparation of a building case for arbitration are allowable in the taxation of legal costs: see *James Longley & Co. Ltd* v. *South West Thames Regional Health Authority* (1983).

It should be pointed out, and should be borne in mind by the architect and the employer, that the contractor's entitlement under clause 4.11 is expressly without prejudice to any other rights or remedies which he may possess. Where the cause of the disturbance of regular progress is a breach of contract by the employer, the contractor may therefore claim damages at common law for breach of contract and thereby include in his claim any matters which the architect or the employer may seek to reject as part of a claim for reimbursement of loss and expense. For example, in the case of *London Borough of Merton* v. *Stanley Hugh Leach* (1985) 32 BLR at pages 107-108, Mr Justice Vinelott in considering clause 24 of JCT 63 (similar in many respects to clause 4.11 and 4.12 of IFC 98) said:

'Moreover there is a clear indication in the contract that the draftsman contemplated that the contractor might have parallel rights to claim compensation under the express terms of the contract and to pursue claims for damages. That arises under clause 24(2) which I have already read and which, of course, expressly provides that the provisions of the conditions are to be without prejudice to other rights and remedies of the contractor. The effect of clause 24(2) (as I understand it) is this. Clause 24(1) specifies grounds upon which the contractor is entitled to make a claim for reimbursement of direct loss or expense for which he would not otherwise be reimbursed by a payment made under the other provisions of the contract. The grounds specified may or may not result from a breach by the architect of his duties under the contract; a claim by the contractor under sub-paragraph (a) will normally, though not perhaps invariably, arise from a failure by the architect to answer with due diligence a proper application by the contractor for instructions, drawings and the like, while a claim by the contractor under sub-clause (b) following a proper instruction requiring the opening up of works under clause 6(3) normally (though not perhaps invariably) will not involve any breach by the architect of any obligation under the contract. In either case the contractor can call on the architect to ascertain the direct loss or expense suffered and to add the loss or expense when ascertained to the contract sum. The contractor will then receive reimbursement promptly and without the expense and delay of a claim for damages. But the contractor is not bound to make an application under clause 24(1). He may prefer to wait until completion of the work and join the claim for damages for breach of the obligation to provide instructions, drawings and the like in good time with other claims for damages for breach of obligations under the contract. Alternatively he can, as I see it, make a claim under clause 24(1) in order to obtain prompt reimbursement and later claim damages for breach of contract, bringing the amount awarded under clause 24(1) into account.'

Global or rolled-up claims

Loss and expense, in order to be reimbursable, must have been caused by a deferment of possession or by one of the clause 4.12 matters. The link between cause and effect must be established. It is sometimes not practicable to relate loss or expense to one specific instance of one specific matter under clause 4.12. If this can be done then it should be. However, there could be a series of events, the interaction of which prevents this approach from working. In the case of *Crosby* v. *Portland Urban District Council* (1967) 5 BLR page 126, a case which concerned the ICE Conditions of Contract, Mr Justice Donaldson (as he then was) said at pages 135–136:

> '... Since, however, the extent of the extra cost incurred depends upon an extremely complex interaction between the consequences of the various denials, suspensions and variations, it may well be difficult or even impossible to make an accurate apportionment of the total extra cost between the several causative events.
> ... so long as the Arbitrator ... ensures that there is no duplication, I can see no reason why he should not recognise the realities of the situation and make individual awards in respect of those parts of individual items of the claim which can be dealt with in isolation and a supplementary award in respect of the remainder of these claims as a composite whole ...'

In suitable cases therefore a global or rolled-up claim may be permissible. It is, however, to be regarded as the exception rather than the rule. Nevertheless, in practice by far the majority of contractors' claims for reimbursement of direct loss and expense are framed, at any rate initially, on this global basis.

In the case of *London Borough of Merton* v. *Stanley Hugh Leach* (1985) this issue was considered again in relation to JCT 63 by Mr Justice Vinelott who said at 32 BLR page 102:

> 'In *Crosby* the arbitrator rolled up several heads of claim arising under different heads and indeed claims for which the contract provided different bases of assessment. The question accordingly is whether I should follow that decision. I need hardly say that I would be reluctant to differ from a judge of Donaldson J's experience in matters of this kind unless I was convinced that the question had not been fully argued before him or that he had overlooked some material provisions of the contract or some relevant authority. Far from being so convinced, I find his reasoning compelling. The position in the instant case is, I think as follows. If application is made (under clause 11(6) or 24(1) or under both sub-clauses) for reimbursement of direct loss or expense attributable to more than one head of claim and at the time when the loss or expense comes to be ascertained it is impracticable to disentangle or disintegrate the part directly attributable to each head of claim, then, provided of course that the contractor has not unreasonably delayed in making the claim and so has himself created

the difficulty, the architect must ascertain the global loss directly attributable to the two causes, disregarding, as in *Crosby*, any loss or expense which would have been recoverable if the claim had been made under one head in isolation and which would not have been recoverable under the other head taken in isolation. To this extent the law supplements the contractual machinery which no longer works in a way in which it was intended to work so as to ensure that the contractor is not unfairly deprived of the benefit which the parties clearly intend he should have.

   ...a rolled up award can only be made in a case where the loss or expense attributable to each head of claim cannot in reality be separated and ... where apart from that practical impossibility the conditions which have to be satisfied before an award can be made have been satisfied in relation to each head of claim.'

This issue of global claims is closely linked to the requirement to properly plead the claim with adequate particulars. In terms of the facts, there should be an analysis of cause and effect which establishes the contractual entitlement to make a claim. Provided this is done, it may then, in terms of the amount of the claim, be possible to claim globally where the loss or expense suffered arises from more than one of the causes particularised.

   A real danger for contractors in advancing a composite financial claim is that it could fail completely if any significant part of the disruption or delay is not established and the court finds no basis for awarding less than the whole amount claimed.

   It is not possible in a book dealing with IFC 98 generally to treat the topic of claims for loss and expense in any greater detail. The reader is referred to specialist works on this topic, e.g. Powell-Smith and Sims: *Building Contract Claims*, third edition. In addition, the following cases dealing with this issue will repay study:

Wharf Properties v. *Eric Cumine Associates* (1991)
ICI v. *Bovis* (1992)
Mid Glamorgan County Council v. *J. Devonald Williams & Partners* (1992)
John Holland Construction & Engineering Pty Ltd v. *Kvaerner R. J. Brown Pty Ltd and Another* (1996)
Bernhard's Rugby Landscapes Ltd v. *Stockley Park Consortium Ltd* (1997)
British Airways Pension Trustees Ltd v. *Sir Robert McAlpine & Sons Ltd* (1994)
Amec Building Ltd v. *Cadmus Investment Co. Ltd* (1996)
Inserco Ltd v. *Honeywell Control Systems* (1998)

Other limits to right of recovery

(i) Loss too remote

Reference has been made earlier (see page 232) to the question of remoteness of damage and the two limbs of the rule in *Hadley* v. *Baxendale* (1854). It seems that similar principles are in practice applied to loss and expense claims. In particular it can be said that the reference to 'direct' loss and expense is treated as a reference to the first limb of the rule in *Hadley* v. *Baxendale*, namely that if loss and expense is to be recoverable, it must be such as to arise naturally in the ordinary course of things. If the loss or expense is attributable to a matter which does not arise in the ordinary course of things but rather is attributable to some unusual or special circumstance, then it falls within the second limb of the rule in *Hadley* v. *Baxendale* and so is not recoverable except in a claim for damages for breach of contract, provided these unusual or special circumstances were in the actual contemplation of the parties as being the foreseeable result of a breach of contract (see earlier page 233 and later page 323).

There are those who might argue that as we are dealing here with an express contractual right to make a claim rather than with a claim for damages for breach of contract, there is no reason to apply the rules as to remoteness of damage which apply in the latter situation. If this is the case, then provided there is a sufficiently direct relationship between the loss and expense caused by disturbance to regular progress and any of the matters listed in clause 4.12, the sum attributable thereto would be recoverable without being subject to the application of the remoteness of damage principle. However, as yet, the courts have not adopted such an approach. See also the possible relevance of remoteness to prolongation costs between the contractor's programmed completion date and the contractual completion date under the heading *(v) Restriction on period of recovery for prolongation costs* (page 244).

(ii) Causation

There must be a sufficiently causal link between the loss and expense for which reimbursement is claimed and the matter under clause 4.12 on which reliance is placed as having caused the disruption to progress and consequentially the loss and expense. If the cause is not sufficiently direct then the claim will fail. For example, the architect may issue an instruction requiring a variation by substituting one specified material for another. The effect of this could be to disturb the regular progress of the works, e.g. a delivery date for the new materials which takes their delivery beyond the date on which the original materials were intended by the contractor to be incorporated into the works. This will properly entitle the contractor to claim.

Supposing, however, that when the materials reach the site they are discovered to be defective. This in turn causes a further period of disturbance to the regular progress of the works while other materials are ordered and obtained. While the loss and expense caused by the original disturbance of progress due to the variation, namely that caused by the delay in delivery, would be reimbursable, the further disturbance of regular progress brought about by reason of the materials being defective would not be sufficiently direct, either at common law or under this clause, to entitle the contractor to recover loss and expense resulting therefrom. While in one sense the contractor can argue that, had it not been for the instruction requiring a variation being issued, those materials would not have been part of the contract and therefore regular progress could not have been affected by them, it is not the *causa causans* (the actual and direct cause) but is only what is called a *causa sine qua non* (an indispensable pre-condition). The real cause of the disturbance of regular progress is the defective materials and not the issue of the instruction requiring a variation. The line is sometimes difficult to draw, but the principle must be kept firmly in mind.

(iii) The duty to mitigate

At common law, following a breach of contract, it is the duty of the aggrieved party to take reasonable steps to mitigate his loss. Subject to what is said below concerning notification from the contractor being required when in his opinion he is likely to incur loss and expense, there is no express requirement in the contract to mitigate the loss caused by one of the matters referred to in clause 4.12, either in relation to the extent of disturbance of regular progress or the financial consequences of it. Even so, it is submitted that there must be a general duty on the contractor to take reasonable steps to mitigate. This may be by reducing the extent to which regular progress is disturbed by the matter in question, or alternatively by limiting the loss and expense which flows from it. If the contractor fails to mitigate the former, then it can be argued that, to that extent, it is not the matter referred to in clause 4.12 which caused the disturbance of regular progress, and if he fails in relation to the latter then it can be argued that, to that extent, not all of the loss and expense is attributable to the disturbance of regular progress as part of it is the result directly of the contractor's failure to mitigate his loss. There are of course limits to the steps which the aggrieved party must take in order to mitigate his loss. It must be a reasonable step to take in all the circumstances. It will certainly not include the expenditure of substantial sums of money.

While the architect or the employer may in appropriate circumstances be able to challenge part of the contractor's claim on the basis of his failure to mitigate, it should also be borne in mind that if the contractor does take reasonable steps to mitigate his loss, and thereby incurs expenditure, then, even if the effect of this is to inadvertently aggravate the loss, this is nevertheless recoverable by him.

(iv) The regular progress of the works must be materially affected

If the disturbance of regular progress is minimal and insignificant, this will not warrant reimbursement of loss and expense.

Regular progress of the works can be materially affected without any overall delay occurring. It is not therefore necessary to establish delay or prolongation in order to found a claim for reimbursement of loss and expense. It could happen that the disturbance to progress is the result of out-of-sequence working brought about by one of the matters referred to in clause 4.12. In this way, for example, the labour force may be less effectively deployed. Certain skilled craftsmen may have to spend longer on a particular task than had been anticipated. The contractor will be involved in additional wage payments. There may be no overall delay to the completion date. Progress, while being disturbed, may not result in delay to completion, e.g. if the activity is not on the critical path.

(v) Restriction on period of recovery for prolongation costs

The contractor may have programmed to complete before the contractual completion date. His pricing will be based on a construction period shorter than the contractual period. If the regular progress of the works is materially affected by a clause 4.12 matter which causes the contractor to overrun his programme, even though not the contract period, he will suffer site-wide preliminary costs, e.g. hutting, supervision, security and general site services which he otherwise would not have incurred. Can he recover these as loss and expense?

If the claim for reimbursement is based on matters falling within clause 4.12.1.2, he cannot recover them as the architect's obligation set out in clause 1.7.2, relating the provision of information to progress of the works, does not apply where that progress would result in completion prior to the contractual completion date. In such a situation the architect's obligation is to provide information to enable the contractor to complete by the contractual completion date and not his earlier programmed completion date: see *Glenlion Construction Ltd* v. *The Guinness Trust* (1987).

If the claim for reimbursement of these prolongation costs arises from any other clause 4.12 matter, the situation may well be different. However, it has been argued that these costs are also irrecoverable in the sense that they are not additional costs at all. It is argued that in tendering against a known contract period it must be assumed that the contractor has priced his preliminaries on the assumption that he may be on site for the whole contract period – see for example the case of *Finnegan Ltd* v. *Sheffield City Council* (1988) and in particular the editorial commentary thereon at 43 BLR page 126.

It is submitted that it cannot be 'deemed' that the contractor's tender includes for all time-related costs for the entire contract period. This would often give the employer the best of both worlds. On the one hand he might be looking towards the lowest tender which in practice will generally be at least partly due to the contractor shortening the construction period. On the other hand, if the employer is responsible for causing the contractor to remain on site beyond that shorter period, he ought not to be able to contend that nothing is due for the period before the contract period has expired.

A possible approach, it is submitted, is to be found in the ordinary principles of remoteness of damage. What could be said at the date of entering into the contract as the likely result of such disturbance to progress of the kind set out in clause 4.12. The knowledge of the employer at the time of entering into the contract is crucial. For instance, if the contractor has provided, with his tender, a realistic programme, showing a construction period shorter than the contract period, it is likely to be foreseeable that prolongation costs will be suffered by an overrun of that shorter period. Whereas, if the information available at the time of entering into the contract, indicates that the contractor intends or expects to be on site for the whole contract period, clearly no prolongation costs will be suffered during that period.

It has to be said, however, that if this approach is the correct one, the situation where a clause 4.12.1.2 matter runs concurrently with one of the other matters in clause 4.12 produces a very difficult situation. If it means that the contractor would not be entitled to recover his prolongation costs between the expiry of his programmed period for completion and the contractual completion date, it might become very tempting for the employer, where the contractor looks as though he could have a claim for prolongation costs under a matter other than that in clause 4.12.1.2, to deprive the architect of information or instructions so as to ensure that the flow of information from the architect is geared to the full contract period plus a short overrun so as to wipe out the contractor's prolongation claim for the period between the contractor's realistic programmed completion date and the contract completion date. On the other hand, a potential problem with this approach is that, dependent as it is on the particular knowledge of the employer at the time of entering into the contract, any claim may fall within the second limb of the rule in *Hadley* v. *Baxendale* (1854) (see earlier page 232) and therefore may be regarded as indirect rather than direct loss and expense and so not recoverable.

The machinery and operation of the clause

The first point to appreciate is that applications for reimbursement of loss and expense are completely unconnected with the granting of extensions of time for completion under clause 2.3. Indeed, IFC 98 usefully separates the clauses whereas their equivalent in JCT 98 is consecutive. It is true that all the matters listed in clause 4.12 are mirrored in clause 2.4, although clause 2.4 also contains other matters. Clauses 2.3 and 2.4 deal with the contractor's right to an extension of time for completion of the works, which relieves him of any liability for liquidated damages. The fact that he is so relieved because of a delay, even if arising out of an event which has a mirror provision in clause 4.12, does not entitle him automatically to claim reimbursement of loss and expense. The delay may not cause a disturbance to regular progress.

In the other direction, it is possible for a claim for reimbursement of loss and expense under clause 4.11 to be made where regular progress is disturbed even though no delay at all is caused and therefore there is no question of an extension of time being granted. Of course, in practice, if there is an extension of time granted owing to delay to completion caused by an event which mirrors one of those listed in clause 4.12, then as a delay to completion almost always involves the contractor in loss and expense, he can use the fact of the grant of an extension of time as compelling evidence that some loss and expense must have been suffered simply by the fact of a delay being involved.

A further point which should be made is that there can be no reimbursement of direct loss and expense where the cause of it falls outside those matters listed in clause 4.12. Simply because the contractor finds that the work is more difficult or more time-consuming, and therefore more expensive, than he had envisaged is not of itself a ground for him to claim.

Clause 4.11 is clearly a mixture of JCT 63 and JCT 98. This cross-fertilisation has not been carried out without leaving in its wake some difficulties of construction brought about by the drafting. These difficulties are mentioned in passing during this discussion of the operation of these clauses and also in the notes to the clause which follow.

The first requirement before clause 4.11 can operate is that the contractor must make a written application. It is a condition precedent to a claim for reimbursement of loss and expense: see *London Borough of Merton* v. *Stanley Hugh Leach* (1985).

The application is the subject of the first paragraph of clause 4.11. Sub-paragraphs (a) and (b) are also relevant. It is in relation to this first paragraph of clause 4.11 that the greatest difficulties in interpretation arise. A comparison of the provisions of JCT 63 (clause 24) and JCT 98 (clause 26) in regard to the contractor's written application helps to illustrate the difficulties.

Under JCT 63 clause 24(1), the written application must be made within a reasonable time of it becoming apparent that the progress of the works has been materially affected.

Under JCT 98 clause 26.1, the written application must be made as soon as it has become, or should reasonably have become, apparent that the regular progress of the works has been or is likely to be materially affected.

Under IFC 98, the words 'within a reasonable time of it becoming apparent' do not appear to refer to the regular progress being materially affected, especially having regard to paragraph (a) of this clause. Indeed, it is difficult to establish to what the word 'it' relates at the end of the second line of the first paragraph of clause 4.11. It probably relates to the occurrence or likely occurrence of loss and expense. Further, JCT 98 goes on to state that the notice is also required where the regular progress of the works is likely to be affected even though it has not been at that date. This extra provision in JCT 98 can clearly be of considerable assistance to the architect in considering ways of preventing or limiting the disturbance of regular progress. It would have been helpful therefore to have such a requirement of advance notice in IFC 98.

Sub-paragraph (a) of clause 4.11 is treated separately rather than being included in one of the matters listed in clause 4.12. One advantage of this is that it avoids the argument which could otherwise be raised that a deferment of possession cannot result in the regular progress of the works being materially affected. To affect progress it may be argued that progress of some sort must have started. If the employer fails to give possession then it could be said that this does not disturb progress at all if the contractor has not even begun work on site.

The next matter to be considered is the contents of the written application made by the contractor. Under JCT 63 there is no express requirement as to the contents of the written application. Under JCT 98, the written application must state that the contractor has incurred, or is likely to incur, loss and expense. In IFC 98, the situation is not clear but provided the word 'it' at the end of the second line of the clause relates to 'has incurred or is likely to incur' it appears to be as in JCT 98, there being a requirement that the contractor should state that he has incurred, or is likely to incur, loss and expense. The clause is certainly not as clear as it could be. Generally the courts will take a liberal approach in relation to the contents of the written application: see *Rees and Kirby* v. *Swansea City Council* (1985).

Having received the notice within the time stated, the architect must form an opinion. Under JCT 63 clause 24(1), the architect's opinion relates to whether or not the loss and expense *has been* incurred. The ascertainment relates back to such loss and expense. Under JCT 98 clause 26.1, the architect's opinion relates to whether the regular progress of the works has been, or is likely to be, materially affected and the ascertainment relates to the loss and expense which has been or *is being* incurred by the contractor. Under IFC 98 the architect's opinion relates to whether the contractor has incurred or *is likely to* incur loss and expense. This is rather odd in that, while the architect's opinion extends to loss and expense which is likely to be incurred by the contractor, the ascertainment carried out by the architect or the quantity surveyor is in relation to loss and expense incurred. Literally, therefore, if the architect is of the opinion that the contractor is likely to incur loss and expense, he can only ascertain or instruct the quantity surveyor to ascertain such loss and expense which has been incurred. Sense can perhaps be made of these words if they are interpreted to mean that the architect, having formed an opinion that loss and expense is likely to be incurred, can then ascertain or instruct the quantity surveyor to ascertain actual loss and expense as and when incurred without the necessity for further written applications by the contractor.

Under IFC 98 (as in JCT 98 but unlike JCT 63) the contractor must support his application with such information required by the architect or the quantity surveyor as is reasonably necessary for the purpose of the operation of the clause. JCT 98 goes on to state (per clause 26.1.3) that the contractor shall also submit on request such details of loss and expense as are reasonably necessary for the ascertainment to be made. Such an express requirement is not to be found in IFC 98 although it is indirectly covered by the requirement referred to above for the contractor to provide such information on request as is reasonably necessary for the purposes of the clause. The degree of information required of the contractor will depend on many factors, including the state of the architect's and quantity surveyor's knowledge of relevant matters, as well as the ease or difficulty faced by the contractor in obtaining the information. The words of Mr Justice Vinelott in *London Borough of Merton* v. *Stanley Hugh Leach* (1985), though relating to the written application under clause 24 of JCT 63, are it is submitted of some relevance to clause 4.11 of IFC 98. He said at 32 BLR page 97:

> 'The contractor must act reasonably: his application must be framed with sufficient particularity to enable the architect to do what he is required to do. He must make his application within a reasonable time: it must not be made so late that, for instance, the architect can no longer form a competent opinion on the matters on which he is required to form an opinion or satisfy himself that the contractor has suffered the loss or expense claimed. But in considering whether the contractor has acted reasonably and with reasonable expedition it must be borne in mind that the architect is not a stranger to the work and may in some cases have a very detailed knowledge of the progress of the work and of the contractor's planning. Moreover, it is always open to the architect to call for further information either before or in the course of investigating a claim. It is possible to imagine circumstances where the briefest and most uninformative notification of a claim would suffice: a case, for instance, where the architect was well aware of the contractor's plans and of a delay in progress caused by a requirement that works be opened up for inspection but where a dispute whether the contractor had suffered direct loss or expense in consequence of the delay had already emerged. In such case the contractor might give a purely formal notice solely in order to ensure that the issue would in due course be determined by an arbitrator when the discretion would be exercised by the arbitrator in the place of the architect.'

The contractor's written application must be made 'within a reasonable time of it becoming apparent'. The difficulties of determining exactly to what 'it' refers have been discussed above. Once that question has been answered, the contractor has a reasonable time to submit his written application to the architect. Failure to notify the architect within a reasonable time may, particularly if the clause can be construed to include an obligation on the contractor to include in his application advance warning of likely future loss and expense, prejudice the architect and therefore the employer in his ability to take avoiding action. However, what is a reasonable time will depend on all the circumstances. In the case of *Tersons Ltd* v. *Stevenage Development Corporation* (1965) Lord Justice Willmer said in relation to the requirement of notice under an early edition of the ICE Conditions of the Contract that:

> '...The contractors must at least be allowed a reasonable time in which to make up their minds. Here the contractors are a limited company and that involves that, in the matter of such importance as that raised by the present case, the relevant intention must be that of the board of management.'

In that case the notice had to be given as soon as practicable but similar considerations will apply to the words 'within a reasonable time'. In the event of the contractor clearly failing to make the written application within a reasonable time, the architect is not empowered to operate the clause. Of course, if this deprives the contractor of his right to claim reimbursement, he may nevertheless pursue an action for damages for breach of contract at common law if the disruption is due to a breach of contract on the part of the employer.

Having received the written application and obtained any further information reasonably required, the architect or the quantity surveyor on his behalf must carry out the ascertainment. If the quantity surveyor is instructed to ascertain the loss and expense it will not be possible subsequently for the architect to ignore that ascertainment. The contract provides that the quantity surveyor, having been instructed by the architect, *shall* carry out the ascertainment of the loss and expense which therefore ceases to be the function of the architect. Once ascertained it shall be added to the contract sum.

The contractor's programme

The contractor's programme of works may be of some assistance both to the contractor in establishing his claim and to the architect in forming an opinion as to whether or not regular progress of the works has been materially affected. However, an agreed programme, while of some evidence, is by no means conclusive as to what regular progress should have been under the contract.

The loss and expense, to be recoverable under clause 4.11, must not be reimbursable by payment under any other provision of the contract. This will avoid the possibility of double payment, e.g. where part of the increased cost of materials or labour is already covered under the fluctuations provisions and therefore paid for in the manner provided elsewhere in the contract.

The matters listed in clauses 4.12.1 to 4.12.10

The wording of the listed matters mirrors the wording of certain of the events listed in clause 2.4. Many of the observations made in relation to the wording of those events in the commentary and notes to clause 2.4 are therefore relevant in considering these listed matters. However it must always be borne in mind that whereas the clause 2.4 events relate to delay in the contractual completion date, the matters referred to in clause 4.12 are related to regular progress of the works being materially affected and, as noted above, these are by no means the same thing. Listed below are the particular events in clause 2.4 to which the reader is referred for a consideration of the words used.

Clause 4.12 matters		*Clause 2.4 events*
4.12.1.1	(information release schedule)	2.4.7.1
4.12.1.2	(late instructions etc.)	2.4.7.2
4.12.2	(opening up and testing)	2.4.6
4.12.3	(work not forming part of this contract carried out by the employer or persons engaged by him)	2.4.8
4.12.4	(employer supplying materials or goods)	2.4.9
4.12.5	(instructions as to postponement)	2.4.5 and 3.15
4.12.6	(failure to give ingress or egress)	2.4.12
4.12.7	(instructions under clauses 1.4, 3.6, 3.3, 3.8)	2.4.5
4.12.8	(approximate quantities not a reasonable forecast)	2.4.15
4.12.9	(compliance or non-compliance with clause 5.7.1)	2.4.17
4.12.10	(contractor suspending performance of obligations pursuant to clause 4.4A)	2.4.18

Apart from the commentary and notes on clause 2.4 to which the reader is referred in relation to the specific matters, one or two further points are made as follows.

Clause 4.12.2 refers to opening up and testing in accordance with clause 3.12. The corresponding event in clause 2.4.6 refers also to clause 3.13.1 dealing with opening up and testing following the discovery of a failure of work or materials to be in accordance with the contract. The question of whether loss and expense is recoverable in respect of testing in such circumstances is discussed in the commentary to clause 3.13 (page 178).

In JCT 63 clause 24(1)(d), the approximate equivalent of what is now contained in clause 4.12.3 of IFC 98 related to *delay* on the part of direct contractors. This reference to the word delay was removed when the equivalent of this provision was inserted into JCT 80, now JCT 98 (see clause 26.2.4). It is proper that loss and expense incurred as a result of disruption to the work carried out by the employer or his direct contractors should be covered even where there has been no delay on their part.

The listed matters in clauses 4.12.1 to 4.12.10 follow quite closely those in clause 26.2 of JCT 98 and represent an improvement over those contained in JCT 63.

NOTES TO CLAUSES 4.11 AND 4.12

[1] '...opinion...'
This opinion is reviewable in adjudication and arbitration proceedings. If the parties agree to litigate it will also be reviewable in the courts: see *Beaufort Developments (NI) Ltd* v. *Gilbert-Ash N.I. Ltd* (1998) overruling *Northern Regional Health Authority* v. *Crouch* (1984).

[2] '...the regular progress...'
As the claiming of loss and expense is tied to the regular progress of the works being materially affected, it is necessary to show that there has been a disturbance

of the regular progress. Before there can be disturbance there must be some form of progress so that if any of the matters listed cause the contractor to be unable to commence progress on the works, this may not be covered by the clause, e.g. failure of the employer to give ingress to or egress from the site of the works over any land, buildings etc. in the possession and control of the employer, in such circumstances that the contractor cannot even commence progress of the works. Although there is a certain logic in this argument, it is considered unlikely to succeed before an adjudicator, an arbitrator or the courts. Further, the employer should appreciate that preventing the contractor from even commencing progress will often be a serious breach of contract so that the employer could well be faced with a substantial claim for damages for breach of contract at common law. If the breach of contract is serious enough, the contractor will also have the right to regard the employer's failure as a repudiation of his obligations under the contract and may then regard the contract as at an end.

In a similar way, loss and expense caused by one of the listed matters after progress has apparently ended could cause difficulties to the contractor in seeking to claim under this clause. It may be that the works have been completed and are awaiting a certificate of practical completion, but that before such a certificate is issued certain tests have to be carried out, e.g. in relation to a heating and ventilation system. If the testing is delayed owing to one of the matters listed in clause 4.12, can it be argued that there has been no disturbance of regular progress as there is nothing left for the contractor to do except await the results of the testing? Again it is submitted that this argument would not succeed. It is submitted that progress can be disturbed by any of the listed matters (with the possible exception of that in clause 4.12.5 (postponement of work)) occurring between the date of possession given in the appendix, or any deferred date allowed under clause 2.2, and the date of practical completion. This uncertainty could have been avoided by including within clause 4.11(b) a reference to the commencement and regular progress of the works being materially affected. (See for example clause 14.1 of NAM/SC.)

[3] '...ascertain...'
This may be regarded as an unfortunate choice of word. Dictionary definitions refer to, for example, 'finding out for certain'. In the case of *Alfred McAlpine Homes North Ltd* v. *Property and Land Contractors Ltd* (1995) Judge Humphrey Lloyd QC said:

> 'Furthermore "to ascertain" means "to find out for certain" and it does not therefore connote as much use of judgment or the formation of an opinion had "assess" or "evaluate" been used. It thus appears to preclude making general assessments as have at times to be done in quantifying damages recoverable for breach of contract.'

This approach has led some employers, and particularly auditors on their behalf, to take the view that if they cannot know for certain that every penny of the contractor's claim for loss and expense has been suffered or incurred, then nothing is due. In other words, if the contractor cannot prove every penny of his claim, he is entitled to nothing in respect thereof as it is not possible under the clause to make estimates. It is submitted that this is taking the matter too far. The architect or quantity surveyor on his behalf must make an ascertainment based on the information available. The lack of information from the contractor, while it may lead to a conservative ascertainment, ought not to lead to a conclusion that no ascertainment at all should be made.

Chapter 8
Statutory obligations etc.

CONTENT

This chapter considers section 5 which deals with statutory obligations, notices, fees and charges together with value added tax, the statutory tax deduction scheme under the Income and Corporation Taxes Act 1988 and obligations falling on the parties as a result of the Construction (Design and Management) Regulations 1994.

SUMMARY OF GENERAL LAW

Statutory requirements in relation to contract works

A building contractor, apart from his contractual obligations to the employer, will also sometimes owe a statutory duty to him and to others under various statutes, e.g. the Defective Premises Act 1972 and possibly the Building Act 1984 (and the Building Regulations enacted pursuant to the powers vested in the Secretary of State under section 1 and schedule 1 thereof). In this chapter we are primarily concerned with duties owed under statute by the contractor to the building owner/employer. However, reference must also be made to the duties falling on both employer and contractor arising out of the Construction (Design and Management) Regulations 1994. It is not every breach of an obligation imposed by statute which will give rise to a civil claim for damages by someone who has been injured or suffered loss or damage by reason of the non-compliance. The injured party must show firstly that the injury suffered was within the ambit of the statute; secondly that the statutory duty imposes a liability to civil action; thirdly that the statutory duty was not fulfilled; and fourthly that the breach of the statutory duty caused the injury.

Where the statute is silent as to a remedy, there is a presumption that the injured party will have a right of action for breach of statutory duty. For a detailed discussion of this topic the reader is referred to leading text books, e.g. *Clerk and Lindsell on Torts*, 17th edition.

So far as the Construction (Design and Management) Regulations 1994 are concerned, made pursuant to section 15 of the Health and Safety at Work etc. Act 1974 ('the Act'), the position is as follows. The Act deals with civil liability in section 47. It provides in section 47(2) that breach of duty imposed by health and safety regulations '... shall, so far as it causes damage, be actionable except insofar as the Regulations provide otherwise'. Turning to the Regulations, Regulation 21 provides that:

'Breach of a duty imposed by these Regulations, other than those imposed by Regulation 10 and Regulation 16(1)(c), shall not confer a right of action in any civil proceedings.'

Regulation 10 requires that every client shall ensure so far as is reasonably practicable that the construction phase of any project does not start unless a health and safety plan has been prepared; Regulation 16(1)(c) provides that the principal contractor shall take reasonable steps to ensure that only authorised persons are allowed into any premises where construction work is being carried out. Accordingly, it follows that so far as civil liability for breach of statutory duty is concerned, it is only contravention of those two provisions which confer a right of action in any civil proceedings.

In relation to contraventions of Regulation 10 it will be necessary for any injured party to demonstrate that the damage to his or her health or safety was as a result of the construction phase starting without a health and safety plan having been prepared.

In relation to Regulation 16(1)(c), it will be necessary for the injured party to demonstrate that the damage to health or safety was caused as a result of the principal contractor not having taken reasonable steps to ensure that only authorised personnel were allowed into premises where construction work was being carried out.

In relation to both liabilities, causation is likely to be a particularly troublesome issue in pursuing any claim.

A brief summary of the Construction (Design and Management) Regulations 1994 will be found later (see page 256).

Duties under building regulations

An issue of particular importance is whether a building contractor owes an independent statutory duty to the employer to comply with the Building Regulations 1991. Section 38 of the Building Act 1984 provides that, subject to the provisions of that section, breach of a duty imposed by building regulations, shall, so far as it causes damage, be actionable except insofar as the regulations provide otherwise. There is provision for the regulations to provide prescribed defences.

It is probable that unless and until section 38 of the Building Act 1984 is brought into force, there is no liability for breach of statutory duty. However the situation is still not absolutely clear. See *Keating on Building Contracts*, 6th edition, page 414, where the relevant cases are summarised in footnotes.

Turning to local authorities, the question of whether a duty of care in the tort of negligence is owed at all in favour of occupiers and users has not been finally determined: see *Murphy* v. *Brentwood District Council* (1990). Even if such a duty is owed, the position is likely to be the same as it is for any builder who owes a duty of care, namely it will be confined to actual personal injury or physical damage to separate property. It does not generally extend to liability in respect of the cost of averting damage to health or safety of person or property.

The employment by building owners of 'approved inspectors' under the 1984 Act on contractual terms is likely to enhance significantly the building owner's position by providing non-occupying building owners with a contractual remedy where none existed before.

Defective dwellings

The Defective Premises Act 1972 imposes duties on persons taking on work for, or in connection with, the provision of dwellings. It may therefore have an application to certain contracts let under IFC 98. Under the Act, the contractor will owe a duty to the employer in the terms of section 1 which says:

> 'A person taking on work for or in connection with the provision of a dwelling (whether the dwelling is provided by the erection or the conversion or enlargement of a building) owes a duty:
> (a) if the dwelling is provided to the order of any person, to that person; and
> (b) without prejudice to paragraph (a) above, to every person who acquires an interest (whether legal or equitable) in the dwelling;
> to see that the work which he takes on is done in a workmanlike or, as the case may be, professional manner with proper materials and so that as regards that work the dwelling will be fit for habitation when completed.'

Once an obligation to carry out work for or in connection with the provision of a dwelling is taken on, section 1 will apply not only in respect of work carried out badly but also where work which should have been done has been left undone: see *Andrews* v. *Schooling and Others* (1991).

A person taking on work will ordinarily include the contractor and also any professional persons, e.g. architects and engineers. Sub-contractors will also be covered.

It is not possible by contract to exclude or restrict the operation of any of the provisions of the Act or any liability arising by virtue of such provisions. However, the Act itself provides a defence. Section 1(2) provides:

> 'A person who takes on any such work for another on terms that he is to do it in accordance with instructions given by or on behalf of that other shall, to the extent to which he does it properly in accordance with those instructions, be treated for the purposes of this section as discharging the duty imposed on him by sub-section (1) above except where he owes a duty to that other to warn him of any defects in the instructions and fails to discharge that duty.'

While the Act will therefore provide the employer with an additional remedy in respect of defective materials or workmanship, it will not apply where materials or goods, which have been specified by the employer, are discovered to be of a kind unsuitable for the purpose for which they are being incorporated into the works. In other words, where the failure is of design or fitness for purpose, then provided the contractor has complied with his obligations under the contract documents, he will not be liable under the Act except where he owes a duty to the employer to warn him of any defects in the instructions and fails to discharge that duty. The duty to warn may well arise in a situation where a reasonably competent and experienced contractor would appreciate that the contract documents or the instructions of the architect were in some way defective.

As this statutory defence is similar in some respects to the operation of clause 5.3 of IFC 98, it is unlikely that that clause will run foul of the prohibition against restricting or excluding liability contained in the Act. Insofar as clause 5.3 (discussed in detail later on page 261) is compatible with the statutory defence, it is not an attempt to restrict or exclude liability under the Act. However, clause 5.3 does not contain the proviso, as does the Act, relating to the duty to warn, and if a duty to warn arises, the contractor's failure to warn in appropriate circumstances will render clause 5.3 void so far as the employer's rights under the Act are concerned.

Any cause of action in connection with a breach of the Act is deemed to have accrued at the time when the dwelling was completed, or where defects in the dwelling have been rectified, from the time when such rectification work was finished so far as that rectified work is concerned – see section 1(5) of the Act. The cause of action for breach of the statutory duty imposed by section 1 of the Act will expire six years from when it is deemed to have accrued.

The Construction (Design and Management) Regulations 1994 ('the Regulations')

Introduction

The Construction (Design and Management) Regulations 1994 ('the Regulations') came into force on 31 March 1995. The Regulations derive their authority principally from section 15 of the Health and Safety at Work etc. Act 1974 ('the Act') which provides that health and safety regulations may be made in respect of a wide range of matters including the carrying out of any process or the carrying out of any operation (schedule 3 para 1(c) of the Act). Breach of the Regulations can raise issues of criminal and civil liability and also having regard to the amendments made to IFC 84, now in IFC 98, in consequence of the Regulations, possible contractual liability.

So far as criminal liability is concerned, by section 33 of the Act '... it is an offence to contravene any health and safety regulations ... or any requirement or prohibition imposed under any such regulations...' (section 33(1)(c)). By section 33(3) of the Act there is provision for a fine on summary conviction not exceeding the 'Prescribed Sum' or on indictment to an unlimited fine. Furthermore, if a breach of the Regulations leads to, for example, a prohibition notice being served and should this not be complied with then this could in turn lead to imprisonment.

The Health and Safety Commission has approved a code of practice (ACOP) entitled 'Managing Construction for Health and Safety'. This gives practical guidance on compliance with the Regulations as they apply to construction projects. The code has special legal status. If a person is prosecuted for breach of the Regulations and it is proved that the ACOP has not been followed, a court will find the defendant at fault unless it can be shown that there has been compliance with the law in some other way.

The Regulations place new duties on clients, clients' agents (where appointed), designers and contractors in order that they should re-think their approach to health and safety which should be taken into account and managed effectively from conception, design and planning through to execution of the works on site and their subsequent maintenance and repair.

Clients

Clients must be reasonably satisfied that they use only competent people as planning supervisors, designers and principal contractors (see below) and must be satisfied that sufficient resources, including time, have been or will be allocated to enable the project to be carried out in compliance with health and safety law. These duties do not apply to domestic householders having construction projects carried out.

Planning supervisor

The Regulations require the appointment of a planning supervisor having overall responsibility for co-ordinating the health and safety aspects of the design and planning phase. He has to ensure that a health and safety plan is prepared and then monitors the health and safety aspects of the design and advises the client on the satisfactory allocation of resources for health and safety. He also prepares a health and safety file.

Designers

Designers must design in a way which avoids, reduces, or controls risks to health and safety as far as is reasonably practicable so that projects they design can be constructed and maintained safely.

Principal contractor

A principal contractor is to be appointed and should take account of the specific requirements of the project when preparing and presenting tenders. He will take over and develop the health and safety plan and co-ordinate the activities of all contractors and sub-contractors to ensure that they comply with relevant health and safety legislation and with the developed health and safety plan.

Other contractors (i.e. other than the principal contractor) are required to co-operate with the principal contractor and provide details on the management and prevention of health and safety risks created by their work.

Health and safety plan

A health and safety plan has to be prepared initially by the planning supervisor. During the pre-construction phase, the plan brings together health and safety information obtained from the client and designers. The health and safety plan during the construction phase will draw on the principal contractor's health and safety policy and assessments etc. The plan will continue to evolve to provide a focus for co-ordination of health and safety as construction progresses.

Health and safety file

A health and safety file must be produced. It effectively amounts to a normal maintenance manual enlarged to alert those who will be responsible for the structure after handover to risks that must be managed when the structure and associated plant is maintained, repaired, renovated or demolished.

Application of the Regulations

The Regulations apply to construction work as defined (Regulation 2(1)). This definition is lengthy and covers any building, civil engineering or engineering construction work, but excludes work concerned with exploration and extraction of mineral resources. It expressly includes such things as construction, alteration, conversion, fitting out, commissioning, renovation, repair, upkeep, redecoration, decommissioning, demolition or dismantling of a structure; preparation for an intended structure including site clearance and investigation (but not a site survey); assembly of prefabricated elements; insulation, commissioning, maintenance, repair or removal of mechanical, electrical, gas, telecommunications, computer or similar services normally fixed within or to a structure.

Regulation 3 deals with the application of the Regulations to construction work. In effect, the substantive Regulations, except for Regulation 13 dealing with the obligation of designers, do not apply to construction work where the client has reasonable grounds for believing that:

(1) The project is not notifiable (it is not notifiable unless the construction work is expected to last more than 30 working days or work is of shorter duration than that but expected to involve more than 500 person days (Regulation 7)); *and*
(2) The largest number of persons at work at any one time carrying out construction work included in the project will be, or as the case may be, is, less than five.

However, even if construction work would be excluded by satisfying the above requirements, the Regulations shall nevertheless apply if and to the extent that the construction work involves demolition or dismantling of a structure.

As noted earlier, generally, the Regulations, except for Regulation 7 (notification of project) and 13 (Designer's duties) do not apply in relation to construction work carried out for a domestic client (Regulation 3(8)).

However, commercial developers building houses are themselves clients to whom the Regulations apply and even if they sell houses before completion they remain clients (Regulation 5).

Furthermore, the Regulations do not apply to or in relation to construction work in respect of which the local authority is the enforcing authority (Regulation 3(4)).

The ACOP usefully produces a brief checking system in order to determine whether the Regulations apply. It is set out in Fig. 8.1.

Is the work minor work in premises normally inspected by the local authority?	YES →	Regulations do not apply

NO ↓

Is the work for a domestic client	YES →	Regulations do not apply, except for Regulation 7 (Notification of project) and Regulation 13 (Requirements on designer)

NO ↓

Is the work 30 days or less involving 4 persons or less on site, and not involving demolition?	YES →	Regulations do not apply except Regulation 13 (Requirements on designer)

NO ↓

Regulations apply

Fig. 8.1   ACOP checking system for application of Regulations.

CONSIDERATION OF THE RELEVANT CLAUSES OF IFC 98

Clauses 5.1 to 5.4

Statutory obligations, notices, fees and charges

5.1   The Contractor shall comply with, and give all notices required by, any statute, any statutory instrument, rule or order or any regulation or byelaw applicable to the Works (hereinafter called 'the Statutory Requirements') and shall pay all fees and charges in respect of the Works legally recoverable from him.
The amount of any such fees or charges (including any rates or taxes other than value added tax) shall be added to the Contract Sum unless they are required by the Specification/Schedules of Work/Contract Bills to be included in the Contract Sum.

Notice of divergence from Statutory Requirements

5.2   If the Contractor finds any divergence between the Statutory Requirements and the Contract Documents or between the Statutory Requirements and any instruction of the Architect/the Contract Administrator he shall immediately give to the Architect/the Contract Administrator a written notice specifying the divergence.

Extent of Contractor's liability for non-compliance

5.3   Subject to clause 5.2 the Contractor shall not be liable to the Employer under this Contract if the Works do not comply with the Statutory Requirements where and to the extent that such non-compliance of the Works results from the Contractor having carried out work in accordance with the Contract Documents or any instruction of the Architect/ the Contract Administrator.

Emergency compliance

5.4   If in any emergency compliance with clause 5.1 requires the Contractor to supply materials or execute work before receiving instructions from the Architect/the Contract Administrator then:

5.4.1 the Contractor shall supply such limited materials and execute such limited work as are reasonably necessary to secure immediate compliance with the Statutory Requirements;

5.4.2 the Contractor shall forthwith inform the Architect/the Contract Administrator thereof; and

5.4.3 the work and materials shall be treated as if they had been executed and supplied pursuant to an Architect's/a Contract Administrator's instruction requiring a Variation issued in accordance with clause 3.6 (*Variations*), provided that the Contractor has informed the Architect/the Contract Administrator in accordance with clause 5.4.2 and the emergency arose because of a divergence between the Statutory Requirements and all or any of the following documents namely: the Contract Drawings or the Specification or the Schedules of Work or the Contract Bills, or any instruction or any drawing or document issued by the Architect/the Contract Administrator under clauses

1.7 (*Further drawings or details*),

3.5 (*Architect's instructions*) or

3.9 (*Levels*).

COMMENTARY ON CLAUSES 5.1 TO 5.4

Clause 5.1 imposes an obligation on the contractor to comply with what are called 'the statutory requirements' and to pay all fees and charges in respect of the works legally recoverable from him.

The question of fees is dealt with differently in IFC 98 when compared with JCT 98. Under JCT 98 the contractor pays the fee or indemnifies the employer in respect of any liability for fees. The amount of these is added to the contract sum unless they have already been included in the contract documents. Under IFC 98 there is no reference to indemnifying the employer. The contractor simply pays those fees which are legally recoverable from him, with the employer paying those legally recoverable from the employer. Those paid by the contractor are recoverable by an addition to the contract sum unless already included for in it.

For a discussion of breaches of statutory requirements involving non-compliance with building regulations, see earlier under 'Summary of general law' page 254.

Clause 5.2 deals with divergences between the statutory requirements and the contract documents or in an instruction of the architect. For some reason, perhaps an oversight in drafting, there is no reference to divergences between the statutory requirements and further drawings or details issued under clause 1.7 or levels under clause 3.9. If the contractor finds any such divergence, he is under a duty to immediately inform the architect by written notice specifying the divergence. From the contractual point of view, the contractor is under no obligation to look for divergences. However, as he may (the position is uncertain) have a potential liability for breach of statutory duty, apart from contract, he should not disregard matters to which such statutory requirements apply, e.g. the building regulations (see earlier under 'Summary of general law' page 254).

Under JCT 98, following notification of a divergence (clause 6.1.3), the architect is expressly required to issue instructions in relation to the divergence. If it requires the works to be varied, any such instruction is to be treated as if it were a variation issued in accordance with the contract. However, under IFC 98 there is no such express requirement on the architect to issue instructions. This may be the result of an oversight in the drafting. In any event it is submitted that, on receipt of a notice from the contractor specifying the divergence, the architect must issue any necessary instructions, no doubt after consultation with the employer where appropriate, to remove the divergence. Failure to do so would be a breach of contract by the employer and would also entitle the contractor to an extension of time under clause 2.4.7.2 and to reimbursement of direct loss and/or expense under clause 4.12.1.2.

Clause 5.3 seeks to exclude the liability of the contractor to the employer 'under this Contract' if the works do not comply with the statutory requirements provided that the non-compliance results from the contractor having carried out the work in accordance with the contract documents or any instruction of the architect. This exclusion of liability is expressed not to apply where the contractor has become aware of a divergence and has failed under clause 5.2 to notify the architect.

This exclusion raises two issues. Firstly, is it effective to reduce or extinguish the contractor's liability to the employer under the contract for breach of the statutory requirements? Secondly, can it affect the independent duty, if any, owed by the contractor to the employer to comply with statutory requirements, e.g. under the building regulations?

As to the first issue, clause 5.3 clearly negatives the express obligation imposed by clause 5.1. However, if, as has been held in the case of *Street and Another* v. *Sibbabridge Ltd and Another* (1980), there is an implied term in the contract involving the contractor in an absolute obligation to comply with building regulations, the exclusion clause is required to satisfy the requirement of reasonableness under the Unfair Contract Terms Act 1977 – see sections 7(1) and 7(3) and the 'Guidelines' for the application of the reasonableness test in schedule 2 to the Act. It is submitted that the exclusion in clause 5.3 is reasonable as the contractor is doing no more than constructing the works in accordance with the particular requirements of the employer (compare paragraph (e) of the Guidelines in schedule 2 of the Act which requires that attention must be given in particular as to 'whether the goods were manufactured, processed or adapted to the special order of the customer'). It is submitted that this exclusion in the clause in favour of the contractor is reasonable despite the fact that from the case of *Acrecrest Ltd* v. *W.S. Hattrell & Partners and London Borough of Harrow* (1983) it would appear that negligence or breach of statutory duty of the architect in failing to design the works to comply with building regulations is not attributable to the employer.

In regard to the second issue, clause 5.3 does not by its terms attempt to exclude liability for any independent duty owed apart from contract. It refers to liability 'under this Contract' being excluded and not any liability apart from the contract itself.

If any loss or damage is suffered by the employer owing to a breach of such an independent duty owed by the contractor, whether depending on fault or not, the indemnity provisions in clause 6.1 will, subject to the provisos contained in that clause, take effect except in relation to damage to the contract works themselves – see clause 6.1.3 and note the case of *Tozer Kemsley Millbourn (Holdings) Ltd* v. *Cavendish Land (Metropolitan) Ltd and Others* (1983).

Clause 5.4 deals with compliance by the contractor with statutory requirements in an emergency. This is a very useful provision enabling the contractor in an emergency to take such limited steps by way of the supply of materials or the execution of work as are necessary to secure immediate compliance with any statutory requirements. The contractor is contractually entitled to do this in the absence of instructions from the architect provided the emergency situation makes compliance necessary before such instructions can be obtained. The contractor must then forthwith inform the architect and the emergency work carried out will be treated as having been done pursuant to an architect's instruction requiring a variation, provided always that the emergency arose because of a divergence between the statutory requirements and the contract documents or any instruction, drawing or document issued by the architect under clauses 1.7, 3.5 or 3.9.

Clause 5.5

Value Added Tax: Supplemental Condition A

5.5   The sum or sums due to the Contractor under article 2 of this Contract shall be exclusive of any value added tax and the Employer shall pay to the Contractor any value added tax properly chargeable by the Commissioners of Customs and Excise on the supply to the Employer of any goods and services by the Contractor under this Contract in the manner set out in Supplemental Condition A. Clause A1.1 of Supplemental Condition A shall only apply where so stated in the Appendix. [o]

To the extent that after the Base Date the supply of goods and services to the Employer becomes exempt from value added tax there shall be paid to the Contractor an amount equal to the loss of credit (input tax) on the supply to the Contractor of goods and services which contribute exclusively to the Works.

---

[o] Clause A1.1 can only apply where the Contractor is satisfied at the date the Contract is entered into that his output tax on all supplies to the Employer under the Contract will be at either a positive or a zero rate of tax.

On and from 1 April 1989 the supply in respect of a building designed for a 'relevant residential purpose' or for a 'relevant charitable purpose' (as defined in the legislation which gives statutory effect to VAT changes operative from 1 April 1989) is only zero rated if the person to whom the supply is made has given to the Contractor a certificate in statutory form: see the VAT leaflet 708 revised 1989. Where a contract supply is zero rated by certificate only the person holding the certificate (usually the Contractor) may zero rate his supply.

Supplemental Conditions

A   **Value Added Tax**

Clause 5.5

Treatment of VAT

A1   In this Supplemental Condition 'tax' means the value added tax introduced by the Finance Act 1972 which is under the care and management of the Commissioners of Customs and Excise (hereinafter called 'the Commissioners'). Supplies of goods and services under this Contract are supplies under a contract providing for periodical payment for such supplies within the meaning of Regulation 26 of the Value Added Tax (General) Regulations 1985 or any amendment or re-enactment thereof.

Alternative provisions to clauses A2 and A3.2 inclusive

A1.1   .1 Where it is stated in the Appendix pursuant to clause 5.5 of the Conditions that clause A1.1 of this Agreement applies clauses A2 to A3.2 inclusive hereof shall not apply unless and until any notice issued under clause A1.1.4 hereof becomes effective or unless the Contractor fails to give the written notice required under clause A1.1.2. Where clause A1.1 applies clauses A1 and A4 to A12 of this Agreement remain in full force and effect.

A1.1   .2 Not later than 7 days before the date for the issue of the first certificate for interim payment the Contractor shall give written notice to the Employer, with a copy to the Architect/the Contract Administrator, of the rate of tax chargeable on the supply of goods and services for which certificates for interim payments and the final certificate are to be issued. If the rate of tax so notified is varied under statute the Contractor shall, not later than 7 days after the date when such varied rate comes into effect, send to the Employer, with a copy to the Architect/the Contract Administrator, the necessary amendment to the rate given in his written notice and that notice shall then take effect as so amended.

A1.1 .3 For the purpose of complying with clause 5.5 for the recovery by the Contractor from the Employer of tax properly chargeable by the Commissioners on the Contractor, an amount calculated at the rate given in the aforesaid written notice (or, where relevant, amended written notice) shall be shown on each certificate for interim payment issued by the Architect/the Contract Administrator and, unless the procedure set out in clause A4 hereof shall have been completed, on the final certificate issued by the Architect/the Contract Administrator. Such amount shall be paid by the Employer to the Contractor or by the Contractor to the Employer as the case may be within the period for payment of certificates set out in clause 4.2 (*certificates for interim payment*) or clause 4.6 (*final certificate*) as applicable.

A1.1 .4 Either the Employer or the Contractor may give written notice to the other with a copy to the Architect/the Contract Administrator, stating that with effect from the date of the notice clause A1.1 shall no longer apply. From that date the provisions of clauses A2 to A3.2 inclusive hereof shall apply in place of clause A1.1 hereof.

Written assessment by Contractor
A2 Unless clause A1.1 applies the Contractor shall not later than the date for the issue of each certificate of interim payment and, unless the procedure set out in clause A4 shall have been completed, for the issue of the final certificate for payment give to the Employer a written provisional assessment of the respective values (less 5% or $2\frac{1}{2}$% applicable thereto: see clauses 4.2 and 4.3) of those supplies of goods and services for which the certificate is being issued and which will be chargeable, at the relevant time of supply under Regulation 26 of the Value Added Tax (General) Regulations 1985 on the Contractor at:

A2.1 a zero rate of tax (Category (i)) and

A2.2 any rate or rates of tax other than zero (Category (ii)).
The Contractor shall also specify the rate or rates of tax which are chargeable on those supplies included in Category (ii), and shall state the grounds on which he considers such supplies are so chargeable.
Employer to calculate amount of tax due – Employer's right of reasonable objection

A3.1 Upon receipt of such written provisional assessment the Employer, unless he has reasonable grounds for objection to that assessment, shall calculate the amount of tax due by applying the rate or rates of tax specified by the Contractor to the amount of the assessed value of those supplies included in Category (ii) of such assessment, and remit the calculated amount of such tax, together with the amount of the certificate issued by the Architect/the Contract Administrator, to the Contractor within the period for payment of certificates set out in clauses 4.2, 4.3 and 4.6.

A3.2 If the Employer has reasonable grounds for objection to the provisional assessment he shall within 3 working days of receipt of that assessment so notify the Contractor in writing setting out those grounds. The Contractor shall within 3 working days of receipt of the written notification of the Employer reply in writing to the Employer either that he withdraws the assessment in which case the Employer is released from his obligation under clause A3.1 or that he confirms the assessment. If the Contractor so confirms then the Contractor may treat any amount received from the Employer in respect of the value which the Contractor has stated to be chargeable on him at a rate or rates of tax other than zero as being inclusive of tax and issue an authenticated receipt under clause A5.

Written final statement – VAT liability of Contractor – recovery from Employer
A4.1 Where clause A1.1 is operated clause A4 only applies if no amount of tax pursuant to clause A1.1.3 has been shown on the final certificate issued by the Architect/the Contract Administrator. After the issue of the certificate under clause 2.10 (*discharge of obligations of Contractor on defects*) the Contractor shall as soon as he can finally so ascertain prepare a written final statement of the respective values of all supplies of goods and services for which certificates have been or will be issued which are chargeable on the Contractor at:

A4.1   .1 a zero rate of tax (Category (i)) and

A4.1   .2 any rate or rates of tax other than zero (Category (ii))
         and shall issue such final statement to the Employer.
         The Contractor shall also specify the rate or rates of tax which are chargeable on the value of those supplies included in Category (ii) and shall state the grounds on which he considers such supplies are so chargeable.
         The Contractor shall also state the total amount of tax already received by the Contractor for which a receipt or receipts under clause A5 have been issued.

A4.2   The Statement under clause A4.1 may be issued either before or after the issue of the final certificate for payment under clause 4.6.

A4.3   Upon receipt of the written final statement the Employer shall, subject to clause A7, calculate the final amount of tax due by applying the rate or rates of tax specified by the Contractor to the value of those supplies included in Category (ii) of the statement and deducting therefrom the total amount of tax already received by the Contractor specified in the statement and shall pay the balance of such tax to the Contractor within 28 days from receipt of the statement.

A4.4   If the Employer finds that the total amount of tax specified in the final statement as already paid by him exceeds the amount of tax calculated under clause A4.3 the Employer shall so notify the Contractor who shall refund such excess to the Employer within 28 days of receipt of the notification, together with a receipt under clause A5 showing the correction of the amounts for which a receipt or receipts have previously been issued by the Contractor.

Contractor to issue receipt as tax invoice
A5   Upon receipt of any amount paid under certificates of the Architect/the Contract Administrator and any tax properly paid under the provisions of clause A2 or clause A1.1 the Contractor shall issue to the Employer a receipt of the kind referred to in Regulation 12(4) of the Value Added Tax (General) Regulations 1985 containing the particulars required under Regulation 13(1) of the aforesaid Regulations or any amendment or re-enactment thereof to be contained in a tax invoice.

Value of supply – liquidated damages to be disregarded
A6.1   If, when the Employer is obliged to make payment under clause A3 or A4 he is empowered under clause 2.7 to deduct any sum calculated at the rate stated in the Appendix as liquidated damages from sums due or to become due to the Contractor under this Contract he shall disregard any such deduction in calculating the tax due on the value of goods and services supplied to which he is obliged to add tax under clause A3 or A4.

A6.2   The Contractor when ascertaining the respective values of any supplies of goods and services for which certificates have been or will be issued under the Conditions in order to prepare the final statement referred to in clause A4 shall disregard when stating such values any deduction by the Employer of any sum calculated at the rate stated in the Appendix as liquidated damages under clause 2.7.

A6.3   Where clause A1.1 is operated the Employer shall pay the tax to which that clause refers notwithstanding any deduction which the Employer may be empowered to make under clause 2.7 from the amount certified by the Architect/the Contract Administrator in a certificate for interim payment or from any amount certified by the Architect/the Contract Administrator as due to the Contractor under the final certificate.

Employer's right to challenge tax claimed by Contractor
A7.1   If the Employer disagrees with the final statement issued by the Contractor under clause A4 he may but before any payment or refund becomes due under clause A4.3 or A4.4 request the Contractor to obtain the decision of the Commissioners on the tax properly chargeable on the Contractor for all supplies of goods and services under this Contract and the Contractor shall forthwith request the Commissioners for such decision. If the Employer disagrees with such decision then, provided the Employer indemnifies and at the option of the Contractor secures the Contractor against all costs

and other expenses, the Contractor shall in accordance with the instructions of the Employer make all such appeals against the decision of the Commissioners as the Employer shall request. The Contractor shall account for any costs awarded in his favour in any appeals for which clause A4 applies.

A7.2 Where, before any appeal from the decision of the Commissioners can proceed, the full amount of the tax alleged to be chargeable on the Contractor on the supply of goods and services under the Conditions must be paid or accounted for by the Contractor, the Employer shall pay to the Contractor the full amount of tax needed to comply with any such obligation.

A7.3 Within 28 days of the final adjudication of an appeal (or of the date of the decision of the Commissioners if the Employer does not request the Contractor to refer such decision to appeal) the Employer or the Contractor, as the case may be, shall pay or refund to the other in accordance with such final adjudication any tax underpaid or overpaid, as the case may be, under the provisions of this Supplemental Condition and the provisions of clause A4.4 shall apply in regard to the provision of authenticated receipts.

Discharge of Employer from liability to pay tax to the Contractor

A8 Upon receipt by the Contractor from the Employer or by the Employer from the Contractor as the case may be, of any payment under clause A4.3 or A4.4 or where clause A1.1 of this Agreement is operated of any payment of the amount of tax shown upon the final certificate issued by the Architect/the Contract Administrator or upon final adjudication of any appeal made in accordance with the provisions of clause A4 and any resultant payment or refund under clause A4.3, the Employer shall be discharged from any further liability to pay tax to the Contractor in accordance with this Supplemental Condition. Provided always that if after the date of discharge under clause A8 the Commissioners decide to correct the tax due from the Contractor on the supply to the Employer of any goods and services by the Contractor under this Contract the amount of any such correction shall be an additional payment by the Employer to the Contractor or by the Contractor to the Employer, as the case may be. The provisions of clause A4 in regard to disagreement with any decision of the Commissioners shall apply to any decision referred to in this proviso.

Award in disputes procedure

A9 If any dispute or difference is referred to adjudication or to arbitration pursuant to article 9A or to legal proceedings then, insofar as any payment awarded in such adjudication or arbitration or legal proceedings varies the amount certified for payment for goods or services supplied by the Contractor to the Employer under this Contract or is an amount which ought to have been so certified but was not so certified, the provisions of this Agreement shall so far as relevant and applicable apply to any such payments.

Arbitration provision excluded

A10 The provisions of article 9A (*Arbitration*) shall not apply to any matters to be dealt with under clause A4.

Employer's right where receipt not provided

A11 Notwithstanding any provisions to the contrary elsewhere in the Conditions the Employer shall not be obliged to make any further payment to the Contractor under the Conditions if the Contractor is in default in providing the receipt referred to in clause A5: provided that clause A11 shall only apply where:

A11.1 the Employer can show that he requires such receipt to validate any claim for credit for tax paid or payable under this Supplemental Condition which the Employer is entitled to make to the Commissioners, and

A11.2 the Employer has
  paid tax in accordance with the provisional assessment of the Contractor under clause A2 of this Agreement unless he has sustained a reasonable objection under clause A3.2 of this Agreement; or
  paid tax in accordance with clause A1.1 of this Agreement.

A12   Where clause 7.4 becomes operative there shall be added to the amount allowable or payable to the Employer in addition to the amounts certified by the Architect/the Contract Administrator any additional tax that the Employer has had to pay by reason of determination under clause 7.1, 7.2 or 7.3 as compared with the tax the Employer would have paid if the determination had not occurred.

COMMENTARY ON CLAUSE 5.5

It is not proposed in this book to deal in any detail with the question of value added tax. The tax was originally introduced under the Finance Act 1972. It has since been significantly amended on a number of occasions, the most recent relevant changes coming into effect on 1 April 1989.

For the most part new buildings as well as works of alteration will attract value added tax at the standard rate. There will however still be some occasions on which the appropriate value added tax is at the zero rate, e.g. domestic residences and buildings for charitable purposes. Accordingly there will also on occasions be situations where part of the work is standard rated and part zero rated. This is considered again below.

Article 2 to the agreement, recitals and articles provides expressly that the contract sum is exclusive of value added tax and the employer must pay to the contractor any value added tax properly chargeable by the Commissioners of Customs and Excise on the supply to the employer of any goods or services by the contractor under the contract and in the manner set out in Supplemental Condition A. If no mention was made of the contract sum being exclusive of VAT it would generally be treated as exclusive in any event, at any rate between those contracting parties who have a knowledge of the practice of the construction industry. This is despite section 19(2) of the Value Added Tax Act 1994 which states that if the supply is for a consideration in money its value shall be taken to be such amount as, with the addition of the value added tax chargeable, is equal to the consideration. In other words the Act assumes that a quoted price is inclusive of value added tax. However in the case of *Tony Cox (Dismantlers) Ltd* v. *Jim 5 Ltd* (1996) it was held that in the construction industry there was a notorious, certain and reasonable custom that prices were quoted exclusive of value added tax and this prevailed. On the other hand, if the contract is with an individual outside of the construction industry and a price is quoted without reference to value added tax then section 19(2) will apply and it will be treated as value added tax inclusive: see *Franks and Collingwood* v. *Gates* (1983) and also *Lancaster* v. *Bird* (1998).

In the event that after the base date the supply of goods and services to the employer becomes exempt from value added tax, this would prejudice the contractor's position in that he would lose the credit of the input tax on the supply to him of goods and services. In such a case the employer must pay the contractor a sum equal to the loss of any such input tax credit.

Prior to 1 April 1989, Supplemental Condition A provided a system of provisional assessment by the contractor whereby each month he had to analyse the supply to ascertain what proportion of the supply was chargeable on him at the standard rate and what proportion was chargeable on him at the zero rate.

With new buildings of a non-residential nature being standard rated, the split between standard and zero rate will now be less frequent though it will still be necessary from time to time, e.g. an extension to a school building which includes a caretaker's flat.

Because it is now likely in many instances that the whole of the supply is either standard rated or zero rated, an alternative mechanism was introduced into Supplemental Condition A. This does not replace the existing system which remains for use where applicable. The alternative system will apply where at the outset the contractor can be confident that the whole of the supply is either standard rated or zero rated. Where there is or may be a combination of these, the previous system of monthly assessments by the contractor will apply with the contractor having to supply the employer, once he is aware of the amount of the certificate, with a provisional assessment of the value on which value added tax is due with the employer having himself to calculate the amount of value added tax thereon.

Whichever system is used there is still provision for issue by the contractor, when he receives the VAT, of a receipt in statutory form (the authenticated receipt) for that VAT for use by the employer as an input credit voucher.

The alternative system

As stated above the alternative mechanism is for use where the contractor is sure that the whole of the supply will either be standard rated or zero rated. In such a case, the appendix entry for clause 5.5 can be completed so that it provides for clause A1.1 of Supplemental Condition A to apply, by deleting 'does not apply' from the appendix entry. The effect of clause A1.1 of Supplemental Condition A can be summarised as follows:

(1) The contractor must notify the employer with a copy to the architect before the issue of the first certificate for interim payment of the appropriate rate of VAT to apply to the tax exclusive amounts shown in payment certificates, both interim and final. If the rate of tax given in this notification is changed, currently either $17\frac{1}{2}$% (standard rate) or nil % (zero rate) then a further notification is required.

(2) Each payment certificate will have shown on it an amount calculated at the rate of tax notified by the contractor. By this means the employer receives certificates showing the VAT exclusive amount as certified by the architect together with an amount of VAT chargeable thereon. He will pay one total sum.

(3) Either party by written notice to the other can revoke the use of clause A1.1 of Supplemental Condition A and thereafter the existing provisional assessment provisions apply. This could happen for instance if variations were introduced resulting in part of the contract works being supplied at the standard rate and part at the zero rate.

(4) The architect is not involved in certifying what rate of VAT is due from the employer since the architect's role is limited to certification of the VAT exclusive amount to which the architect then simply by a matter of arithmetic applies the percentage VAT notified by the contractor.

Clause 5.6

Statutory tax deduction scheme: Supplemental Condition B

5.6   Where at the Base Date the Employer was a 'contractor', or where at any time up to the issue and payment of the final certificate for payment the Employer becomes a 'contractor' for the purposes of the statutory tax deduction scheme referred to in Supplemental Condition B, that Condition shall be operated.

B   **Statutory tax deduction scheme [II]**
Supplemental Condition B
Clause 5.6
Definitions

B1.1   In this clause 'the Act' means the Income and Corporation Taxes Act 1988;
'the Regulations' means the Income Tax (Sub-Contractors in the Construction Industry) Regulations 1993 S.I. No. 743;
'contractor' means a person who is a contractor for the purposes of the Act and the Regulations;
'evidence' means such evidence as is required by the Regulations to be produced to a 'contractor' for the verification of a 'sub-contractor's' tax certificate;
'statutory deduction' means the deduction referred to in S.559(4) of the Act or such other deduction as may be in force at the relevant time;
'sub-contractor' means a person who is a sub-contractor for the purposes of the Act and the Regulations;
'tax certificate' is a certificate issuable under S.561 of the Act.

[II] The application of the Tax Deduction Scheme and these provisions is explained in JCT Practice Note 8.

Provision of evidence – tax certificate

B2.1   Not later than 21 days before the first payment becomes due under clause 4.2 or after the Employer becomes a 'contractor' as referred to in clause 5.6 the Contractor shall: either

B2.1.1 provide the Employer with the evidence that the Contractor is entitled to be paid without the statutory deduction;
or

B2.1.2 inform the Employer in writing, and send a duplicate copy to the Architect/the Contract Administrator, that he is not entitled to be paid without the statutory deduction.

B2.2   If the Employer is not satisfied with the validity of the evidence submitted in accordance with clause B2.1.1 hereof, he shall within 14 days of the Contractor submitting such evidence notify the Contractor in writing that he intends to make the statutory deduction from payments due under this Contract to the Contractor who is a 'sub-contractor' and give his reasons for that decision. The Employer shall at the same time comply with clause B5.1.

Uncertificated Contractor obtains tax certificate

B3.1   Where clause B2.1.2 applies, the Contractor shall immediately inform the Employer if he obtains a tax certificate and thereupon clause B2.1.1 shall apply.

Expiry of tax certificate

B3.2   If the period for which the tax certificate has been issued to the Contractor expires before the final payment is made to the Contractor under this Contract the Contractor shall, not later than 28 days before the date of expiry: either

B3.2.1 provide the Employer with evidence that the Contractor from the said date of expiry is entitled to be paid for a further period without the statutory deduction in which case the provisions of clause B2.2 hereof shall apply if the Employer is not satisfied with the evidence;
or

B3.2.2 inform the Employer in writing that he will not be entitled to be paid without the statutory deduction after the said date of expiry.

Cancellation of tax certificate

B3.3 The Contractor shall immediately inform the Employer in writing if his current tax certificate is cancelled and give the date of such cancellation.

Vouchers

B4 The Employer shall as a 'contractor', in accordance with the Regulations send promptly to the Inland Revenue any voucher which, in compliance with the Contractor's obligations as a 'sub-contractor' under the Regulations, the Contractor gives to the Employer.

Statutory deduction – direct costs of materials

B5.1 If at any time the Employer is of the opinion (whether because of the information given under clause B2.1.2 or of the expiry or cancellation of the Contractor's tax certificate or otherwise) that he will be required by the Act to make a statutory deduction from any payment due to be made the Employer shall immediately so notify the Contractor in writing and require the Contractor to state not later than 7 days before each future payment becomes due (or within 10 days of such notification if that is later) the amount to be included in such payment which represents the direct cost to the Contractor and any other person or materials used in carrying out the Works.

B5.2 Where the Contractor complies with clause B5.1 he shall indemnify the Employer against loss or expense caused to the Employer by any incorrect statement of the amount of direct cost referred to in that clause.

B5.3 Where the Contractor does not comply with clause B5.1 the Employer shall be entitled to make a fair estimate of the amount of direct cost referred to in that clause.

Correction of errors

B6 Where any error or omission has occurred in calculating or making the statutory deduction the Employer shall correct that error or omission by repayment to, or by deduction from payments to, the Contractor as the case may be subject only to any statutory obligation on the Employer not to make such correction.

Relation to other Clauses of the Contract

B7 If compliance with this Supplemental Condition involves the Employer or the Contractor in not complying with any other provisions of this Contract, then the provisions of this Supplemental Condition shall prevail.

Disputes or differences – application of relevant Procedures

B8 The relevant procedures applicable under this Contract to the resolution of disputes or differences shall apply to any dispute or difference between the Employer and the Contractor as to the operation of this Supplemental Condition except where the Act or the Regulations or any other Act of Parliament or statutory instrument, rule or order made under an Act of Parliament provide for some other method of resolving such dispute or difference.

COMMENTARY ON CLAUSE 5.6

It is not proposed to deal in this book with the provisions of the statutory tax deduction scheme. Reference should be made to the Income and Corporation Taxes Act 1988, together with the regulations made thereunder, and to text books on the subject. Further assistance can be obtained from publications issued by the Inland Revenue and also from JCT Practice Note 8.

The Inland Revenue's new Construction Industry Scheme comes into force on 1 August 1999. It introduces significant changes to the existing system. It imposes much more stringent requirements on sub-contractors seeking certification to enable them to be paid gross, without deduction of income tax. Many main contractors will be classified as sub-contractors as many employers will find themselves falling within a widened category of 'contractor' under the scheme. One very important point to appreciate is that even those sub-contractors seeking payment net of tax rather than gross will still need to hold a registration card, without which they are not entitled to be paid at all.

Accordingly, IFC 98 will be subject to amendment by the tribunal to cater for the new scheme. It is believed that these amendments will include transitional provisions to be brought into contracts already existing on 1 August 1999 if the parties to the contract agree to this.

Clause 5.7

Provisions for use where the Appendix states that all the CDM Regulations apply
5.7.1   The Employer shall ensure:
that the Planning Supervisor carries out all the duties of a planning supervisor under the CDM Regulations; and
where the Contractor is not the Principal Contractor, that the Principal Contractor carries out all the duties of a principal contractor under the CDM Regulations.

5.7.2   Where the Contractor is and while he remains the Principal Contractor, the Contractor shall comply with all the duties of a principal contractor set out in the CDM Regulations; and in particular shall ensure that the Health and Safety Plan has the features required by regulation 15(4) of the CDM Regulations. Any amendment by the Contractor to the Health and Safety Plan shall be notified to the Employer, who shall where relevant[1] thereupon notify the Planning Supervisor and the Architect/the Contract Administrator.

5.7.3   Clause 5.7.3 applies from the time the Employer pursuant to article 6 appoints a successor to the Contractor as the Principal Contractor. The Contractor shall comply at no cost to the Employer[2] with all the reasonable requirements of the Principal Contractor to the extent that such requirements are necessary for compliance with the CDM Regulations; and, notwithstanding clause 2.3, no extension of time shall be given in respect of such compliance.

5.7.4   Within the time reasonably required[3] in writing by the Planning Supervisor to the Contractor, the Contractor shall provide, and shall ensure that any sub-contractor, through the Contractor, provides, such information to the Planning Supervisor or, if the Contractor is not the Principal Contractor, to the Principal Contractor as the Planning Supervisor reasonably requires for the preparation, pursuant to regulations 14(d), 14(e) and 14(f) of the CDM Regulations, of the health and safety file required by the CDM Regulations.

COMMENTARY ON CLAUSE 5.7

It is not proposed to deal in this book in detail with the provisions of the Construction (Design and Management) Regulations 1994 (the CDM Regulations). A brief summary of these is to be found earlier in this chapter (see page 256). Those interested in finding out more about the Regulations are recommended to read the Health and Safety Commission Approved Code of Practice *Managing Construction for Health and Safety*, no. L.54.

The CDM Regulations are defined in clause 8.3 to include any remaking or amendment. Clause 8.3 also defines 'Health and Safety Plan', 'Planning Supervisor' and 'Principal Contractor' by reference to their meaning in the CDM Regulations.

Clause 5.7 applies where the appropriate item in the appendix to IFC 98 states that all of the CDM Regulations apply rather than only Regulations 7 (notification) and 13 (duties on designers).

The employer in effect contractually guarantees that the planning supervisor, who is to be appointed by the 'client' (the employer or the employer's agent) under the Regulations, carries out all of the duties of the planning supervisor under the Regulations and also similarly that if the contractor is not the principal contractor appointed by the 'client' under the Regulations, the principal contractor will likewise carry out all the duties of a principal contractor under the Regulations.

This is a potentially significant and onerous obligation falling on the employer. However, as between the contractor and the employer, it is entirely reasonable that it should be the employer who is responsible for appointing the planning supervisor and principal contractor, to make sure that they comply with their statutory duties. If they do not so comply then the employer will be in breach of contract and liable for damages. In addition, if the employer in either ensuring compliance or in failing to ensure compliance causes the contractor to be delayed in completing the works or causes the regular progress of the works to be materially affected, the contractor will obtain an appropriate extension of time under clause 2.4.17 and reimbursement of direct loss and/or expense suffered under clause 4.12.9.

Finally, if the employer fails pursuant to the conditions to comply with the requirements of the CDM Regulations then this can lead to the contractor determining his own employment under clause 7.9.1(d). It is worth noting, however, that it is only where the employer fails to comply with the requirements of the CDM Regulations that the determination can take place. Therefore, if the employer is not the 'client' under the Regulations and instead appoints someone else as the 'client', then if that client fails to comply with the client's obligations under the Regulations, that will not involve a failure by the employer to fulfil his obligations under the Regulations, provided the 'client' was properly appointed (the employer must be reasonably satisfied that the agent or other person appointed to act as client has the competence to perform the duties imposed on a client by the Regulations – Regulation 4(2)).

Generally speaking the principal contractor will also be the contractor under the IFC 98 contract so that clause 5.7.1 will often only apply so far as the planning supervisor is concerned. However there will be occasions where the contractor under IFC 98 is not the principal contractor, in which case the whole of clause 5.7.1 will be relevant.

Bearing in mind the employers' obligations under clause 5.7.1, it is strongly advisable for the employer to ensure that there are adequate terms of engagement between himself and the 'client' (if different), and between himself and the planning supervisor and principal contractor (where not the contractor under IFC 98), including appropriate indemnity clauses in the employer's favour should a failure by the client, planning supervisor or principal contractor to carry out their duties under the Regulations involve the employer in liability whether to the contractor or to anyone else.

Some of the main duties of client, planning supervisor and principal contractor are referred to in the brief summary of the Regulations earlier in this chapter (see page 256). In addition, more is said concerning the provision of a health and safety file, and the obligation of the contractor to provide adequate information to the planning supervisor in order that he may prepare it, as a condition of the contractor achieving practical completion, in the commentary on clauses 2.9 and 2.10 (see page 106).

Where the contractor is the principal contractor under the Regulations, he must comply with all of the duties of a principal contractor set out in the Regulations. Some of these are mentioned earlier in this chapter (see page 257). Clause 5.7.2 makes specific reference to the contractor's duty to ensure that a health and safety plan is provided having the features required by Regulation 15(4). This regulation provides that the principal contractor is to take such measures as is reasonable for a person in his position to take to ensure that the health and safety plan contains, until the end of the construction phase, the following features:

(1)   Arrangements for the project which ensure so far as reasonably practicable the health and safety of all persons at work carrying out construction work and all persons who may be affected by the work of such persons taking account both of the risks involved in construction work and any activities which could affect health and safety;
(2)   Sufficient information about the arrangements for the welfare of persons at work to enable any contractor (i.e. in this context, any sub-contractor or other direct contractor of the employer) to understand how they can comply with requirements placed on them in respect of welfare by or under the relevant statutory provisions.

If the health and safety plan is amended by the contractor, he is to notify the employer who in turn where relevant is to notify the planning supervisor and the architect. It is quite likely, perhaps even inevitable, that during the course of a construction project, the health and safety plan will be amended numerous times to deal with the construction process as it takes place. This may be the result of variation instructions changing the design or specification or the conditions under which the work is to be carried out; the proposed methods of working of sub-contractors of which the contractor may be unaware until some time during the construction period; or the contractor's or sub-contractor's decision to change methods of working, e.g. using a cradle instead of scaffolding. Some may question whether the employer as 'client' under the Regulations or otherwise needs to know of each and every such amendment. Under the Regulations, it is the principal contractor who takes over responsibility for the health and safety plan from the planning supervisor who prepares it in outline. The employer is not responsible either for the plan itself or for ensuring that it is complied with by the contractor or sub-contractors. This contractual obligation falling on the contractor is not something which is mirrored in the Regulations themselves.

If the contractor while principal contractor under the Regulations does not comply with all of his duties thereunder, this will be a breach of contract by the contractor which can lead to the determination of his employment by the employer under clause 7.2.1(e).

If the contractor is, at the outset of the contract, appointed as the principal contractor pursuant to the Regulations and subsequently is replaced, he does of course remain as the contractor under this contract. In such a case he has an obligation to comply at no cost to the employer with all the reasonable requirements of the appointed principal contractor to the extent that such requirements are necessary for compliance with the Regulations. The contractor cannot therefore recover loss and expense incurred as a result of such compliance. Further, the contractor will not be entitled to an extension of time if that compliance causes delay to completion.

Clause 5.7.4 provides that within the time reasonably required in writing by the planning supervisor the contractor must provide and ensure that sub-contractors provide information to the planning supervisor or, if the contractor is not also the principal contractor, then to the principal contractor, which the planning supervisor reasonably requires for the preparation, pursuant to Regulations 14(d), 14(e) and 14(f), of the health and safety file. These Regulations require the preparation by the planning supervisor of a health and safety file which must include adequate information affecting the health and safety of persons at work, and information which it is reasonably foreseeable will be necessary to ensure health and safety of persons at work, and its hand over to the client.

While the planning supervisor retains responsibility for the health and safety file throughout the course of a project, the principal contractor and other contractors have a duty to provide relevant information to enable the planning supervisor to review, amend or add to the health and safety file.

The file needs to contain design information and also information added during the construction phase. The Health and Safety Commission approved code of practice (ACOP) *Managing Construction for Health and Safety* provides in its appendix 5 a list of typical information which may be included in a health and safety file. It refers to:

(1) 'As built' drawings and plans along with design criteria
(2) General details of the construction methods and materials used
(3) Details of equipment and maintenance facilities
(4) Maintenance procedures and requirements
(5) Manuals produced by specialist contractors and suppliers which outline operating and maintenance procedures and schedules for plant and equipment
(6) Details on the location and nature of utilities and services, including emergency and fire-fighting systems.

For further guidance reference may be made to the National Joint Consultative Committee for Building Procedure Note 16.

It can be seen therefore that a comprehensive amount of information may have to be supplied before there is sufficient for compliance. It will be a contentious area.

It will be recalled (see clause 2.9) that there must be sufficient compliance with this obligation before the architect certifies practical completion.

NOTES TO CLAUSE 5.7

[1] '...who shall where relevant...'
Presumably it will be relevant where action is likely to be required on the part of the planning supervisor or the architect. However, the employer will not neces-

sarily be in a position to know if this is the case. As a precaution employers may well therefore supply copies of all amendments notified to them to both planning supervisor and architect. Alternatively, employers may well insert a requirement somewhere in the contract documentation for the contractor to notify amendments directly to the planning supervisor and the architect to remove the administrative burden from the employer.

[2] '... at no cost to the Employer...'
This is interesting. Clause 5.7.1 provides that the employer is to ensure that where the contractor is not also the principal contractor, the principal contractor carries out all of his duties under the Regulations. Yet, the requirement here in clause 5.7.3 is for the contractor at no cost to the employer to comply with all reasonable requirements of the principal contractor. So, for example, Regulation 19(1)(a) requires a contractor to co-operate with the principal contractor so far as is necessary to enable each of them to comply with their duties under the relevant statutory provisions. And, under Regulation 19(1)(b) the contractor must so far as reasonably practicable promptly provide the principal contractor with any information which might affect the health or safety of any person at work carrying out the construction work or anyone else likely to be affected by that work or which might justify a review of the health and safety plan. What then if the principal contractor requires information for such a purpose from the contractor? The effect may be that the works are delayed or progress is disrupted, e.g. the principal contractor may require information to give to the planning supervisor which is necessary in connection with a health and safety assessment in respect of a variation in design emanating from the architect.

   Under clause 5.7.1 in conjunction with 2.4.17 and 4.12.9, it would appear that if delay or disruption is caused, the contractor can obtain an extension of time and reimbursement of loss and expense; whereas, under clause 5.7.3, he must comply at no cost to the employer and it is further stated that no extension of time shall be given in respect of such compliance. The only way to make sense of this appears to be to construe clause 5.7.3 as a limitation on the contractor's rights under clauses 2.4.17 and 4.12.9 to the effect that the delay or disruption is not in fact being caused by the compliance or non-compliance by the employer in ensuring that the principal contractor carries out his duties under the Regulations, but rather that the delay or disruption is caused by the contractor complying with a reasonable requirement of the principal contractor under clause 5.7.3. However, such a view can only be tentatively put forward.

[3] '... the time reasonably required...'
What amounts to 'the time reasonably required' is bound to be the subject of differences of view from time to time. Presumably, if the time given is not reasonable the contractor need not provide the information unless and until a reasonable time is given. If a reasonable time is given and this involves the contractor in delay or disruption, what is the position? Is the delay or disruption the result of the planning supervisor complying or not complying with his duties as planning supervisor under the Regulations so as to bring in clause 2.4.17 (extension of time) and clause 4.12.9 (loss and expense), or is the delay or disruption caused by the contractor complying with the requirement? Perhaps if the contractor provides

the information within the time reasonably required then he will get his extension of time and reimbursement of loss and expense, whereas if he takes longer than a reasonable time the cause of the delay and disruption will be attributable to the contractor rather than the planning supervisor complying with his duties under the Regulations. However, the position is certainly not clear cut.

Chapter 9
Injury, damage and insurance

CONTENT

This chapter considers section 6, which deals with indemnities and insurance. It covers personal injury and injury or damage to property arising out of the carrying out of the contract works, and indemnities and insurances related thereto. It also deals with loss or damage to the contract works themselves and insurances in respect of this.

SUMMARY OF GENERAL LAW

(A) Loss or damage to the contract works

As a general principle, subject to relatively few exceptions, a contractor's obligation to complete the works means that he will be obliged, at his own cost, to repair or reinstate the contract works should they be damaged or destroyed before completion. The main exception to this is in relation to damage or destruction which is sufficiently fundamental to cause the contract to become frustrated in law, in which case the contractor will be excused further performance. For a brief discussion on the doctrine of frustration see page 33. In practice, a building contract will often specify expressly who is to bear the risks in the event of loss or damage to the works before completion. If the words used are sufficiently clear, the contract can enable a party whose negligence has caused the loss or damage to nevertheless impose the responsibility for such loss or damage on the other contracting party. A case in point is *Farr* v. *The Admiralty* (1953).

Facts:

The plaintiffs agreed to build a destroyer wharf on behalf of the defendants. The contract provided that the plaintiff should be responsible for, and should make good, any loss or damage arising from any cause whatsoever. A vessel belonging to the defendants collided with and damaged the works as a result of negligent navigation.

Held:

The words 'any cause whatsoever' included negligent navigation of a ship by the defendant's employee and the plaintiffs were accordingly liable under the contract to make good the damage at their own cost.

A frequent means by which the risk of loss or damage to the contract works is imposed on one of the contracting parties, even if caused by the other, is by the use of the phrase 'sole risk'. Used in the right circumstances this will not only relieve a negligent party from responsibility for the loss or damage caused by his own negligence but will impose the risk on the innocent party. There can be very good reasons for this related to insurance of the works. A good example of this is to be found in the House of Lords' case of *Scottish Special Housing Association* v. *Wimpey Construction (UK) Ltd* (1986).

Facts:

This case concerned a JCT 63 Contract (1977 Revision). Clause 20(c) dealing with insurance of the works applied. That clause provided that the existing structures together with the contract work itself 'shall be at the sole risk of the Employer as regards loss or damage by fire...'. In the course of carrying out work on the modernisation of 128 houses in Edinburgh, one of the houses was damaged by fire. The question which came before the House of Lords was whether, assuming for the point of argument that the contractor had been negligent, he was liable to the employer for the damage resulting from the fire.

Held:

The employer was to bear the sole risk of damage by fire including fire caused by the negligence of the contractor or his sub-contractors. This view of clause 20(c) was reinforced by a consideration of clause 18(2) which provided:

> 'Except for such loss or damage as is at the risk of the Employer under clause ... 20(c) ... the Contractor shall be liable for and shall indemnify the Employer against, any expense, liability, loss, claim...'

In holding that the contractor was not responsible for his own negligence in damaging the works, the House of Lords followed the earlier English case of *James Archdale & Co. Ltd* v. *Comservices Ltd* (1954).

While it may seem odd that a contractor should be excused from liability even in respect of his own negligence, this is understandable when the insurance background to many of these contractual provisions is considered. The contractual insurance provisions in JCT 63 and indeed in IFC 84 prior to Amendment 1 of November 1986 gave three options:

(1) The contractor to insure for the benefit of both contractor and employer. This applied to new building works and if this option was adopted, even if the risk which materialised was due to negligence on the part of either contractor or employer, as they would both be covered by the policy of insurance, it was thought that as a result of this there was no need to provide for one party accepting the sole risk of loss or damage. The main purpose behind including a provision that one party should take the sole risk of loss or damage to the works was so that the insurers, in respect of such loss or damage, having paid out under an insurance claim could not seek recompense from the negligent party. In other words, any right of subrogation which the insurers had was

removed by the use of such a phrase. Clearly this elimination of the right of subrogation was generally beneficial as the existence of any subrogation rights would have led both parties to take out insurance cover in respect of negligent damage to the works which would in insurance terms be both inefficient and uneconomical.

If a works policy is taken in joint names then as both contractor and employer are protected by the policy there should be no opportunity for subrogation rights to apply. However, it was held by the Court of Appeal in the case of *Surrey Heath Borough Council* v. *Lovell Construction Ltd and Another* (1990) that even in such a case the liability and indemnity provisions extended to all damage to the employer's property, including the works, even if it was damage which was the subject of a joint names insurance. In so deciding they overruled Judge Fox-Andrews QC who, at first instance, held that the overall scheme of liability, indemnity and insurance provisions produced a result whereby the contractor was not liable for negligently damaging the works.

(2) The employer to insure (except if a local authority) for the benefit of both employer and contractor. Again, as both parties were covered it was thought that it would not matter if there was negligence as no rights of subrogation could arise. However, as the *Surrey Heath* case referred to above indicates, the employer may still be able to sue the contractor. Quite what effect this has when the contractor is covered by the same insurance policy is unclear.

(3) In relation to existing structures and work done to them, no one was required to insure and the employer was to take the sole risk. Here, the employer might well have taken out insurance cover and if the employer then claimed under his own insurance policy, the purpose of placing the 'sole risk' with the employer was to prevent the insurer paying out the employer and seeking by way of subrogation to claim against the negligent contractor.

The current IFC 98 insurance provisions are dealt with in the commentary on clauses 6.3 to 6.3C (see page 301).

The insertion of a 'sole risk' provision in relation to loss or damage to contract works in a main contract, expressed to cover sub-contractors also, has had the effect of removing any duty of care owed by the sub-contractor to the employer, thus relieving the sub-contractor from liability for negligently damaging the contract works: see *Norwich City Council* v. *Paul Clarke Harvey and Another* (1988); also *Welsh Health Technical Services Organisation* v. *Haden Young Ltd and IDC* (1987). Even where the 'sole risk' technique is not used, it may be that looked at as a whole the liability, indemnity and insurance provisions reveal an intention that the sub-contractor should owe no duty of care: see *Ossory Road (Skelmersdale) Ltd* v. *Balfour Beatty Building Ltd and Others* (1993), though this absence of duty may be limited to damage of the type to be covered by the insurance and would not therefore prevent a duty of care arising in respect of uncovered consequential losses: see *Kruger Tissue (Industrial) Ltd* v. *Frank Galliers Ltd and Others* (1998).

However, in a case involving JCT 80 – *British Telecommunications Plc* v. *James Thomson & Sons (Engineers) Ltd* (1998) it was held that a sub-contractor did owe a duty of care in the tort of negligence – see also a similar decision in *National Trust*

v. *Haden Young Ltd* (1994), a case based on the JCT Minor Works Form of Contract containing different but similar provisions. Finally, see also *London Borough of Barking and Dagenham* v. *Stamford Asphalt Co. Ltd and Others* (1997).

If a party is required to bear the risk rather than the sole risk, this will not absolve the other party from liability in respect of that other party's negligence. Such a clause, being akin to an exclusion clause, will be interpreted strictly against the party relying on it: see *Dorset County Council* v. *Southern Felt Roofing Co. Ltd* (1989).

The main contract may expressly provide that sub-contractors are to be covered by the works policy or alternatively that any subrogation rights against them in favour of the insurers are waived, for example clause 6.3.3 of IFC 98.

The existence of insurance cover against loss or damage to the works is of potential benefit to both parties. For the contractor, it guarantees a fund out of which he can meet his primary obligation to complete the contract works notwithstanding their damage or destruction. For the employer, it ensures that his right to have the works completed at no additional cost to him, despite their damage or destruction, is not merely a legal remedy with no substance, as it could be if the contractor was both uninsured and impecunious.

The extent of the insurance cover required will depend on the express terms of the contract. For example, it may require what is often termed 'all risks' cover or alternatively it may be more limited covering only the most frequent and potentially serious risks, e.g. fire and flood.

Very often the contract will contain exceptions to the contractor's obligation to insure in respect of certain risks. Generally these arise either because insurance cover is unobtainable and/or because there exists a statutory compensation scheme, e.g. loss or damage due to nuclear explosions; or because the loss or damage has been caused by the employer or by someone for whom he is, for this purpose, responsible, e.g. the architect's design failure where design does not form part of the contractor's obligation.

(B) Third party claims: indemnities and insurance

Most standard forms of building contract will contain express clauses dealing with third party claims, i.e. third party claims from the employer's point of view, arising out of the carrying out of the contract works. In other words, the situation in which a stranger to the contract claims against the employer in respect of a wrong done to the stranger in the form of personal injury or physical damage to his property arising out of the carrying out of the works by the contractor. Typically the express provision will require the contractor in certain circumstances to indemnify the employer against any loss, damage, claims, proceedings, costs etc. for which the employer becomes responsible to the third party. This will generally be coupled with an obligation on the contractor to insure himself against liability in respect of such claims. That liability – and accordingly the insurance to cover it – may well extend to sub-contractors employed by a main contractor.

These indemnity provisions have tended to be construed by the courts very strictly against the party seeking to rely on them, especially where the provision would have the effect of indemnifying one party to the contract in respect of his own negligence. This is so even where the contract contains equivalent cross-indemnities so that either party can benefit or suffer: see *E.E. Caledonia Ltd* v. *Orbit Valve Co.* (1993). Furthermore, a person dealing as consumer – i.e. if he does not make the contract in the course of a business whereas the other party does – cannot be made to indemnify the other party in respect of that other party's liability for negligence or breach of contract, except in so far as such indemnity satisfies the requirement of reasonableness – see sections 4 and 12 of the Unfair Contract Terms Act 1977. See also the Unfair Terms in Consumer Contracts Regulations 1994, particularly paragraph 4 which provides that any term which is contrary to the requirements of good faith and causes a significant imbalance in the parties' rights and obligations to the detriment of the consumer will not be binding. Neither the Act nor the Regulations purport to control indemnity provisions in non-consumer transactions so that the case law on this topic prior to the Act is still very relevant.

In *Walters* v. *Whessoe Ltd and Shell Refining Co. Ltd* (1960) the Court of Appeal had to consider an indemnity clause in a contract between the first and second defendants following a finding that the second defendants had been negligent. The second defendants sought to recover the damages payable by them to the plaintiffs from the first defendants under an indemnity clause, despite the fact that the second defendants had themselves been at fault when an industrial accident caused the death of an employee of the first defendants. The clause required that the first defendants:

> 'shall indemnify and hold Shell [the second defendants] their servants and agents free and harmless against all claims arising out of the operations being undertaken by [the first defendants] in pursuance of this contract or order or incidental thereto ...'.

The court held that the reference to 'all claims' did not indemnify Shell against the results of their own negligence. Lord Justice Sellars said:

> 'It is well established that indemnity will not lie in respect of loss due to a person's own negligence or that of his servants unless adequate and clear words are used or unless the indemnity could have no reasonable meaning or application unless so applied.'

Lord Justice Devlin said:

> 'It is now well established that if a person obtains an indemnity against the consequences of certain acts, the indemnity is not to be construed so as to include the consequences of his own negligence unless those consequences are covered either expressly or by necessary implication. They are covered by necessary implication if there is no other subject matter upon which the indemnity could operate.'

In *Canada Steamship Lines Ltd* v. *The King* (1952), a Privy Council case, Lord Morton of Henryton set out the approach of the courts to indemnity (and exclusion) clauses as follows:

> '(1) If the clause contains language which expressly exempts the person in whose favour it is made (hereafter called "the *proferens*") from the con-

sequence of the negligence of his own servants, effect must be given to that provision...

(2) If there is no express reference to negligence, the court must consider whether the words used are wide enough, in their ordinary meaning, to cover negligence on the part of the servants of the *proferens*...

(3) If the words used are wide enough for the above person, the court must then consider whether "the head of damage may be based on some ground other than negligence", to quote again Lord Greene in the *Alderslade* case. The "other ground" must not be so fanciful or remote that the *proferens* cannot be supposed to have desired protection against it, but subject to this qualification, which is no doubt to be implied from Lord Greene's words, the existence of a possible head of damage other than that of negligence is fatal to the *proferens* even if the words are *prima facie* wide enough to cover negligence on the part of his servants.'

This test was subsequently applied in the case of *Smith* v. *South Wales Switchgear Co. Ltd* (1978) and again by the Court of Appeal in *Dorset County Council* v. *Southern Felt Roofing Co. Ltd* (1989). However, the wording may of course be such as to make it clear that one contracting party is taking the risk of loss or damage even if caused by the other contracting party's negligence: see *Farr* v. *The Admiralty* (1953) (page 276).

Where the indemnity provision does not expressly or by implication cover losses consequent on a person's own negligence, if the loss is due in part to the fault of the contracting party against whom an indemnity is being sought, and in part due to the fault of the party seeking to enforce the indemnity, then, unless the indemnity provision expressly provides for an apportionment to be made, the indemnity provision will fail in its entirety: see *A.M.F. (International) Ltd* v. *Magnet Bowling Ltd and G.P. Trentham* (1968).

The application of the statutory limitation periods in connection with actions on indemnities should be noted. The limitation period under the Limitation Act 1980 will operate from the date when the cause of action on the indemnity accrued, i.e. most probably when the person claiming the indemnity actually incurred a liability, e.g. when the third party obtains judgment against him. In the case of *County and District Properties Ltd* v. *C. Jenner & Sons and Others* (1976) Mr Justice Swanwick said:

'...an indemnity against a breach, or an act, or an omission, can only be an indemnity against the harmful consequences that may flow from it, and I take the law to be that the indemnity does not give rise to a cause of action until those consequences are ascertained.'

If this is a correct statement of the law, in a typical third party claim, while it is the date of the damage to the third party plaintiff which will start time running against the defendant, if that defendant in turn seeks to rely on an indemnity against another defendant, time will not start to run until the first defendant's liability has been ascertained. Only then will his cause of action under the indemnity provisions accrue. However, a contrary view – that the cause of action accrues, and therefore time begins to run, as soon as the original loss, damage, claim etc. has been incurred, suffered or made – has received some support as a result of the House of Lords' case of *Scott Lithgow Ltd* v. *Secretary of State for Defence* (1989).

As a matter of convenience and expediency, a defendant can ask the court to join into the proceedings between the plaintiff and the defendant the person from whom an indemnity is being sought so that if any liability does attach to the defendant, the question of the indemnity can be considered immediately thereafter – see the Civil Procedure Rules Part 19. The use of contractual indemnities has a number of other advantages or disadvantages depending on whether the party concerned is the beneficiary under an indemnity or the giver of it. For example, if the act in respect of which the indemnity is given is a negligent act, the indemnity may be capable of extending to the recovery of purely economic loss, even though this may not otherwise have been recoverable against the negligent party. Additionally, the ordinary contractual rules as to remoteness of damage may not apply to the like extent in relation to indemnities which may be worded so as to expressly cover all loss, expense etc. resulting from the act complained of, whether or not the nature of that loss or expense is too remote in terms of ordinary contractual principles as to the measure of damages.

CONSIDERATION OF THE RELEVANT CLAUSES OF IFC 98

INTRODUCTION

As originally published, IFC 84 contained indemnity and insurance clauses which were virtually identical to those contained in JCT 63 and subsequently in JCT 80. These clauses, which have become very familiar to those connected with the building industry, were then thoroughly reviewed and substantially revised, the revisions being introduced in almost identical form in most forms of building contract published by the Tribunal. Amendment 1 to IFC 84 covering amendments to the 'insurance and related liability provisions' was published in November 1986. These amended provisions form the basis for the insurance provisions in IFC 98.

Unfortunately, many professionals who are involved with building contracts do not, in general, give this subject the attention it deserves. Their reluctance to become too deeply involved probably stems from the fact that they consider insurance to be secondary in importance to the main task of designing and constructing the works. Lack of interest may also stem from the fact that there is insufficient standardisation in the scope of insurance cover and in policy wordings and, although there has been some movement towards simplification in recent years, the drafting of many policies remains somewhat complex. However, the importance of the subject to the employer and, indeed, to all those connected with the building industry cannot be overstressed.

Some working knowledge of the subject is essential. Furthermore, the employer or the employer's insurance brokers must be involved if there is any doubt at all and probably involved as a matter of routine, when the project is in any way complex, in order to ensure that there is adequate insurance cover.

Clause 6.1

Injury to persons and property and indemnity to Employer

6.1.1 The Contractor shall be liable for, and shall indemnify the Employer against, any expense, liability, loss, claim or proceedings whatsoever arising under any statute or at common law in respect of personal injury to or the death of any person whomsoever arising out of[1] or in the course of or caused by the carrying out of the Works, except to the extent that the same is due to any act or neglect[2] of the Employer or of any person for whom the Employer is responsible including the persons employed or otherwise engaged by the Employer to whom clause 3.11 refers.

6.1.2 The Contractor shall be liable for, and shall indemnify the Employer against, any expense, liability, loss, claim or proceedings in respect of any loss, injury or damage whatsoever to any property real or personal in so far as such loss, injury or damage arises out of or in the course of or by reason of the carrying out of the Works and to the extent that the same is due to any negligence[3], breach of statutory duty, omission or default of the Contractor, his servants or agents or of any person employed or engaged upon or in connection with the Works or any part thereof, his servants or agents or of any other person who may properly be on the site upon or in connection with the Works or any part thereof, his servants or agents, other than the Employer or any person employed, engaged or authorised by him or by any local authority or statutory undertaker executing work solely in pursuance of its statutory rights or obligations. This liability and indemnity is subject to clause 6.1.3 and, where clause 6.3C.1 is applicable, excludes loss or damage to any property required to be insured thereunder[4] caused by a Specified Peril.

6.1.3 Subject to clause 6.1.4, the reference in clause 6.1.2 to 'property real or personal' does not include the Works, work executed and/or Site Materials up to and including the date of issue of the certificate of Practical Completion or up to and including the date of determination of the employment of the Contractor (whether or not the validity of that determination is disputed) under clauses 7.2 to 7.4 or clause 7.9 or 7.10 or 7.13 or, where clause 6.3C applies, under clause 6.3C.4.3 or clauses 7.2 to 7.4 or clause 7.9 or 7.10 or 7.13, whichever is the earlier.

6.1.4 If clause 2.11 has been operated then, in respect of the relevant part and as from the relevant date, such relevant part shall not be regarded as 'the Works' or 'work executed' for the purpose of clause 6.1.3.

COMMENTARY ON CLAUSE 6.1

Injury to persons

For a summary of the general law in relation to indemnities see page 279.

Clause 6.1.1 deals with the indemnity of the employer by the contractor where the employer suffers any expense, liability, loss or claim whatsoever arising under any statute or at common law in respect of personal injury or death of any person arising out of or which is caused by the carrying out of the works. If the employer is held liable to a third party either under statute or at common law, the contractor must indemnify the employer in respect of such a claim but not 'to the extent' that the claim may be due to the act or neglect of the employer or any person for whom the employer is responsible. An apportionment can therefore be made.

Clearly, if the loss sustained by the employer is entirely due to his own 'act or neglect' or the act or neglect of those for whom he is responsible, then no liability rests with the contractor. If, however, the employer is, or those for whom he is responsible are, only partly at fault, then the words 'except to the extent that' mean that liability will be apportioned between the parties, the degree of apportionment reflecting their relative liability. The question of apportionment was considered in the case of *Barclays Bank Plc* v. *Fairclough Building Ltd* (1993) at first instance (not affected on appeal) by Judge Havery QC who said:

> 'The clause offers no guidance how one should assess the extent to which a death or personal injury is due to an act or omission of the employer. It seems that culpability is irrelevant. In my judgment, the wording of the clause does not justify the court in apportioning liability on the basis provided for in section 1 of the (Law Reform (Contributory Negligence) Act 1945)...'.

Accordingly, section 1 of the 1945 Act which effectively deals with apportionment on the basis of fault is not appropriate.

The onus of proving that the employer should accept any part of the liability rests with the contractor. Generally, the employer would not be responsible for independent contractors appointed by him providing reasonable care was taken in their selection. However, it is expressly provided that the employer is responsible for such contractors – see also the second paragraph to clause 3.11 of IFC 98.

It might be thought that the architect is a person for whom the employer is responsible in this connection. In *Acrecrest* v. *Hattrell and London Borough of Harrow* (1982), the Court of Appeal held that the negligence of the architect in failing to ensure compliance of the works with building regulations should not be attributable to the building owners. It might appear therefore that in certain circumstances negligence on the part of the architect which leads to a third party claim for damages in respect of personal injury against the employer could fall within clause 6.1.1, enabling the employer to obtain an indemnity from the contractor on the basis that the architect was not a person for whom the employer is responsible. However, such a contention is, it is submitted, open to doubt, particularly as in the *Barclays Bank* case referred to above, Judge Havery QC, when dealing with a case on the IFC 84 Conditions held (unaffected on appeal) that the employer's damages should be reduced on the basis that negligence on the part of the architect could form the basis of a contributory negligence claim by the contractor. In this case, the contractor was held liable to the employer for the negligent manner in which a sub-sub-contractor had cleaned an asbestos roof using the high pressure jetting method, which resulted in a slurry being formed which, when it dried, allowed asbestos dust to enter the employer's building.

The judge said:

> 'I find that the Plaintiff, through its Supervising Officer, should have informed itself that the high pressure jetting method was to be used; that it ought to have been aware of the dangers associated with that method, namely the possibility of the ingress into the building of water containing asbestos, with the danger that ensues when the water dries out. I find that the Supervising Officer could and ought to have secured that proper precautions were carried out, and failed to do so...
>
>   In my judgment, the Defendant is the party primarily at fault; the fault of the Plaintiff is a failure to prevent the Defendant from committing that fault. But since the Plaintiff was the Architect (as well as the employer) under the Con-

tract, in my judgment its responsibility for the damage is considerable. This is not a case where the Plaintiff could reasonably leave the implementation of the works to the Defendant, even though the latter was an experienced and reputable contractor.

In my judgment having regard to the Plaintiff's share in the responsibility for the damage, it is just and equitable to reduce the damages by 40%. The Defendant is therefore liable for 60% of such damages as may be proved.'

While in this case, the architect was employed within the plaintiff's organisation so that clearly the plaintiff was responsible for its employee, it is likely, it is submitted, that the position would have been the same even if the architect was not an employee of the employer.

Injury or damage to property

Clause 6.1.2 deals with the indemnity of the employer by the contractor against any expense, liability, loss, claim or proceedings in respect of any loss, injury or damage to property other than personal injury which is covered under clause 6.1.1. While the words 'loss' and 'damage' have a reasonably precise meaning, the use of the term 'injury' is not so precise and there is consequently a question as to whether it could be held to apply to, say, interference with the reasonable access by third parties to their property or to a similar situation where the use of a third party's building is seriously affected. There is a possibility that it could also extend to such matters as the infringement of a patent. There is no specific clause in IFC 98 dealing with the infringements of patents and like rights as there is in JCT 98 (clause 9).

The basis of indemnity under this clause differs from that provided under clause 6.1.1. The contractor is only liable to indemnify the employer under this clause to the extent that the loss can be shown to be due to the negligence, breach of statutory duty, omission or default of the contractor or those for whom he is responsible as defined by the clause. This will certainly include sub-contractors, sub-sub-contractors and so on, as well as suppliers. It does not include a local authority or statutory undertaker executing work solely pursuant to its statutory rights or obligations, though they may well be included if they have any con-tractual relationship with the contractor.

If the injury or damage is not due to such negligence, breach of statutory duty, omission or default, this indemnity is of no effect and the loss will fall on the employer, unless the employer can recover all or some of the loss from the con-tractor in some other way under the general law, e.g. in contribution proceedings where the employer suffers loss as a result of being sued by someone in respect of some tortious act of the contractor which does not involve any negligence, breach of statutory duty, omission or default such as nuisance or trespass – see section 1 Civil Liability (Contribution) Act 1978.

The use of the words 'to the extent that' means that even if the employer is partly at fault the indemnity will still apply, but the loss will be apportioned between the parties, the degree of apportionment reflecting the relative respon-sibility of the contractor. See on this the *Barclays Bank* case referred to earlier (page 284).

By clause 6.1.3, the injury or damage to 'any property real or personal' covered by clause 6.1 does not apply to the works themselves or to materials on site, both of which are covered by works insurance under clause 6.3A, 6.3B or 6.3C to which reference is made later. Nevertheless, if part of the works have reached practical completion, for instance in the case of partial possession where optional clause 2.11 is implemented, then the indemnity provided by clause 6.1.2 will be applicable to that part of the works (clause 6.1.4). However, it must be remembered that the indemnity is limited to loss resulting from any negligence, breach of statutory duty, omission or default of the contractor, or those for whom he is responsible. The employer will therefore still need to consider insuring that part of the works which has reached practical completion even if he has no contractual obligation to do so since any loss may well fall outside the indemnity provisions of this clause.

Clause 6.1.2 is expressed to be subject not only to clause 6.1.3, but also to clause 6.3C.1 which imposes on the employer an obligation to insure existing structures for which he is responsible in the joint names of himself and the contractor against loss or damage from specified perils. The indemnity by the contractor in favour of the employer will not therefore extend to these existing structures if the loss or damage to them has been caused by one or more of the specified perils. The wording of clause 6.1.2 was revised by the Tribunal in Amendment 10 to IFC 84 issued July 1996 as the previous wording failed to achieve its purpose of preventing the contractor's indemnity extending to existing structures where the loss or damage was caused by a specified peril: see *Ossory Road (Skelmersdale) Ltd* v. *Balfour Beatty Building Ltd and Others* (1993). Hopefully the revised wording which provides that the indemnity excludes loss or damage to any property required to be insured under 6.3C.1 now makes the position absolutely clear.

It is interesting to note with regard to clause 6.1.1 (injury to persons), that the contractor only indemnifies the employer to the extent that the injury or death is not due to any act or neglect of the employer or those for whom he is responsible; whereas here, in clause 6.1.2 (loss injury or damage to property), the position is reversed and the contractor's indemnity only applies to the extent that the injury or damage arises due to the negligence, breach of statutory duty, omission or default of the contractor, or those for whom he is responsible. The onus of proving that the contractor should accept any part of the liability therefore rests with the employer.

NOTES TO CLAUSES 6.1.1 TO 6.1.3

[1] '. . . arising out of . . .'
These words were considered in the case of *Richardson* v. *Buckinghamshire County Council and Others* (1971) in connection with the ICE Conditions of Contract 4th Edition, clause 22(1), which is an indemnity clause. The plaintiff had been injured when he fell from his motorcycle at the point of some road works which were the subject of the contract. The defendant local authority successfully resisted a claim from the motorcyclist but the costs which were incurred in so doing were irrecoverable from the plaintiff as he was legally aided. The local authority therefore sought to recover these costs from the contractor under the wording quoted above, contending that the costs were incurred as a result of a claim arising out of

the construction of the contract works. It was held that the local authority was not entitled to an indemnity from the contractor as the motorcyclist's claim was not one which arose out of, or in consequence of, the construction of the works. This clearly makes good sense, for otherwise a contractor could find himself having to indemnify an employer in respect of the employer's costs in defending all manner of specious or far-fetched claims. As the motorcyclist could not establish that his injuries arose out of the construction of the road works, it could not be said that the costs arose out of the construction of the works.

[2] '...except to the extent that the same is due to any act or neglect...'
This exception appears to extend to those situations where the employer is not at fault in a culpable sense – see the *Barclays Bank* case dealt with earlier (page 284).

[3] '...negligence...'
In this context this means negligence in law, i.e. where there has been a breach of a legal duty of care rather than just a careless act: see *Anthony Callaghan and Thomas Welton (Trading as R. W. Construction)* v. *Hewgate Construction Ltd* (1995).

[4] '...excludes loss or damage to any property required to be insured thereunder...'
This exclusion in favour of the contractor does not extend to economic loss flowing from the physical loss or damage to the existing structures, e.g. loss of profits or increased cost of working in a damaged structure. The employer can still sue the contractor in respect of such losses where they flow from the contractor's negligence: see *Kruger Tissue (Industrial) Limited* v. *Frank Galliers Limited and Others* (1998) CLJ vol. 14, issue 6, at page 437.

Clauses 6.2.1 to 6.2.3

Insurance against injury to persons or property
6.2.1 Without prejudice to his obligation to indemnify the Employer under clause 6.1 the Contractor shall take out and maintain insurance which shall comply with clause 6.2.1 in respect of claims arising out of his liability referred to in clauses 6.1.1 and 6.1.2.
The insurance in respect of claims for personal injury to or the death of any person under a contract of service or apprenticeship with the Contractor, and arising out of and in the course of such person's employment, shall comply with all relevant legislation. For all other claims to which clause 6.2.1 applies the insurance cover: [p]
– shall indemnify the Employer in like manner to the Contractor but only to the extent that the Contractor may be liable to indemnify the Employer under the terms of this Contract; and
– shall be not less than the sum stated in the Appendix for any one occurrence or series of occurrences arising out of one event. [q]

---

[p] It should be noted that the cover granted under public liability policies taken out pursuant to clause 6.2.1 may not be co-extensive with the indemnity given to the Employer in clauses 6.1.1 and 6.1.2: for example each claim may be subject to the excess in the policy and cover may not be available in respect of loss or damage due to gradual pollution.
[q] The Contractor may, if he so wishes, insure for a sum greater than that stated in the Appendix.

6.2.2 As and when he is reasonably required to do so by the Employer the Contractor shall send to the Architect/the Contract Administrator for inspection by the Employer documentary evidence that the insurances required by clause 6.2.1 have been taken out and are being maintained, but at any time the Employer may (but not unreasonably or vexatiously) require to have sent to the Architect/the Contract Administrator for inspection by the Employer the relevant policy or policies and the premium receipts therefor.

6.2.3   If the Contractor defaults in taking out or in maintaining insurance as provided in clause 6.2.1 the Employer may himself insure against any liability or expense which he may incur arising out of such default and a sum or sums equivalent to the amount paid or payable by him in respect of premiums therefor may be deducted by him from any monies due or to become due to the Contractor under this Contract or such amount may be recoverable by the Employer from the Contractor as a debt.

COMMENTARY ON CLAUSES 6.2.1 TO 6.2.3

While clauses 6.1.1 and 6.1.2 establish an indemnity on the part of the contractor in favour of the employer, clause 6.2.1 requires the contractor also to insure in respect of any claims arising out of the contractor's liability under clause 6.1. Without that insurance the indemnity given by the contractor may well be worthless if he is financially unable to meet his liabilities under the indemnity provisions. However, even when the contractor's liabilities under clause 6.1.1 and 6.1.2 are adequately covered by insurance, employers should remember the limited nature of that indemnity. Claims arising out of any act or neglect of the employer or of any person for whom the employer is responsible are excluded from the indemnity against injury or death to the persons under clause 6.1.1 and, in the case of injury or damage to property under clause 6.1.2, the contractor is only liable if the claim arises out of his negligence, breach of statutory duty, omission or default. A separate policy taken out by the employer to protect him against claims not covered by the indemnity may therefore need to be considered.

Even where a claim made against the employer is covered by the indemnity provisions of clause 6.1.1 or 6.1.2, and where the employer has satisfied himself that the contractor's insurance under clause 6.2.1 is adequate, it would be prudent for the employer to ensure that the contractor has immediately notified his insurers if there is a claim or the possibility of a claim against him under clause 6.1.

In respect of claims for personal injury to or death of the contractor's own employees, the cover provided by the policy must comply with all relevant employment legislation. For all other claims which may arise from injury or death to third parties under clause 6.1.1 or injury or damage to property under clause 6.1.2, the insurance cover must be not less than the figure stated in the appendix for any one occurrence or series of occurrences arising out of one event. It may be prudent to obtain the advice of the employer's insurers or insurance brokers before deciding on the amount to be inserted. There is a footnote to clause 6.2.1 noting that the contractor may of course, if he wishes, insure for a sum greater than that stated in the appendix. Often the contractor's annual public liability policy will be for a sum greater than that required by the appendix. Should it be less, this issue will need to be addressed and resolved before the contract is entered into.

The indemnity which the contractor offers by this clause only extends to that for which the contractor may be liable to indemnify the employer under the terms of the contract.

Under clause 6.2.2, the contractor has an obligation to send to the architect, documentary evidence, for inspection by the employer, that the insurance provisions of clause 6.2.1 have been complied with, and that the policies are being maintained. Care should be taken when checking the evidence submitted by the contractors since policies frequently include limitation or exclusion clauses and there is always a possibility that the polices may not comply with the obligations of the contractor under the indemnity provisions. Where there is any doubt it would be advisable for the employer to seek the advice of his insurers or insurance brokers.

There is a further footnote to clause 6.2.1 pointing out that the contractor's public liability policy may not be co-extensive with the indemnity given under clauses 6.1.1 and 6.1.2, e.g. claims may be subject to an excess payable by the contractor or cover may not be available in respect of certain loss or damage, including that due to gradual pollution. Again, in such cases, this issue should be addressed and resolved before the contract is entered into.

While the indemnity provided by the contractor under clauses 6.1.1 and 6.1.2, and the insurance provisions under clause 6.2.1 covering those indemnities, are in respect of claims arising out of the carrying out of the works, the Tribunal recommend contractors and sub-contractors not to terminate their insurance cover at the date of practical completion. The Tribunal suggests that the cover remains until expiry of any defects liability period or the date of making good of defects and possibly until the issue of a final certificate (see the Tribunal's Practice Note 22 page 34/35).

The right given to the employer under clause 6.2.3 to take out insurance in default of the contractor insuring is a useful one. The employer needs to be sure that the contractor will be able to financially withstand any claim against him. A failure by the contractor to insure while being a breach of contract and entitling the employer to sue, will be a worthless right against an impecunious contractor without the express right to take out the necessary insurance cover. Should that become necessary, the employer can recover the cost by a deduction from any monies due to the contractor or, failing that, such amount may be recovered from the contractor as a debt.

Finally, it should be noted that by clause 6.2.5, it is made clear that the indemnities and insurances referred to do not apply and are not required in respect of any damage, loss or injury caused to the works or site materials or work executed, the site or any property, if it is by reason of an 'Excepted Risk'. This is defined in clause 8.3 and deals with the effect of ionising radiations or contamination by radioactivity from nuclear fuel etc. or pressure waves caused by aircraft or other aerial devices.

Clauses 6.2.4 and 6.2.5

Insurance – liability etc. of Employer

6.2.4   Where it is stated in the Appendix that the insurance to which clause 6.2.4 refers may be required by the Employer the Contractor shall, if so instructed by the Architect/the Contract Administrator, take out a policy of insurance[r] in the names of the Employer and the Contractor for such amount of indemnity as is stated in the Appendix in respect of any expense, liability, loss, claim or proceedings which the Employer may incur or sustain by reason of injury or damage to any property caused by collapse, subsidence, heave, vibration, weakening or removal of support or lowering of ground water arising out of or in the course of or by reason of the carrying out of the Works excepting injury or damage:

.1 for which the Contractor is liable under clause 6.1.2;

.2 attributable to errors or omissions in the designing of the Works;

.3 which can reasonably be foreseen to be inevitable having regard to the nature of the work to be executed and the manner of its execution;

.4 which it is the responsibility of the Employer to insure under clause 6.3C.1 (if applicable);

.5 to the Works and Site Materials brought on to the site of this Contract for the purpose of its execution except in so far as any part or parts thereof are the subject of a certificate of Practical Completion;

.6 arising from any consequence of war, invasion, act of foreign enemy, hostilities (whether war be declared or not), civil war, rebellion or revolution, insurrection or military or usurped power;

.7 directly or indirectly caused by or contributed to by or arising from the Excepted Risks;

.8 directly or indirectly caused by or arising out of pollution or contamination of buildings or other structure or of water or land or the atmosphere happening during the period of insurance; save that this exception shall not apply in respect of pollution or contamination caused by a sudden identifiable, unintended and unexpected incident which takes place in its entirety at a specific moment in time and place during the period of insurance provided that all pollution or contamination which arises out of one incident shall be considered for the purpose of this insurance to have occurred at the time such incident takes place;

.9 which results in any costs or expenses being incurred by the Employer or in any other sums being payable by the Employer in respect of damages for breach of contract except to the extent that such costs or expenses or damages would have attached in the absence of any contract.

Any such insurance as is referred to in clause 6.2.4 shall be placed with insurers to be approved by the Employer, and the Contractor shall send to the Architect/the Contract Administrator for deposit with the Employer the policy or policies and the receipts in respect of premiums paid.

The amounts expended by the Contractor to take out and maintain the insurance referred to in clause 6.2.4 shall be added to the Contract Sum.

If the Contractor defaults in taking out or in maintaining the Joint Names Policy as provided in clause 6.2.4 the Employer may himself insure against any risk in respect of which the default shall have occurred.

---

[r] A policy of insurance taken out for the purposes of clause 6.2.4 should not have an expiry date earlier than the end of the defects liability period named in the Appendix.

Excepted Risks

6.2.5 Notwithstanding the provisions of clauses 6.1.1, 6.1.2 and 6.2.1, the Contractor shall not be liable either to indemnify the Employer or to insure against any personal injury to or the death of any person or any damage, loss or injury caused to the Works or Site Materials, work executed, the site, or any property, by the effect of an Excepted Risk.

COMMENTARY ON CLAUSES 6.2.4 AND 6.2.5

Clause 6.2.4 covers the situation where the employer may incur or sustain any expense, liability, loss, claim or proceedings due to injury or damage to property, arising out of or in the course of or by reason of the carrying out of the works and which is caused by collapse, subsidence, heave, vibration, weakening or removal of support or lowering of ground water. A typical example would be a claim made by the owner of an adjoining property due to damage to his property arising from piling, demolition or from basement excavation in connection with the works.

Whether insurance is necessary in order to protect the employer against such risks will clearly depend on the situation of the site and the works. Close proximity to other buildings will represent a greater risk than building on a green field site where it is unlikely that insurance would be required. Obviously the extent to which injury or damage is likely to be caused to adjoining property will also depend on the nature of the works themselves.

It is important to remember that the indemnity provided by the contractor to the employer under clause 6.1.2 against injury or damage to property other than the works and the insurance under clause 6.2.1 to cover that indemnity, only extends to injury or damage due to the negligence, breach of statutory duty or omission or default of the contractor. The indemnity to the employer does not therefore cover injury or damage arising from any other cause. In some construction situations, that risk can be considerable and the purpose of clause 6.2.4 is therefore to provide insurance cover for both employer and contractor where such cover is deemed to be necessary.

Where such insurance cover is required by the employer, this is achieved by an entry in the appendix to IFC 98 which also provides for the level of indemnity to be stated in respect of any one occurrence or series of occurrences arising out of one event, or alternatively, for an aggregate amount. Clearly, appropriate instructions from the employer and advice from his insurers or insurance brokers should be sought.

The need for such insurance cover was highlighted by the case of *Gold* v. *Patman and Fotheringham Ltd* (1958) which led to a provision similar to clause 6.2.4 being included in JCT 63 and subsequently in JCT 80.

Facts:

A clause in an earlier RIBA form of building contract required the contractor to indemnify the employer against claims in respect of damage to property, provided that the contractor or his sub-contractor was guilty of negligence or default. The contractor was also required to insure as might be required in the bills of quantities. The bills of quantities required the contractor to insure adjoining properties against subsidence or collapse. A specialist piling sub-contractor, without any negligence or default, caused damage to adjoining properties during the piling operation. The contractor had insured himself, but not the employer, against such damage. The employer contended that the contractor was in breach of contract in failing to insure for the employer's benefit.

Held:

The wording of that clause required the contractor to insure himself and not the employer.

Where it is considered by the employer that there is a risk, then clause 6.2.4 provides for the architect to instruct the contractor to take out a policy of insurance in the names of the employer and the contractor to indemnify the employer against claims and losses as defined by the clause.

No action is required from the contractor until he receives an instruction from the architect to obtain a quotation for the insurance cover from insurers approved by the employer. When the employer has approved the quotation the architect should then instruct the contractor to take out the insurance.

It is important that the cover provided conforms to the requirements of this clause and that the policy and premium receipt are sent to the architect for deposit with the employer. Because the greatest risk may well stem from the early work, it is important that the architect ensures that the policy is in place before the contractor commences work on site. If the contractor fails to insure then the employer may do so in which case the premium is clearly not then added to the contract sum which would otherwise be the case.

The insurance cover provided by this clause is by no means all-embracing and the considerable number of specific exceptions should be noted, particularly those to be found in .3, .5, .8 and .9 of clause 6.2.4. While all the exceptions will clearly add to the employer's vulnerability, that which excludes damage 'which can reasonably be foreseen to be inevitable having regard to the nature of the work to be executed and the manner of its execution' could clearly be the cause of considerable dispute. For that reason it may well be prudent for the employer to cover the risk as part of his own insurances.

Clause 6.2.5 is commented on earlier at page 289.

Clauses 6.3 to 6.3C

Insurance of the Works – alternative clauses [s]

6.3.1   Clause 6.3A or clause 6.3B or clause 6.3C shall apply whichever clause is stated to apply in the Appendix.

---

[s] Clause 6.3A is applicable to the erection of a new building where the Contractor is required to take out a Joint Names Policy for All Risks Insurance for the Works and clause 6.3B is applicable where the Employer has elected to take out such Joint Names Policy. Clause 6.3C is to be used for alterations of or extensions to existing structures under which the Employer is required to take out a Joint Names Policy for All Risks Insurance for the Works and also a Joint Names Policy to insure the existing structures and their contents owned by him or for which he is responsible against loss or damage thereto by the Specified Perils.

Definitions

6.3.2   In clauses 6.3A, 6.3B, 6.3C and, so far as relevant, in other clauses of the Conditions the following phrases shall have the meanings given below:
All Risks Insurance: [t]
means insurance which provides cover against any physical loss or damage to work executed and Site Materials and against the reasonable cost of the removal and disposal of debris and of any shoring and propping of the Works which results from such physical loss or damage but excluding the cost necessary to repair, replace or rectify
1   property which is defective due to
     .1   wear and tear,
     .2   obsolescence,
     .3   deterioration, rust or mildew;
2   any work executed or any Site Materials lost or damaged as a result of its own defect in design, plan, specification, material or workmanship or any other work executed which is lost or damaged in consequence thereof where such work relied for its support or stability on such work which was defective; [u]

---

[t] The definition of 'All Risks Insurance' in clause 6.3.2 is intended to define the risks for which insurance is required. Policies issued by insurers are not standardised and there will be some variation in the way the insurance for those risks is expressed. See also Practice Note 22 and Guide, Part A.

[u] In any policy for 'All Risks Insurance' taken out under clauses 6.3A, 6.3B or 6.3C.2 cover should not be reduced by the terms of any exclusion written in the policy beyond the terms of

paragraph 2; thus an exclusion in terms 'This Policy excludes all loss of or damage to the property insured due to defective design, plan, specification, materials or workmanship' would not be in accordance with the terms of those clauses and of the definition of 'All Risks Insurance'. Cover which goes beyond the terms of the exclusion in paragraph 2 may be available though not standard in all policies taken out to meet the obligation in clauses 6.3A, 6.3B or 6.3C.2; and leading insurers who underwrite All Risks cover for the Works have confirmed that where such improved cover is being given it will not be withdrawn as a consequence of the publication of the terms of the definition in clause 6.3.2 of 'All Risks Insurance'.

3　loss or damage caused by or arising from
   .1　any consequence of war, invasion, act of foreign enemy, hostilities (whether war be declared or not), civil war, rebellion, revolution, insurrection, military or usurped power, confiscation, commandeering, nationalisation or requisition or loss or destruction of or damage to any property by or under the order of any government *de jure* or *de facto* or public, municipal or local authority;
   .2　disappearance or shortage if such disappearance or shortage is only revealed when an inventory is made or is not traceable to an identifiable event;
   .3　an Excepted Risk (as defined in clause 8.3);
and if the Contract is carried out in Northern Ireland:
   .4　civil commotion;
   .5　any unlawful, wanton or malicious act committed maliciously by a person or persons acting on behalf of or in connection with any unlawful association; 'unlawful association' shall mean any organisation which is engaged in terrorism and includes an organisation which at any relevant time is a proscribed organisation within the meaning of the Northern Ireland (Emergency Provisions) Act 1973; 'terrorism' means the use of violence for political ends and includes any use of violence for the purpose of putting the public or any section of the public in fear.

Joint Names Policy:
means a policy of insurance which includes the Employer and the Contractor as the insured and under which the insurers have no right of recourse against any person named as an insured, or, pursuant to clause 6.3.3, recognised as an insured thereunder.

Sub-contractors – benefit of Joint Names Policies – Specified Perils
6.3.3　The Contractor where clause 6.3A applies, and the Employer where either clause 6.3B or clause 6.3C applies shall ensure that the Joint Names Policy referred to in clause 6.3A.1 or clause 6.3A.3 or the Joint Names Policy referred to in clause 6.3B.1 or in clause 6.3C.1 and 6.3C.2 shall
   – either provide for recognition of each sub-contractor referred to in clause 3.3 as an insured under the relevant Joint Names Policy
   – or include a waiver by the relevant insurers of any right of subrogation which they may have against any such sub-contractor
in respect of loss or damage by the Specified Perils to the Works and Site Materials where clause 6.3A or clause 6.3B or clause 6.3C.2 applies and, where clause 6.3C.1 applies, in respect of loss or damage by the Specified Perils to the existing structures (which shall include from the relevant date any relevant part to which clause 2.11 refers) together with the contents thereof owned by the Employer or for which he is responsible; and that this recognition or waiver shall continue up to and including the date of issue of any certificate or other document which states that the Sub-Contract Works of such a sub-contractor are practically complete or the date of determination of the employment of the Contractor (whether or not the validity of that determination is contested) under clauses 7.2 to 7.4 or clause 7.9 or 7.10 or 7.13 or, where clause 6.3C applies, under clause 6.3C.4.3 or clauses 7.2 to 7.4 or clause 7.9 or 7.10 or 7.13, whichever is the earlier. The provisions of clause 6.3.3 shall apply also in respect of any Joint Names Policy taken out by the Employer under clause 6.3A.2 or by the Contractor under clause 6.3B.2 or under clause 6.3C.3.
Except in respect of the Joint Names Policy referred to in clause 6.3C.1 (or the Joint Names Policy referred to in clause 6.3C.3 taken out by the Contractor in respect of a default by the Employer under clause 6.3C.1) the provisions of clause 6.3.3 in regard to recognition or waiver shall apply to sub-contractors as referred to in clause 3.2.

Such recognition or waiver for such sub-contractors shall continue up to and including the date of issue of any certificate or other document which states that the Sub-Contract Works of such a sub-contractor are practically complete or the date of determination of the employment of the Contractor as referred to in this clause 6.3.3 whichever is the earlier.

Erection of new buildings – All Risks Insurance of the Works by the Contractor [s]

6.3A.1   The Contractor shall take out and maintain a Joint Names Policy for All Risks Insurance for cover no less than that defined in clause 6.3.2 [t] [v] for the full reinstatement value of the Works (plus the percentage, if any, to cover professional fees stated in the Appendix) and shall (subject to clause 2.11) maintain such Joint Names Policy up to and including the date of issue[1] of the certificate of Practical Completion or up to and including the date of determination of the employment of the Contractor under clauses 7.2 to 7.4 or clause 7.9 or 7.10 or 7.13 (whether or not the validity of that determination is contested), whichever is the earlier.

Where the Employer's status for VAT purposes is exempt or partially exempt the full reinstatement value to which this clause 6.3A.1 refers shall be inclusive of any VAT on the supply of the work and materials referred to in clause 6.3A.4.3 for which the Contractor is chargeable by the Commissioners.

---

[v] In some cases it may not be possible for insurance to be taken out against certain of the risks covered by the definition of 'All Risks Insurance'. This matter should be arranged between the Parties prior to entering into the Contract and either the definition of 'All Risks Insurance' given in clause 6.3.2 amended or the risks actually covered should replace this definition; in the latter case clause 6.3A.1, clause 6.3A.3 or clause 6.3B.1, whichever is applicable, and other relevant clauses in which the definition 'All Risks Insurance' is used should be amended to include the words used to replace the definition.

Single policy – insurers approved by Employer – failure by Contractor to insure

6.3A.2   The Joint Names Policy referred to in clause 6.3A.1 shall be taken out with insurers approved by the Employer and the Contractor shall send to the Architect/the Contract Administrator for deposit with the Employer that Policy and the premium receipt therefor and also any relevant endorsement or endorsements thereof as may be required to comply with the obligation to maintain that Policy set out in clause 6.3A.1 and the premium receipts therefor. If the Contractor defaults in taking out or in maintaining the Joint Names Policy as required by clauses 6.3A.1 and 6.3A.2 the Employer may himself take out and maintain a Joint Names Policy against any risk in respect of which the default shall have occurred and a sum or sums equivalent to the amount paid or payable by him in respect of premiums therefor may be deducted by him from any monies due or to become due to the Contractor under this Contract or such amount may be recoverable by the Employer from the Contractor as a debt.

Use of annual policy maintained by Contractor – alternative to use of clause 6.3A.2

6.3A.3   .1 If the Contractor independently of his obligations under this Contract maintains a policy of insurance which provides (inter alia) All Risks Insurance for cover no less than that defined in clause 6.3.2 for the full reinstatement value of the Works (plus the percentage, if any, to cover professional fees stated in the Appendix) then the maintenance by the Contractor of such policy shall, if the policy is a Joint Names Policy in respect of the aforesaid Works, be a discharge of the Contractor's obligation to take out and maintain a Joint Names Policy under clause 6.3A.1. If and so long as the Contractor is able to send to the Architect/the Contract Administrator for inspection by the Employer as and when he is reasonably required to do so by the Employer documentary evidence that such a policy is being maintained then the Contractor shall be discharged from his obligation under clause 6.3A.2 to deposit the policy and the premium receipt with the Employer but on any occasion the Employer may (but not unreasonably or vexatiously) require to have sent to the Architect/the Contract Administrator for inspection by the Employer the policy to which clause 6.3A.3.1 refers and the premium receipts therefor. The annual renewal date, as supplied by the Contractor, of the insurance referred to in clause 6.3A.3.1 is stated in the Appendix.

6.3A.3   .2 The provisions of clause 6.3A.2 shall apply in regard to any default in taking out or in maintaining insurance under clause 6.3A.3.1.

Loss or damage to Works – insurance claims – Contractor's obligations – use of insurance monies

6.3A.4 .1 If any loss or damage affecting work executed or any part thereof or any Site Materials[2] is occasioned by any one or more of the risks covered by the Joint Names Policy referred to in clause 6.3A.1 or clause 6.3A.2 or clause 6.3A.3 then, upon discovering the said loss or damage, the Contractor shall forthwith give notice in writing both to the Architect/the Contract Administrator and to the Employer of the extent, nature and location thereof.

6.3A.4 .2 The occurrence of such loss or damage shall be disregarded in computing any amounts payable to the Contractor under or by virtue of this Contract.

6.3A.4 .3 After any inspection[3] required by the insurers in respect of a claim under the Joint Names Policy referred to in clause 6.3A.1 or clause 6.3A.2 or clause 6.3A.3 has been completed the Contractor with due diligence shall restore such work damaged, replace or repair any such Site Materials which have been lost or damaged, remove and dispose of any debris and proceed with the carrying out and completion of the Works.

6.3A.4 .4 The Contractor, for himself and for all sub-contractors referred to in clauses 3.2 and 3.3 who are, pursuant to clause 6.3.3, recognised as an insured under the Joint Names Policy referred to in clause 6.3A.1 or clause 6.3A.2 or clause 6.3A.3, shall authorise the insurers to pay all monies from such insurance in respect of the loss or damage referred to in clause 6.3A.4.1 to the Employer. The Employer shall pay all such monies (less only the amount properly incurred by the Employer in respect of professional fees[4] but not exceeding the amount arrived at by applying the percentage to cover professional fees stated in the Appendix to the amount of the monies so paid excluding any amount included therein for professional fees) to the Contractor by instalments under certificates of the Architect/the Contract Administrator issued at the intervals to which clause 4.2 refers.

6.3A.4 .5 The Contractor shall not be entitled to any payment in respect of the restoration, replacement or repair of such loss or damage, and (when required) the removal and disposal of debris other than the monies received under the aforesaid insurance.

Erection of new buildings – All Risks Insurance of the Works by the Employer [s]

6.3B.1 The Employer shall take out and maintain a Joint Names Policy for All Risks Insurance for cover no less than that defined in clause 6.3.2 [t] [v] for the full reinstatement value of the Works (plus the percentage, if any to cover professional fees stated in the Appendix) and shall (subject to clause 2.11) maintain such Joint Names Policy up to and including the date of issue of the certificate of Practical Completion or up to and including the date of determination of the employment of the Contractor under clauses 7.2 to 7.4 or clause 7.9 or 7.10 or 7.13 (whether or not the validity of that determination is contested), whichever is the earlier. Where the Employer's status for VAT purposes is exempt or partially exempt the full reinstatement value to which this clause refers shall be inclusive of any VAT on the supply of the work and materials referred to in clause 6.3B.3.3 for which the Contractor is chargeable by the Commissioners.

Failure of Employer to insure – rights of Contractor

6.3B.2 Except where the Employer is a local authority the Employer shall, as and when reasonably required to do so by the Contractor, produce documentary evidence and receipts showing that the Joint Names Policy required under clause 6.3B.1 has been taken out and is being maintained. If the Employer defaults in taking out or in maintaining the Joint Names Policy required under clause 6.3B.1 then the Contractor may himself take out and maintain a Joint Names Policy against any risk in respect of which a default shall have occurred and a sum or sums equivalent to the amount paid or payable by him in respect of the premiums therefor shall be added to the Contract Sum.

Loss or damage to Works – insurance claims – Contractor's obligations – payment by Employer

6.3B.3 .1 If any loss or damage affecting work executed or any part thereof or any Site Materials[5] is occasioned by any one or more of the risks covered by the Joint Names Policy referred to in clause 6.3B.1 or clause 6.3B.2 then, upon discovering the said loss or damage, the Contractor shall forthwith give notice in writing both to the Architect/the Contract Administrator and to the Employer of the extent, nature and location thereof.

6.3B.3 .2 The occurrence of such loss or damage shall be disregarded in computing any amounts payable to the Contractor under or by virtue of this Contract.

6.3B.3 .3 After any inspection required by the insurers in respect of a claim under the Joint Names Policy referred to in clause 6.3B.1 or clause 6.3B.2 has been completed the Contractor with due diligence shall restore work damaged, replace or repair any Site Materials which have been lost or damaged, remove and dispose of any debris and proceed with the carrying out and completion of the Works.

6.3B.3 .4 The Contractor, for himself and for all sub-contractors referred to in clauses 3.2 and 3.3 who are, pursuant to clause 6.3.3, recognised as an insured under the Joint Names Policy referred to in clause 6.3B.1 or clause 6.3B.2, shall authorise the insurers to pay all monies from such insurance in respect of the loss or damage referred to in clause 6.3B.3.1 to the Employer.

6.3B.3 .5 The restoration, replacement or repair of such loss of damage and (when required) the removal and disposal of debris shall be treated as if they were a Variation required by an instruction of the Architect/the Contract Administrator under clause 3.6.

Insurance of existing structures – Insurance of Works in or extensions to existing structures [s]

6.3C.1 The Employer shall take out and maintain a Joint Names Policy in respect of the existing structures (which shall include from the relevant date any relevant part to which clause 2.11 refers) together with the contents thereof owned by him or for which he is responsible, for the full cost of reinstatement, repair or replacement of loss or damage due to one or more of the Specified Perils [w] up to and including the date of issue of the certificate of Practical Completion or up to and including the date of determination of the employment of the Contractor under clause 6.3C.4.3 or clauses 7.2 to 7.4 or clause 7.9 or 7.10 or 7.13 (whether or not the validity of that determination is contested), whichever is the earlier. The Contractor, for himself and for all sub-contractors referred to in clause 3.3 who are, pursuant to clause 6.3.3, recognised as an insured under the Joint Names Policy referred to in clause 6.3C.1 or clause 6.3C.3, shall authorise the insurers to pay all monies from such insurance in respect of loss or damage to the Employer [x]. Where the Employer's status for VAT purposes is exempt or partially exempt the full cost of reinstatement, repair or replacement of loss or damage to which the clause refers shall be inclusive of any VAT chargeable on the supply of such reinstatement, repair or replacement.

---

[w] In some cases it may not be possible for insurance to be taken out against certain of the Specified Perils or the risks covered by the definition of 'All Risks Insurance'. This matter should be arranged between the parties prior to entering into the Contract and either the definition of Specified Perils and/or All Risks Insurance given in clauses 6.3 and 8.3 amended or the risks actually covered should replace the definition; in the latter case clause 6.3C.1 and/or clause 6.3C.2 and other relevant clauses in which the definitions 'All Risks Insurance' and/or 'Specified Perils' are used should be amended to include the words used to replace the definition.
[x] Some Employers e.g. tenants, may not be able to fulfil the obligations in clause 6.3C.1. If so, clause 6.3C.1 should be amended accordingly.

Works in or extensions to existing structures – All Risks Insurance – Employer to take out and maintain Joint Names Policy

6.3C.2 The Employer shall take out and maintain a Joint Names Policy for All Risks Insurance (as defined in clause 6.3.2) [t] [w] for the full reinstatement value of the Works (plus the percentage, if any, to cover professional fees stated in the Appendix) and shall (subject to clause 2.11) maintain such Joint Names Policy up to and including the date of issue of the certificate of Practical Completion or up to and including the date of determination of the employment of the Contractor under clause 6.3C.4.3 or clauses 7.2 to 7.4 or clause 7.9 or 7.10 or 7.13 (whether or not the validity of such determination is contested), whichever is the earlier.
Where the Employer's status for VAT purposes is exempt or partially exempt the full

reinstatement value to which this clause refers shall be inclusive of any VAT on the supply of the work and materials referred to in clause 6.3C.4.4 for which the Contractor is chargeable by the Commissioners.

Failure of Employer to insure – rights of Contractor

6.3C.3   Except where the Employer is a local authority the Employer shall, as and when reasonably required to do so by the Contractor, produce documentary evidence and receipts showing that the Joint Names Policy required under clause 6.3C.1 or clause 6.3C.2 has been taken out and is being maintained. If the Employer defaults in taking out or in maintaining the Joint Names Policy required under clause 6.3C.1 the Contractor may himself take out and maintain a Joint Names Policy against any risk in respect of which the default shall have occurred and for that purpose shall have such right of entry and inspection as may be required to make a survey and inventory of the existing structures and the relevant contents. If the Employer defaults in taking out or in maintaining the Joint Names Policy required under clause 6.3C.2 the Contractor may take out and maintain a Joint Names Policy against any risk in respect of which the default shall have occurred. A sum or sums equivalent to the premiums paid or payable by the Contractor pursuant to clause 6.3C.3 shall be added to the Contract Sum.

Loss or damage to Works – insurance claims – Contractor's obligations – payment by Employer

6.3C.4   .1 If any loss or damage affecting work executed or any part thereof or any Site Materials is occasioned by any one or more of the risks covered by the Joint Names Policy referred to in clause 6.3C.2 or clause 6.3C.3 then, upon discovering the said loss or damage, the Contractor shall forthwith give notice in writing both to the Architect/the Contract Administrator and to the Employer of the extent, nature and location thereof.

6.3C.4   .2 The occurrence of such loss or damage shall be disregarded in computing any amounts payable to the Contractor under or by virtue of this Contract;

6.3C.4   .3 The Contractor, for himself and for all sub-contractors referred to in clauses 3.2 and 3.3 who are, pursuant to clause 6.3.3, recognised as an insured under the Joint Names Policy referred to in clause 6.3C.2 or clause 6.3C.3, shall authorise the insurers to pay all monies from such insurance in respect of the loss or damage referred to in clause 6.3C.4 to the Employer.

6.3C.4   .4 If it is just and equitable so to do the employment of the Contractor under this Contract may within 28 days of the occurrence of such loss or damage be determined at the option of either party by notice by special delivery[6] or recorded delivery from either party to the other. Within 7 days[7] of receiving such a notice (but not thereafter) either party may invoke the relevant procedures under this Contract relevant to the resolution of disputes or differences in order that it may be decided whether such determination is just and equitable.
Upon the giving or receiving by the Employer of such a notice of determination, or where the relevant procedures referred to in this clause 6.3C.4.4 have been invoked and the notice of determination has been upheld, the provisions of clause 7.11 shall apply except the words in clause 7.11.3(d) 'and any direct loss and/or damage caused to the Contractor by the determination'.

6.3C.4   .5 If no notice of determination is served under clause 6.3C.4.4 or, where the relevant procedures referred to in clause 6.3C.4.4 have been invoked and the notice of determination has not been upheld, then
after any inspection required by the insurers in respect of a claim under the Joint Names Policy referred to in clause 6.3C.2 or clause 6.3C.3 has been completed, the Contractor with due diligence shall restore such work damaged, replace or repair any such Site Materials which have been lost or damaged, remove and dispose of any debris and proceed with the carrying out and completion of the Works; and
the restoration, replacement or repair of such loss or damage and (when required) the removal and disposal of debris shall be treated as if they were a Variation required by an instruction of the Architect/the Contract Administrator under clause 3.6.

AMENDMENT TC/94/IFC – ISSUED APRIL 1994

Clause 6.3.2

Definitions
After the definition of 'All Risks Insurance' insert as additional definitions:
terrorism:
means any act of any person acting on behalf of or in connection with any organisation with activities directed towards the overthrowing or influencing of any government *de jure* or *de facto* by force or violence.
terrorism cover:
means insurance provided under a Joint Names Policy to which clause 6.3A, clause 6.3B and clause 6.3C refer for physical loss or damage to work executed and Site Materials and to an existing structure and/or its contents due to fire or explosion caused by terrorism.

Clause 6.3A

Erection of new buildings – All Risks Insurance of the Works by the Contractor
After clause 6.3A.4.5 insert:
Terrorism cover – non-availability
6.3A.5   .1 If the insurers named in the Joint Names Policy notify the Contractor or the Employer (the 'Insurer's Notification') that, with effect from a date stated by the insurers (the 'Effective Date'), terrorism cover will cease and will no longer be available the Contractor shall immediately so inform the Employer or the Employer shall immediately so inform the Contractor.

Employer's options
6.3A.5   .2 The Employer, after receipt of the Insurer's Notification but before the Effective Date, shall notify the Contractor in writing:
either
    .2.1   that on and from the Effective Date clause 6.3A.5.3 shall apply in respect of physical loss or damage to work executed and/or Site Materials due to fire or explosion caused by terrorism;
    or
    .2.2   that on a date stated by the Employer in his notice (which date shall be after the date of the Insurer's Notification and on or before the Effective Date) the employment of the Contractor under this Contract shall be and is determined; and that upon such determination the provisions of this Contract which require any further payment to the Contractor shall not apply and the provisions of clauses 7.15, 7.16, 7.17 and 7.18 (except clause 7.18.5) shall thereupon apply.

6.3A.5   .3 Where clause 6.3A.5.2.1 applies then if work executed or Site Materials suffer physical loss or damage due to fire or explosion caused by terrorism the Contractor shall with due diligence restore such work damaged, replace or repair any such Site Materials which have been lost or damaged, remove and dispose of any debris and proceed with the carrying out of the Works; and the restoration, replacement or repair of such loss or damage and (when required) the removal and disposal of debris shall be treated as if they were a Variation required by an instruction of the Architect/the Contract Administrator under clause 3.6. The Employer shall not reduce any amount payable to the Contractor pursuant to clause 6.3A.5.3 by reason of any act or neglect of the Contractor or of any sub-contractor which may have, or is alleged by the Employer to have, contributed to the physical loss or damage to which this clause refers.

Premium rate changes – terrorism cover
6.3A.5   .4.1   If the rate on which the premium is based for terrorism cover required under the Joint Names Policy to which clause 6.3A.1 or clause 6.3A.3.1 refer is varied at any renewal of the cover the Contract Sum shall be adjusted by the net amount

of the difference in the premium paid by the Contractor as compared to the premium that would have been paid but for the change in the rate.

6.3A.5 .4.2 Where the Employer is a local authority the Employer may, in lieu of any adjustment of the Contract Sum under clause 6.3A.5.4.1, instruct the Contractor not to renew the terrorism cover under the Joint Names Policy to which clause 6.3A.1 or clause 6.3A.3.1 refer; and state that from the Effective Date the provisions in clause 6.3A.5.3 shall apply if work executed and/or Site Materials suffer physical loss or damage by fire or explosion caused by terrorism.

Clause 6.3B

Erection of new buildings – All Risks Insurance of the Works by the Employer
After clause 6.3B.2 insert:
Terrorism – cover certificate
6.3B.2 .1 Where the Employer is a local authority, as and when reasonably required by the Contractor to do so the Employer shall produce to the Contractor a copy of the cover certificate issued by the insurer named in the Joint Names Policy to which clause 6.3B.1 refers and which certifies that terrorism cover is being provided under that Policy.

After clause 6.3B.3.5 insert:
Terrorism cover – non-availability
6.3B.4 .1 If the insurers named in the Joint Names Policy notify the Employer or the Contractor (the 'Insurer's Notification') that, with effect from a date stated by the insurers (the 'Effective Date'), terrorism cover will cease and will no longer be available the Employer shall immediately so inform the Contractor or the Contractor shall immediately so inform the Employer.

Employer's options
6.3B.4 .2 The Employer, after receipt of the Insurer's Notification but before the Effective Date, shall notify the Contractor in writing:
either
.2.1   that on and from the Effective Date clause 6.3B.4.3 shall apply in respect of physical loss or damage to work executed and Site Materials due to fire or explosion caused by terrorism;
or
.2.2   that on a date stated by the Employer in his notice (which date shall be after the date of the Insurer's Notification and on or before the Effective Date) the employment of the Contractor shall be and is determined; and that upon such determination the provisions of this Contract which require any further payment to the Contractor shall not apply and the provisions of clauses 7.15, 7.16, 7.17 and 7.18 (except clause 7.18.5) shall thereupon apply.

6.3B.4 .3 Where clause 6.3B.4.2.1 applies then if work executed or Site Materials suffer physical loss or damage due to fire or explosion caused by terrorism the Contractor shall with due diligence restore such work damaged, replace or repair any such Site Materials which have been lost or damaged, remove and dispose of any debris and proceed with the carrying out of the Works; and the restoration, replacement or repair of such loss or damage and (when required) the removal and disposal of debris shall be treated as if they were a Variation required by an instruction of the Architect/the Contract Administrator under clause 3.6. The Employer shall not reduce any amount payable to the Contractor pursuant to clause 6.3B.4.3 by reason of any act or neglect of the Contractor or of any sub-contractor which may have, or is alleged by the Employer to have, contributed to the physical loss or damage to which this clause refers.

Insurance of existing structures – Insurance of Works in or extensions to existing structures
After clause 6.3C.1 insert:

Terrorism cover – existing structures and contents – non-availability – Employer's options

6.3C.1   A.1 If the insurers named in the 6.3C.1 Policy notify the Employer or the Contractor (the 'Insurer's Notification') that, with effect from a date stated by the insurers (the 'Effective Date'), terrorism cover under the 6.3C.1 Policy will cease and will no longer be available, the Employer shall immediately so inform the Contractor or the Contractor shall immediately so inform the Employer. The Employer, after receipt of the Insurer's Notification but before the Effective Date, shall notify the Contractor in writing; either

    A.1.1   that on and from the Effective Date clause 6.3C.1A.2 shall apply if loss or damage occurs to the structures and/or the contents due to fire or explosion caused by terrorism;

    or

    A.1.2   that on a date stated by the Employer in his notice (which date shall be after the date of the Insurer's Notification and on or before the Effective Date) the employment of the Contractor under this Contract shall be and is determined.

6.3C.1   A.2 Where clause 6.3C.1A.1.1 applies, the Employer shall continue to require the Works to be carried out notwithstanding that the existing structures and/or the contents thereof owned by him or for which the Employer is responsible suffer loss or damage due to fire or explosion caused by terrorism; provided that clause 6.3C.1A.2 shall not be construed so as to impose an obligation on the Employer to reinstate the existing structure after such loss or damage caused by terrorism.

6.3C.1   A.3 Where under clause 6.3C.1A.1.2 the employment of the Contractor under this Contract is determined, then upon such determination the provisions of this Contract which require any payment to the Contractor shall not apply and the provisions of clauses 7.15, 7.16, 7.17 and 7.18 (except clause 7.18.5) shall thereupon apply.

Clause 6.3C.3

After clause 6.3C.3 insert:

Terrorism – cover certificate

6.3C.3   .1 Where the Employer is a local authority, as and when reasonably required by the Contractor to do so the Employer shall produce to the Contractor a copy of the cover certificate issued by the insurers named in the Joint Names Policies to which clauses 6.3C.1 and 6.3C.2 refer and which certify that terrorism cover is being provided under each Policy.

After clause 6.3C.4.4.2 [sic] 6.3C.4 insert:

Terrorism cover – non-availability

6.3C.5   .1 If the insurers named in the Joint Names Policy notify the Employer or the Contractor (the 'Insurer's Notification') that, with effect from a date stated by the insurers (the 'Effective Date'), terrorism cover will cease and will no longer be available the Employer shall immediately so inform the Contractor or the Contractor shall immediately so inform the Employer.

6.3C.5   .2 The Employer, after receipt of the Insurer's Notification but before the Effective Date, shall notify the Contractor in writing: either

    .2.1   that on and from the Effective Date clause 6.3C.5.3 shall apply in respect of physical loss or damage to work executed and Site Materials due to fire or explosion caused by terrorism;

    or

    .2.2   that on a date stated by the Employer in his notice (which date shall be after the date of the Insurer's Notification and on or before the Effective Date) the employment of the Contractor shall be and is determined; and that upon such determination the provisions of this Contract which require any further payment to the Contractor shall not apply and the provisions of clauses 7.15, 7.16, 7.17 and 7.18 (except clause 7.18.5) shall thereupon apply.

6.3C.5   .3 Where clause 6.3C.5.2.1 applies then if work executed or Site Materials suffer physical loss or damage due to fire or explosion caused by terrorism the Contractor shall with due diligence restore such work damaged, replace or repair any such Site Materials which have been lost or damaged, remove and dispose of any debris and proceed with the carrying out of the Works; and the restoration, replacement or repair of such loss or damage and (when required) the removal and disposal of any debris shall be treated as if they were a Variation required by an instruction of the Architect/ the Contractor Administrator under clause 3.6. The Employer shall not reduce any amount payable to the Contractor pursuant to clause 6.3C.5.3 by reason of any act or neglect of the Contractor or of any sub-contractor which may have, or is alleged by the Employer to have, contributed to the physical loss or damage to which this clause refers.

COMMENTARY ON CLAUSES 6.3 TO 6.3C

(A) The nature and extent of the insurance required

Before commenting on the detailed provisions, it is useful to comment on the general principles on which the clauses covering the insurance of the works is based. The insurance provisions include three methods by which insurance of the works can be arranged. In the case of the erection of a new building this can be done by either the contractor insuring the works and site materials (clause 6.3A) or by the employer insuring (6.3B). Where the works are for the alteration or extension to an existing building there is no option for the contractor to insure and the policy must be taken out by the employer under clause 6.3C. Clauses 6.3A, 6.3B and 6.3C are therefore mutually exclusive clauses and an entry in the appendix must indicate which is to apply. It should be noted that in the case of an existing building, the employer also has a further obligation as far as insurance is concerned, and that is to insure the remainder of the building together with its contents.

In all cases, irrespective of whether the contractor or the employer is to insure, the works themselves must be covered by what is called a 'Joint Names Policy' which is a policy which includes both the contractor and the employer as the insured. 'Joint Names Policy' is defined in clause 6.3.2 as 'a policy of insurance which includes the Employer and the Contractor as the insured and under which the insurers have no right of recourse against any person named as an insured, or, pursuant to clause 6.3.3, recognised as an insured thereunder'. The result is that the insurer is prevented from exercising any possible subrogation rights as between the employer and the contractor. The right of subrogation, put in its simplest terms, means that the insurer (i.e. the insurance company or under-writer), on paying out the claim, is entitled to stand in the shoes of the insured and seek recovery of all or part of the loss from anyone having a legal liability in respect of the loss or damage to the works, e.g. where the loss or damage was caused by someone's negligence. If for instance the insurance policy was taken out in the name of the employer alone, and the loss or damage was suffered by reason of the contractor's wrongful act or omission, the insurance company could settle the claim on behalf of the employer and then, standing in the employer's shoes, could seek recovery from the contractor. Insurance in the joint names of contractor and employer as defined protects both from this type of situation.

It should be noted that the conditions provide for 'All Risks' insurance cover for the works irrespective of whether the contractor or the employer insures and irrespective of whether the works are for new construction or are works of alteration or an extension to an existing building. 'All Risks' cover is defined by clause 6.3.2 as insurance which 'provides cover against any physical loss or damage to work executed and Site Materials'. It must also extend to cover the reasonable cost of removal and disposal of debris and any shoring or propping of the works which results from the physical loss or damage. It must be for the full reinstatement value of the works plus any percentage for professional fees stated in the appropriate appendix entry. The words used therefore do not include consequential loss.

Express exclusions include damage to any part of the works resulting from a defect in that part's design, plan, specification, material or workmanship. This exclusion extends to other parts of the works which are lost or damaged in consequence where such other part relied for its support or stability on the defective work. Resulting damage, for example, by fire, to other parts of the works not so dependent will still therefore be covered. Other exclusions include the following:

- Property which is defective due to wear and tear, obsolescence, deterioration, rust or mildew
- Consequences of war, invasion, act of foreign enemy etc
- Disappearance or shortage if only revealed when an inventory is made or is not traceable to an identifiable event

and if the contract is carried out in Northern Ireland then also:

- Civil commotion
- Any unlawful wanton or malicious act committed maliciously in connection with any unlawful association.

There are also the 'Excepted Risks' which are also excluded – see clause 8.3 for the definition of these.

Finally, there may be an exception in respect of cover for terrorist activities if the optional Amendment TC/94 issued April 1994 is incorporated into the contract.

Optional Amendment TC/94

[Authors Note: At the time of publication of this book, Optional Amendment TC/ 94 is being reviewed to bring it into line, as appropriate, with IFC 98, although the changes will be very few, if any. A decision will also be made as to whether to consolidate it into IFC 98 or keep it as a discrete amendment.]

Very briefly, the background and purpose to this optional Amendment is as follows.

In 1992, reinsurers indicated to the insurance industry that they would not reinsure in respect of the risk of loss or damage due to fire and explosion caused by terrorism. Insurers informed the Government that they could not therefore cover damage to commercial and industrial buildings resulting from terrorism. The Government agreed to act as insurers of last resort and a new method of providing terrorism cover was introduced. Policies accordingly now exclude terrorism cover and then bring it back upon payment of a standard premium fixed for all policies according to graded risk zones in the UK. These premiums are then paid into a reinsurance pool administered by a company formed by the Government, Pool Reinsurance Co. Ltd.

However, the Government retained a right to terminate its agreement with Pool Reinsurance Co. Ltd to act as an insurer of last resort. As a result, there remains a possibility that during the course of the works insurers could remove terrorism cover. Optional Amendment TC/94 has therefore been produced. If it is incorporated as part of the contract, and during the course of the works insurers withdraw terrorism cover, the employer is given two options in relation to the contract works and two, slightly different options, in relation to existing structures.

The contract works

The employer can either:

- Himself pay the contractor in respect of any necessary restoration, replacement or repair; *or*
- Terminate the contractor's employment and, in effect, pay him for work carried out up to that date.

Note: Where the contractor insures and insurance premiums in respect of terrorism cover increase, the increase is added to the contract sum. If however the employer is a local authority, it has the option not to pay the premium increase but instead to let terrorism cover lapse and to itself pay for any necessary restoration, replacement or repair.

Existing structures

Where the terrorism cover is in respect of existing structures and their contents, if terrorism cover ceases then the employer may either:

- Notify the contractor that even if loss or damage occurs to the existing structures, the contractor is nevertheless to carry on with the contract works; *or*
- Terminate the contractor's employment and in effect pay him for work carried out up to that date.

For anyone who needs more information in relation to this optional provision, the Tribunal has produced a *Guide to Terrorism Cover* (April 1994).

In the case of clause 6.3A the contractor can fulfil his obligation to insure through a suitable existing annual policy, in which case its annual renewal date must be stated in the appendix.

The insurance in respect of the works includes 'Site Materials' and this is defined in clause 8.3 as meaning all unfixed materials and goods delivered to, placed on or adjacent to the works and intended for incorporation therein. As such they clearly exclude items such as formwork or scaffolding, the insurance for which would be entirely a separate matter for the contractor.

Cover does not extend to off-site materials and goods, even though these may have become the property of the employer – see clause 1.11. In such a case the contractor takes the risk of loss or damage, though the employer may still consider it worthwhile to insure them, at any rate if the contractor has not.

When arranging insurance it is important to ensure that the cover provided by the all risks policy conforms to the definition included in the conditions. It may be that cover is obtainable which goes beyond that required in the definition of all risks insurance in clause 6.3.2 (see footnote [u]), in which case leading insurers have confirmed that such extended cover will not be withdrawn as a consequence of the more limited definition contained in clause 6.3.2. If there are any doubts about whether the cover being obtained meets the definition, the insurer should be asked to confirm that the cover provided is no less than that defined in the conditions. Where the cover provided by a policy clearly falls short of that definition, discussions should be held with the insurers with the aim of amending or supplementing the policy in order to provide that required degree of cover. However, if the parties agree to accept a lower degree of cover, the conditions should be amended accordingly, although it is clearly desirable that the required degree of cover remains unaltered – see footnote [v] to clauses 6.3A.1 and 6.3B.1 and footnote [w] to clauses 6.3C.1 and 6.3C.2.

It may be that while the employer does not want the contractor to insure under clause 6.3A, neither does he wish to insure himself under clause 6.3B. Similarly if the work is in, or an extension to, an existing structure, the employer may not wish to insure the works and/or the existing structures. If this is so, clearly amendments will have to be made. In the case of JCT 98, the Tribunal has provided in Practice Note 22 appendices C and D – Model Clauses 22E to 22K which cover such a situation, as well as modifying the employer's obligations in relation to excesses. Provided care is taken to update and tailor them for the particular requirements, these model clauses may readily be adapted for use with IFC 98.

Where the works are for the alteration and extension to an existing structure the employer has an obligation to insure the existing structure together with its contents against risks termed 'Specified Perils'. This degree of cover is less than that provided for in an all risks policy. 'Specified Perils' means 'fire, lightning, explosion, storm, tempest, flood, bursting or overflowing of water tanks, apparatus or pipes, earthquake, aircraft and other aerial devices or articles dropped therefrom, riot and civil commotion, but excluding 'Excepted Risks' – see clause 8.3 for definition.

Differences between all risks cover and specified perils should be noted, the most important of which are that the former does cover whereas the latter does not cover matters such as vandalism, subsidence, impact, malicious damage and theft.

These differences between the two types of cover can of course be crucial. For example, in the case of *Computer and Systems Engineering plc* v. *John Lelliott (Ilford)* (1990), which was concerned with JCT 80 prior to its insurance provisions for the works changing from specified perils to all risks, the insurance was required to cover 'flood, bursting or overflowing of water tanks, apparatus or pipes...'. The sub-contractor negligently dropped a purlin which fractured a sprinkler pipe causing damage by the consequent escape of water. The question for consideration by the court was whether or not the escape of water as a result of the bursting of a pipe caused by a negligently created external impact was covered by the clause. The Court of Appeal held that reference in the required cover to flood involved a natural phenomenon of some kind or at least some form of abnormal occurrence. Bursting and overflowing both contemplated a rupture of the pipe due to internal causes. Accordingly, the damage concerned was not covered within the definition of the perils.

Had the cover in this case been of the all risks type, it would no doubt have provided protection.

The cover provided must be for the 'full reinstatement value of the Works plus the percentage, if any, to cover professional fees stated in the Appendix'. The term 'full reinstatement value' is intended to cover the cost of any necessary demolition and the removal of debris together with the cost of reinstating the works and the replacement of any materials or goods on site which may have been lost or damaged. As it is common practice to provide cover for the amount of the contract sum, this should always be kept under review, particularly for a contract with a relatively long contract period or during periods of high inflation. For existing structures and contents, the cover must be for the full cost of reinstatement. This does not oblige the employer to insure in respect of economic loss consequent upon physical loss or damage to the existing structures or contents. Accordingly, the contractor is not covered for such loss in the joint names policy. If the physical loss or damage has been caused by the contractor's negligence he will be liable to the employer for such economic loss: see *Kruger Tissue (Industrial) Limited* v. *Frank Galliers Limited and Others* (1998) mentioned earlier (see page 287). Whether the cover is in respect of full reinstatement value or the full cost of reinstatement, if the employer's status for VAT purposes is exempt or partially exempt, the level of cover must be such as to include any VAT chargeable on the supply of work associated with the reinstatement etc.

Excesses

As the conditions stipulate that the all risks policy in respect of the works which is to be taken out by either contractor or employer is to be for the full reinstatement value, the conditions do not expressly permit policy excesses to apply. In practice, however, they are likely to do so. Excesses occur where the terms of a policy stipulate that the first stated amount of any claim will not be met by the insurers. Where excesses apply, payment by the insurers will therefore fall short of the full reinstatement value and that will clearly be a breach of the insurance provisions by the party responsible for obtaining the cover. It is therefore the party who has a contractual obligation to insure who will be responsible for meeting any cost arising from those excesses. It is a matter which should be carefully considered since excesses may be substantial.

Inflation cost of remaining work

If any of the risks against which the insurance cover is taken, materialise, it is highly probable that reinstatement will cause delay to overall completion of the works. This will mean that the outstanding work which had not been completed at the time when the loss or damage occurred, will have to be completed later than would otherwise have been the case. If the contract sum is subject to fluctuation provisions by virtue of clause 4.9(b), the increased cost is likely to be for the most part catered for. However, the situation could be very different if it is a fixed price contract.

If clause 6.3A applies, the cost of the increase in finishing off the outstanding work will fall on the contractor. The contractor may therefore wish to take special steps to ensure that this risk is covered by some endorsement to the policy or by a separate policy. If clause 6.3B or 6.3C applies, then, by virtue of clause 6.3B.3.5 or 6.3C.4.5 (last paragraph) the increased cost of completing the outstanding work may be covered by the treatment of the contractor's obligation to reinstate following an inspection of the loss or damage as a variation, in which case it is submitted that clause 3.7.9 may apply so as to entitle the contractor to have the outstanding work treated as though it were itself the subject of a variation instruction. This is considered again later as is the question of extensions of time and loss and expense – see page 308. Accordingly, if clause 6.3B or 6.3C applies, the employer may well wish to cover the increased cost for which he may be responsible in relation to the completion of outstanding work by means of additional cover to the all risks policy or by way of a separate policy.

Duration of cover

The insurance cover must be maintained up to and including the date of the issue of the certificate of practical completion, except to the extent that clause 2.11 (partial possession) is operated under which part of the works can be taken over by the employer before practical completion of the whole works; or up to and including the date of determination of the contractor's employment under the contract (even if the validity of that determination is contested), whichever is the earlier (see clauses 6.3A.1, 6.3B.1 and 6.3C.2).

(B) Approvals, inspections of policies and evidence of insurance cover

Where clause 6.3A applies (contractor responsible for insuring), the insurer selected by the contractor is subject to the approval of the employer. However, if the contractor has opted to utilise any existing and adequate all risks policy, the employer's approval to the insurer is not required.

If the policy is obtained specifically for the particular contract by the contractor, he must send to the architect for deposit with the employer, the policy and premium receipts including any relevant endorsements.

If risks are covered by the contractor's annual all risks policy and if the contractor sends to the architect for inspection by the employer as and when reasonably required, documentary evidence that the policy is being maintained, e.g. a broker's certificate may suffice, the contractor is discharged from the need to deposit the policy itself and the premium receipts with the employer, though the employer may still call for inspection of the policy and premium receipts provided this is not done unreasonably or vexatiously. If the contractor does utilise his own all risks policy, its renewal date must be stated in the appendix.

If clause 6.3B or 6.3C apply (employer responsible for insuring), there is no provision for the insurer to be approved by the contractor. Unless the employer is a local authority, however, he must, as and when reasonably required by the contractor, produce documentary evidence and receipts showing that a policy has been taken out and is being maintained as required.

The express exception for local authorities is not justifiable. While no doubt one reason for insuring, that is having a fund available to the employer to finance reinstatement, is not likely to be critical for a local authority, this could apply equally to any other public undertaking. Further, as the employer, even if a local authority, is bound to insure, and as the contractor has a right to get the cover if the employer defaults, it would have been sensible and logical for the evidence to be required even where the employer is a local authority.

If clause 6.3C applies, the same requirements falling on the employer in connection with evidence of insurance extend also to the employer's insurance policy relating to the existing structures and contents.

(C) Default in insuring

If the party upon whom the obligation to insure, fails to do so, the other party may take out the required cover. If it is the contractor who has defaulted, the employer can recover any premiums paid by deduction from monies due or to become due or can claim it as a debt. If it is the contractor who insures as a result of the employer's failure to do so, the premiums paid by the contractor are to be added to the contract sum. Where existing structures and their contents are concerned, the contractor is given a right of access for the purpose of inspecting, carrying out a survey and making an appropriate inventory of the contents.

(D) Occurrence of loss or damage

Where any loss or damage is occasioned to the works by one or more of the insured risks, the contractor must forthwith upon discovery, give notice in writing to both the architect and the employer of the extent, nature and location of the damage.

After an inspection by the insurers has taken place, the contractor is obliged to reinstate and proceed with the carrying out of the remainder of the works. However, if the loss or damage relates to works in, or extensions to, existing structures, the contractor will not have such an obligation if either the employer or the contractor has, within 28 days of the occurrence of the loss or damage, determined the employment of the contractor on the basis that it is just and equitable to do so – see clause 6.3C.4.4 and 6.3C.4.5 – and provided further that no objection is taken by the other party within seven days of the notice of determination (or if one is taken where an adjudicator or an arbitrator nevertheless upholds the determination). In such a case the contractor will be entitled to the benefits but subject to the obligations which apply following a determination of the contractor's employment under clause 7.11 (dealt with later – page 359), except that any direct loss or damage caused by the determination otherwise recoverable under clause 7.11.3(d) is excluded.

Clearly, where existing structures are concerned, this is a sensible provision as it could well be inappropriate to expect the contractor to reinstate the works, e.g. where the existing structure has itself been destroyed.

In all cases the loss or damage is disregarded in computing any amounts payable to the contractor under or by virtue of the contract. The contractor must on his and his sub-contractor's (where recognised as an insured – see clause 6.3C.4.3) behalf, authorise the insurers to pay the insurance money over to the employer. The method by which the contractor gets paid for the reinstatement is then as follows:

- In the case of 6.3A (new buildings – contractor to insure) the employer will pay the contractor the insurance monies under separate certificates of the architect. These sums are not included in the contract sum. However, the employer is entitled to deduct any amount properly incurred in respect of professional fees subject to an upper limit equal to the percentage for professional fees stated in the appendix. If there is any shortfall between the insurance monies and the cost of reinstatement it is the contractor who must meet the difference from his own resources.
- In the case of clause 6.3B (new buildings – employer to insure) or 6.3C (works in or extensions to existing buildings – employer to insure) the reinstatement is treated as a variation required by an instruction of the architect under clause 3.6. The instruction will then be valued in accordance with the valuation rules in clause 3.7, so that any risk of a shortfall in the insurance monies will effectively fall on the employer.

The cost of completing the outstanding work

As noted earlier (see page 305) there is every chance that work yet to be done at the time of the occurrence of the loss or damage will, due to the resultant delays, cost more to complete than would otherwise have been the case. In the case of clause 6.3A this cost will fall on the contractor, although if the contract is subject to fluctuations this will clearly assist him. In the case of clause 6.3B or 6.3C, it must at least be arguable that any increased cost in finishing off can be recovered by the contractor either under clause 3.7.9 and/or as loss and expense under clause 4.12.7. Further, some or all of the extra cost may in any event be recoverable if the contract includes a fluctuations clause.

If the appropriate means of recovery is by way of seeking reimbursement for direct loss and/or expense, certain points should be noted. The loss and expense must be attributable to the variations. Where clause 6.3B applies, the variation will be treated as having been instructed after completion of inspection by or on behalf of the insurers. Clearly, loss or expense could have been incurred between the loss or damage to the works being occasioned and the variation instruction being treated as having been given. There seems no way for the contractor to recover this. In the case of clause 6.3C, the appropriate date for the variation being treated as having been instructed, seems to be either the date of completion of the inspection by or on behalf of the insurers, or the decision on the issue of whether a determination is just and equitable where either party seeks to determine the contractor's employment under clause 6.3C.4.4 and 6.3C.4.5. Any adjudicator's or arbitrator's decision could of course take a considerable time. The risks for the contractor here may therefore be considerable.

Whichever party bears the increased cost of completing the outstanding work, it should be possible to include for this either as part of the full reinstatement value under the all risks policy or alternatively by way of some special endorsement or even as a separate policy.

(E) Extensions of time

So far as an extension of time is concerned, it is tentatively submitted that the position is as follows:

- If clause 6.3A applies the contractor will obtain an extension of time for any delay to completion resulting from the occurrence of loss or damage provided the loss or damage was due to a specified peril – clause 2.4.3. Accordingly, if the loss or damage is due to, for example, impact, subsidence or malicious damage, the contractor will not be entitled to an extension of time.
- Where clause 6.3B or 6.3C applies the contractor will similarly be entitled to an extension of time under clause 2.4.3 in relation to specified perils. However, once the architect's instruction is treated as being issued, prompting the contractor's obligation to reinstate, the appropriate extension of time event will switch to clause 2.4.5.

(F) Determination of contractor's employment

Quite apart from the 'just and equitable' ground for determination of the contractor's employment referred to earlier (see page 307), if as a result of the loss or damage there is a suspension of the whole or substantially the whole of the uncompleted works for a period of not less than three months, then, provided the loss or damage was occasioned by a specified peril, either the contractor or the employer may determine the contractor's employment – see clause 7.13.1. The contractor cannot utilise this provision if the loss or damage was brought about by his or his sub-contractor's or agent's negligence – clause 7.13.2. This provision and the consequence of a determination under it are dealt with in greater detail in Chapter 7 (page 364).

(G) Works insurance and the position of sub-contractors

Under clause 6.3.3, the benefit of the joint names policy covering the works taken out by either the contractor or the employer is extended to both domestic and named sub-contractors, but only in respect of specified perils and not to the extent of cover provided by all risks insurance. This is achieved by the policy either providing for recognition of sub-contractors as an insured party or by the inclusion in the policy of a waiver of any rights of subrogation which the insurers may have against any sub-contractor.

The same benefit must also be extended to named sub-contractors in respect of the insurance of existing buildings and contents taken out by the employer under clause 6.3C. It should be noted that in this case the cover is not extended to domestic sub-contractors. This being so, it might be thought that if the loss or damage was caused by a domestic sub-contractor's negligence, the employer could sue the domestic sub-contractor in tort. However, there appear to be conflicting decisions on this: see for example *Norwich City Council* v. *Harvey* (1989); *Ossory Road (Skelmersdale) Ltd* v. *Balfour Beatty Building Ltd and Others* (1993), all suggesting that the domestic sub-contractor does not owe a duty of care; and *National Trust* v. *Haden Young Ltd* (1994); *London Borough of Barking and Dagenham* v. *Stamford Asphalt Co. Ltd and Others* (1997); *British Telecommunications Plc* v. *James Thomson & Sons (Engineers) Ltd* (1998) suggesting that the sub-contractor does owe such a duty. This last mentioned case, being a House of Lords' decision, carries considerable weight and should now resolve this particular issue.

The recognition of a waiver for sub-contractors under this clause continues up to and including the date of the issue of any certificate or other document which states that the sub-contractor's works are practically complete, or the date of determination of the employment of the contractor, whichever is the earlier.

NOTES TO CLAUSES 6.3 TO 6.3C

[1] '... up to and including the date of issue...'
Note that it is not the date of practical completion but the date of the issue of the certificate that is important. This is clearly necessary as otherwise there could be an uninsured period if the date of practical completion is back-dated, as it often is, to a date before the date of the issue of the certificate itself.

[2] '...loss or damage affecting ... any Site Materials...'
The employer may or may not own the materials or goods on site. Perhaps only in the world of insurance do we find a situation in which on the one hand the contractor could be liable to replace materials or goods owned by the employer (if the employer has paid for them) and yet on the other, the employer can be responsible for the cost of replacing materials or goods still owned by the contractor – clause 6.3B and 6.3C.

[3] 'After any inspection...'
The risk of loss or damage brought about by an insured peril will generally fall on the contractor who is contractually responsible for completing the works and will therefore, in pursuance of this obligation, have to reinstate the damaged part as necessary. It might therefore be thought that the obligation to reinstate ought not to depend on any inspection by or on behalf of the insurers being completed. If for some reason the insurers are able to avoid the claim in circumstances where there has been no breach or default by the contractor, it would appear that the obligation to reinstate never arises. The insurance cover ought primarily to be for the protection of the contractor and effective insurance cover should not be a condition precedent to his obligation to reinstate loss or damage to the works.

On the other hand, it is right that the obligation to reinstate etc. should not operate immediately loss or damage has been incurred. The insurance policy will almost certainly provide for the insurers to have the right to inspect the damage before restoration work begins and it would be unreasonable to require the contractor to begin such work before adequate time for an inspection has elapsed.

[4] '...less only the amount properly incurred by the Employer in respect of professional fees...'
Note that while the obligation on the employer to insure includes a requirement to insure not only for the full reinstatement value of the works but also the percentage, if any, to cover professional fees stated in the appendix, when it comes to a pay out, the employer's right to deduct from the insurance monies before paying them over to the contractor is limited to the actual cost properly incurred in professional fees. If the employer pays less in professional fees than the amount in the appendix then it is the lesser sum, provided it is properly incurred, that can be claimed and not the whole of the sum stated in the appendix.

[5] '...or any Site Materials...'
Presumably, this includes *all* unfixed materials and goods, even if they are prematurely or not properly brought on to site and indeed even if they are not properly protected by the contractor against weather and other casualties. It might be thought harsh for the employer (or his insurers) to meet the cost of replacing materials or goods which perhaps ought not to have been on the site at all. The contractor still owns them and is likely to have them covered under some appropriate policy in any event. The position is even more startling if the loss or damage is restricted to site materials, e.g. accidental damage or theft when all or part of the cost of replacement will be met by the employer due to the inevitable excess on the policy. However, it is no doubt a matter of convenience or expediency to include all site materials, which avoids any argument about premature delivery or inadequate protection.

[6] '...special delivery...'
This has now replaced registered post.

[7] 'Within 7 days...'
The party objecting to a determination of the contractor's employment must invoke the relevant dispute resolution procedure within seven days if it is to be disputed. The dispute resolution procedures include adjudication. This clause therefore suggests that unless the agreed party gives notice of adjudication within seven days of the notice of determination, it is too late to challenge it. Clause 9A setting out the adjudication procedure is intended to comply with the requirements of section 108 of the Housing Grants, Construction and Regeneration Act 1996. If it does not then it is invalid and the Scheme for Construction Contracts (England and Wales) Regulations 1998 will apply. Section 108(2)(a) provides that a contractual scheme must 'enable a party to give notice at any time of his intention to refer a dispute to adjudication'. It appears a party cannot do so under this clause outside the seven day period.

Does this render the whole of the IFC 98 adjudication procedure invalid as not complying with the Act? This would be most unfortunate. The requirement for a very short time limit is obviously sensible when the issue relates to termination of the contractor's employment under the contract. The problem arises because of the wording of section 108(2)(a) which lacks the necessary refinement to cover situations where a time limit is clearly sensible and highly desirable. This point has also been discussed earlier in relation to a time limit in clause 3.13.2 (dispute regarding architect's instructions following discovery of non-complying work – see page 181).

It might be possible to overcome this problem by introducing some sort of conclusiveness provision in the same manner as the 28 day time limit in relation to the final certificate (see clause 4.7.1).

Clause 6.3D

Insurance for Employer's loss of liquidated damages – clause 2.4.3

6.3D.1   Where it is stated in the Appendix that the insurance to which clause 6.3D refers may be required by the Employer then forthwith after the Contract has been entered into the Architect/the Contract Administrator shall either inform the Contractor that no such insurance is required or shall instruct the Contractor to obtain a quotation for such insurance. This quotation shall be for an insurance on an agreed value basis [y] to be taken out and maintained by the Contractor until the date of Practical Completion and which will provide for payment to the Employer of a sum calculated by reference to clause 6.3D.3 in the event of loss or damage to the Works, work executed, Site Materials, temporary buildings, plant and equipment for use in connection with and on or adjacent to the Works by any one or more of the Specified Perils and which loss or damage results in the Architect/the Contract Administrator giving an extension of time under clause 2.3 in respect of the Event in clause 2.4.3. The Architect/the Contract Administrator shall obtain from the Employer any information which the Contractor reasonably requires to obtain such quotation. The Contractor shall send to the Architect/the Contract Administrator as soon as practicable the quotation which he has obtained and the Architect/the Contract Administrator shall thereafter instruct the Contractor whether or not the Employer wishes the Contractor to accept that quotation and such instruction shall not be unreasonably withheld or delayed. If the Contractor is instructed to accept the quotation the Contractor shall forthwith take out and maintain the relevant policy and send it to the Architect/the Contract Administrator, for deposit with the Employer, together with the premium receipt therefor and also any relevant endorsement or endorsements thereof and the premium receipts therefor.

6.3D.2   The sum insured by the relevant policy shall be a sum calculated at the rate stated in the Appendix as liquidated damages for the period of time stated in the Appendix.

6.3D.3   The payment in respect of this insurance shall be calculated at the rate referred to in clause 6.3D.2 (or any revised sum produced by the application of clause 2.11) for the period of any extension of time finally given by the Architect/the Contract Administrator as referred to in clause 6.3D.1 or for the period of time stated in the Appendix, whichever is the less.

6.3D.4   The amounts expended by the Contractor to take out and maintain the insurance referred to in clause 6.3D.1 shall be added to the Contract Sum. If the Contractor defaults in taking out or in maintaining the insurance referred to in clause 6.3D.1 the Employer may himself insure against any risk in respect of which the default shall have occurred.

---

[y] The adoption of an agreed value is to avoid any dispute over the amount of the payment due under the insurance once the policy is issued. Insurers on receiving a proposal for the insurance to which clause 6.3D refers will normally reserve the right to be satisfied that the sum referred to in clause 6.3D.2 is not more than a genuine pre-estimate of the damages which the Employer considers, at the time he enters into the Contract, he will suffer as a result of any delay.

COMMENTARY ON CLAUSE 6.3D

Introduction

In the event of the progress of the works being delayed due to loss or damage caused by any one or more of the specified perils, the contractor will be entitled to an extension of time under clause 2.4.3. It should be noted that the contractor's entitlement to an extension of time is limited to the events included under the heading of specified perils as defined in clause 8.3 and not to the wider scope of all risks as defined in clause 6.3.2. It is quite likely that the contractor will be entitled to an extension of time even if the specified peril was brought about by his own negligence.

The award to the contractor of an extension of time and the fixing of a later completion date will result in the employer being unable to claim liquidated damages under clause 2.7 for the period of the delay.

Some employers amend clause 2.4.3 to qualify its application where the contractor has been negligent. However, it is the architect who has the duty to determine whether the contractor is entitled to an extension of time and it may be unwise and perhaps unfair to require or expect the architect to decide whether there has been negligence. Accordingly the Tribunal itself has not gone down this path. Instead, clause 6.3D has been introduced to compensate the employer for his lost liquidated damages by way of insurance with the added benefit that the insurance monies are payable irrespective of whether the specified peril was the result of the contractor's negligence.

This clause therefore provides a means whereby the employer can require the contractor to arrange insurance to either fully or partially compensate him for the lost liquidated damages.

The operation of clause 6.3D

An entry in the appendix should indicate whether or not insurance under this clause may be required. Immediately after the contract is entered into the employer must decide whether the contractor is to be required to obtain a quotation and the architect must either inform the contractor that no such insurance is required or must instruct the contractor to obtain a quotation on an agreed value basis for the employer to approve. The architect must then instruct the contractor whether the quotation is to be accepted. If it is, the contractor must forthwith take out and maintain the policy for the employer's benefit, sending it to the architect for deposit with the employer together with premium receipts.

If the appendix entry provides that insurance under this clause may be required then the period of delay for which the employer will require cover should also be entered in the appendix. While the employer's intention will be to obtain reimbursement for the full amount of liquidated damages, the insurers will wish to be satisfied that the amount covered by the policy is no more than a genuine pre-estimate of the employer's loss resulting from any delay. The number of weeks delay which the insurers are prepared to cover may be less than the employer might have wished.

The nature and extent of cover

Clause 2.4.3 which provides for an extension of time where loss or damage is caused by one of the specified perils, is not limited by its terms to loss or damage to the works themselves. Consequently, the scope of loss or damage for which the clause 6.3D cover against lost liquidated damages is applicable, includes loss or damage to the works, work executed, site materials, temporary buildings, plant and equipment for use in connection with and on or adjacent to the works.

It has been said above that in the event of loss or damage brought about by one of the specified perils, there must be an extension of time pursuant to clause 2.4.3. However, it has already been indicated earlier (page 309), that where clauses 6.3B or 6.3C apply, the obligation to reinstate, once it arises, is deemed to be a variation so that as from that time, 2.4.3 would no longer, it is submitted, be appropriate and clause 2.4.5 would take over. This could therefore very significantly restrict the benefit of any cover under a clause 6.3D policy wherever the employer is responsible for insuring the works.

If clause 2.11 (partial possession) has been operated, the effect of the employer taking part of the works into possession will be that the rate of liquidated damages will be proportionately reduced by the value of the works taken over in proportion to the contract sum. In these circumstances, presumably some proportionate refund of insurance premiums may be possible.

Finally, before any sum becomes payable under a clause 6.3D policy, it is necessary to know the amount of any extension of time awarded by the architect in clause 2.4.3. Extensions awarded prior to practical completion are based on estimates. Following practical completion, the architect may, up to 12 weeks thereafter, make a more considered and final decision. Does this mean that the monies under the policy will not be paid out against any estimate and that the employer will have to wait and see if the architect makes any adjustments to extensions of time previously given or makes new extensions, within the 12 weeks following practical completion?

The cover provided by a clause 6.3D policy covers much of the same ground as that already available, particularly to private employers, if they have business interruption cover. As such, clause 6.3D will be of little assistance to them. Further, since its introduction in 1986, this form of insurance has not been popular. On balance it adds nothing to the previous situation where the employer could in any event have obtained this type of cover had he wished to. Expressly incorporating it within the contract does not really add much.

Clause 6.3FC

Joint Fire Code – compliance
Application of clause
6.3FC   .1 Clause 6.3FC applies where it is stated in the Appendix that the Joint Fire Code applies.

Compliance with Joint Fire Code
6.3FC   .2.1 The Employer shall comply with the Joint Fire Code and ensure such compliance by his servants or agents and by any person employed, engaged or authorised by him upon or in connection with the works or any part thereof other than the

Contractor and the persons for whom the Contractor is responsible pursuant to clause 6.3FC.2.2.

6.3FC .2.2 The Contractor shall comply with the Joint Fire Code and ensure such compliance by his servants or agents or by any person employed or engaged by him upon or in connection with the Works or any part thereof their servants or agents or by any other person who may properly be on the site upon or in connection with the Works or any part thereof other than the Employer or any person employed, engaged or authorised by him or by any local authority or statutory undertaker executing work solely in pursuance of its statutory rights or obligations.

Breach of Joint Fire Code – Remedial Measures

6.3FC .3.1 If a breach of the Joint Fire Code occurs and the insurer under the Joint Names Policy in respect of the Works specifies by notice the remedial measures he requires ('the Remedial Measures') and the time by which such Remedial Measures are to be completed ('the Remedial Measures Completion Date') the Contractor shall ensure that the Remedial Measures are carried out, where relevant in accordance with the instructions of the Architect/the Contract Administrator, by the Remedial Measures Completion Date.

6.3FC .3.2 If the Contractor, within 7 days of receipt of a notice specifying the Remedial Measures, does not begin to carry out or thereafter fails without reasonable cause regularly and diligently to proceed with the Remedial Measures then the Employer may employ and pay other persons to carry out the Remedial Measures; and, subject to clause 6.3FC.4, all costs incurred in connection with such employment may be withheld and/or deducted by him from any monies due or to become due to the Contractor or may be recoverable from the Contractor by the Employer as a debt.

Indemnity

6.3FC .4 The Contractor shall indemnify the Employer and the Employer shall indemnify the Contractor in respect of the consequences of a breach of the Joint Fire Code to the extent that these consequences result from a breach by the Contractor or by the Employer of their respective obligations under clause 6.3FC.

Joint Fire Code – amendments

6.3FC .5 If after the Base Date the Joint Fire Code is amended and the Joint Fire Code as amended is, under the Joint Names Policy, applicable to the Works, the net extra cost, if any, of compliance by the Contractor with the amended Joint Fire Code shall be added to the Contract Sum.

COMMENTARY ON CLAUSE 6.3FC

Where insurance is required in respect of construction work, many insurers require those taking out the insurance cover (often of course both employer and contractor jointly) to comply with The Joint Code of Practice on the Protection from Fire of Construction Sites and Buildings Undergoing Renovation (the Joint Fire Code) – see clause 8.3 for full definition.

Any failure to abide by the code could lead to insurance being withdrawn and this of course would mean one or other of the parties to IFC 98 being in breach of their insurance obligations. The code may not apply in respect of small contracts, so an appendix item is included in IFC 98 to indicate whether or not the code applies. Further, additional requirements are set out in the code in relation to a 'Large Project', and again an appendix entry indicates whether or not this is so.

This clause requires both parties to comply with the code and to ensure that those for whom they are responsible also comply. If there is a breach of the code and the insurer requires remedial measures and a time period within which they are to be completed, the contractor must ensure that they are carried out, where relevant, in accordance with the architect's instructions. If the contractor, within seven days of receiving the insurer's notice specifying the remedial measures, does not begin to carry them out or, if he does begin to carry them out, fails without reasonable cause to regularly and diligently proceed with them, the employer may employ others to carry out the work and may withhold or deduct the costs from the contractor or may recover them as a debt. The appropriate notice of deduction or withholding will be required – see clauses 4.2.3(b), 4.3(c) and 4.6.1.3.

Cross indemnities are provided in respect of the consequences of a breach by either party of the code or of their respective obligations under this clause.

If after the base date (stated in the appendix) the code is amended and is, under the relevant insurance policy, applicable, as amended, to the works, then the net extra cost, if any, of compliance by the contractor is to be added to the contract sum.

Chapter 10
Determination

CONTENT

This chapter considers section 7, which deals with the determination of the employment of the contractor by the employer, by the contractor himself or automatically by operation of the contract.

SUMMARY OF GENERAL LAW

(A) Introduction

Contractual obligations may come to an end in a number of ways. In the ordinary course of things this will most commonly be brought about by the parties performing all their contractual obligations or promises.

However, contractual obligations can be determined in a number of other ways, e.g. by accord and satisfaction (mutual discharge by agreement not to require performance of outstanding obligations); by the unilateral release by one party of the others obligations; by frustration of the contract; by breach (a sufficiently serious breach of contract by one party might entitle the other party to treat himself as discharged from any further performance); or by contractual terms providing for discharge in certain events.

In this summary we are concerned only with the discharge of contractual obligations in accordance with the express conditions of the contract itself and with the discharge of contractual obligations brought about by breach of contract by one of the parties.

(B) Provision for discharge of the contract itself

Building contracts often make provision for one or both of the parties to have the right in certain situations either to bring the contract itself to an end or (and the distinction might be slight) to bring the employment of the contractor under the contract to an end.

At common law, a contracting party will only be able to regard himself as discharged from any further performance of his contractual obligations by a particularly serious breach of contract by the other party. The inclusion therefore of an express provision for determination has the advantage to the party exercising the right that it can permit such a discharge of further obligations for less serious breaches and even for an event which does not amount to a breach of contract at all (although its consequences might), e.g. administrative receivership or a voluntary arrangement for a composition of debts. However, in certain circumstances a term of this nature would have to be shown to be fair and reasonable by virtue of the provisions of the Unfair Contract Terms Act 1977 (section 3).

There might be very good reasons for determining the employment of the contractor rather than bringing the contract as a whole to an end. It might be desirable for one or both parties that the contract survives and that express provisions can be called into play in the event of the employment of the contractor being determined. Such express provisions may cover: the right of the contractor to payment for work done but not paid for; rights over goods, materials, equipment and plant on site; and recovery of loss or damage incurred by the employer or the contractor as the case may be, as a result of the determination of the employment of the contractor.

If a breach of contract occurs which is both sufficiently serious to be regarded as an intention by the party in breach no longer to be bound by the contract, i.e. a repudiation, and which is accepted as a repudiation by the other party, the effect of the parties' actions is to bring an end to their obligations as to future performance. On one view this might be seen as having brought the contract to an end, because the primary purpose of the contract is no longer to be achieved, but in fact the contract will remain in existence in the sense that a remedy in damages for breach of contract will survive which will be calculated by reference to the terms of the contract, and certain express provisions of the contract (e.g. arbitration and, possibly, adjudication provisions) will remain operative for the purposes of regulating the parties' residual rights.

It is perhaps a matter of debate whether or not the inclusion in the contract of express provision for determination takes away either party's right to treat the contract as discharged by reason of repudiation, in the absence of a clear statement to that effect in the contract. In *Architectural Installation Services Limited* v. *James Gibbons Windows Limited* (1989) Judge Bowsher QC held that clear words were necessary to exclude common law rights, and that the words 'without prejudice to other rights' were not therefore necessary to preserve such rights. However, the Court of Appeal in *Lockland Builders Limited* v. *John Kim Rickwood* (1995) apparently accepted that although there were no words expressly excluding common law rights in the contract in question, as a matter of implication those rights were excluded in the face of an express mechanism regulating the parties' rights in the circumstances that had occurred.

The express right to determine either the contract or the contractor's employment thereunder is often made subject to the service of one or more notices and any such notices must be clear and unambiguous in order to be valid. Further, a failure to comply with any procedural requirements, e.g. timetable or mode of service, might prevent the notice from taking effect. More is said about this topic later in this chapter (page 331).

A wrongful determination of the contractor's employment by the employer might well itself amount to a repudiation of contract by the employer giving the contractor the right to treat the contract as at an end and to sue for damages for breach of contract.

A determination of the contractor's employment by the employer which is challenged by the contractor can create great difficulties if the contractor is unwilling to leave the site. The employer/landowner will no doubt want to engage another contractor to finish off the work. The continued presence on site of the original contractor will almost certainly make this impossible. The contractor's right to enter the site is likely to be by reason of an express or implied contractual licence to do so. The important question is whether or not, and in what circumstances, this licence is revocable. An important case in this regard is *London Borough of Hounslow* v. *Twickenham Garden Developments* (1970).

Facts:

The defendants were employed by the London Borough of Hounslow to carry out the sub-structure works on a site in Middlesex. The contract was based on JCT 63 and by clause 25(1) it was provided that if the contractor should make default, *inter alia*, in failing to proceed regularly and diligently with the work, the architect could give notice by registered post or recorded delivery specifying the default, and if the contractor continued or repeated the default the council could by notice by registered post or recorded delivery forthwith determine the employment of the contractor. The council purported to do this and the contractor contested the validity of the council's action, stating that it regarded the service of such a notice as a repudiation of the contract by the council which it, the contractor, elected not to accept.

The council unsuccessfully tried to obtain possession of the site and issued a writ claiming damages for trespass and seeking an injunction restraining the contractor from trespassing on the site. The council argued that it was entitled – irrespective of the validity or invalidity of the notices given pursuant to clause 25 – to evict the contractor and resume possession of its own property. The contractor claimed to be able to insist on performing the contract.

Held:

There was an implied term of the contract that the council would not revoke the contractor's licence to enter the site while the contract period was still running other than in accordance with the contract. The court would not grant the council the injunction requested to expel the contractor from site because it had not been decided at that stage whether the contractor's employment had been validly determined by the council.

The effect of this decision is to produce a legal stalemate. The decision has been criticised – see *Hudson's Building and Engineering Contracts*, 11th edition, at page 1298, *Keating on Building Contracts*, 6th edition, at page 294 and earlier in this book (page 68).

The criticism is well deserved. It has been suggested that the *Hounslow* case was decided at a time when a plaintiff could only obtain an interim injunction if he had shown a prima facie case in his favour. In the subsequent case of *American Cyanamid Company* v. *Ethicon Limited* (1975) the test applied by the House of Lords was concerned with the balance of convenience as to whether or not an injunction should be granted, and then an injunction would only generally be granted if damages were not an adequate remedy. Even so, the decision in the case of

Vonlynn Holdings v. *T. Flaherty* (1988) concerning a JCT 80 contract provides limited support for the *Hounslow* case. The position under the ICE Conditions of Contract (where the operation by the employer of the forfeiture provisions under clause 63 is dependent on the certified opinion of the engineer) is very different, the contractor being bound to vacate the site notwithstanding that the engineer's opinion is challenged by the contractor: see *Tara Civil Engineering Limited* v. *Moorfield Developments Limited* (1989).

Under IFC 98, the contractor is expressly required to give up possession of the site in the event of the employer determining the contractor's employment under the contract. As has been suggested in Chapter 4, the express provision is likely to be regarded as requiring the contractor to give up possession even if the validity of the determination is challenged.

(C) Discharge by breach

One party to a contract may, by reason of the other party's breach, be entitled to treat himself as discharged from any obligation to further perform his contractual promises under the contract, and treat that other party's breach as being a repudiation by him of his contractual obligations. The innocent party will additionally have a right to claim damages consequent on the breach.

However, while any breach of contract can give rise to a claim for damages, it is not every breach of contract which will entitle the innocent party to a discharge from further liability to perform his own contractual promises. The default must be of a particularly serious or fundamental nature. Broadly speaking, it will be sufficiently serious to justify a discharge in three situations:

(1) In the case of a renunciation by one party of his contractual liabilities where that party by his words or conduct evinces an intention – whether due to unwillingness or inability – no longer to continue his part of the contract.
(2) Where by breach or default one party has rendered himself unable to perform his outstanding contractual obligations, e.g. if a contract requires the builder to be a member of the National House Building Council and to obtain an appropriate NHBC certificate and he is removed from the register kept by that body during the course of the contract.
(3) In the case of a very serious failure of performance by one party which will discharge the other. The failure must be of a fundamental nature going to the root of the contract.

These three situations may be described as repudiation or repudiatory events (sometimes perhaps less accurately as repudiatory breaches).

Once a repudiation has taken place, the innocent party must elect whether or not to accept the repudiation and this must be clear and unequivocal and must be communicated to the other contracting party. Once made, it cannot be withdrawn. The requirement that the election must be clear, unequivocal and communicated should not, however, be construed as a requirement that any particular form of election is required. In *Vitol S.A.* v. *Norelf Limited* (1996) the House of Lords recognised that non-performance of an obligation arising under a contract might suffice. In giving the unanimous opinion of the House, Lord Steyn gave the following example:

'Postulate the case where an employer at the end of a day tells a contractor that he, the employer, is repudiating the contract and that the contractor need not return the next day. The contractor does not return the next day or at all. It seems to me that the contractor's failure to return may, in the absence of any other explanation, convey a decision to treat the contract as at an end. Another example may be an overseas sale providing for shipment on a named ship in a given month. The seller is obliged to obtain an export licence. The buyer repudiates the contract before loading starts. To the knowledge of the buyer the seller does not apply for an export licence with the result that the transaction cannot proceed. In such circumstances it may well be that an ordinary businessman, circumstanced as the parties were, would conclude that the seller was treating the contract as at an end.'

The innocent party may, if he wishes, treat the contract as continuing if this is still a possibility, despite the existence of repudiatory conduct. This will not prevent him from claiming damages while at the same time continuing with his contractual obligations.

If one party mistakenly treats an event as amounting to a repudiation by the other party and purports to accept it as such, this will often in itself create a repudiation which the wrongly accused party might be forced to accept. Indeed, in practice it is not unusual for both sides to argue that they have accepted a repudiation by the other.

However, if a party refuses to perform the contract, giving an inadequate or wrong reason, he may be able to justify his conduct if he subsequently discovers that at the time there existed another good reason entitling him to refuse further performance.

Once a repudiation has been accepted by the innocent party, both parties are excused from further performance of their primary obligations under the contract. Instead, secondary obligations are imposed on the guilty party, namely to pay monetary compensation for non-performance: see *Photo Production Limited* v. *Securicor Transport Limited* (1980).

Breach of a term of a contract which is not in itself sufficiently serious to amount to a repudiation could become so if it persists, especially after a notice from the innocent party requesting proper performance. Furthermore, the degree of wilfulness of a breach of contract might be a relevant factor in displaying an intention to no longer be bound by the contract terms.

Examples of serious breaches of contract amounting in the particular circumstances of the case to a repudiation include: the prolonged failure by the employer to hand over the site to the contractor (*Carr* v. *J. A. Berriman (Property) Limited* (1953)); and a failure by the contractor to obtain the required performance bond (*Swartz & Son (Property) Limited* v. *Wolmaransstad Town Council* (1960)).

It has been suggested that a contractual term for due expedition by the contractor of the construction etc. of the contract works must be implied in building contracts, if not expressed, and that the repeated failure by the contractor to proceed with due expedition after notice by the employer will entitle the employer to treat the contractor's failure as a repudiation: see *Hudson's Building and Engineering Contracts*, 11th edition, page 1123 *et seq*. However, the decision at first instance of Mr Justice Staughton in the case of *Greater London Council* v. *Cleveland*

Bridge & Engineering Co Limited and Another (1984) has thrown some doubt on the extent and nature of such a term. In the context of IFC 98, however, this debate is of academic interest only because of the express obligation so to proceed. On the question of the construction of the express term to proceed regularly and diligently: see *West Faulkner Associates* v. *London Borough of Newham* (1994). See also note [3] to clauses 2.1 and 2.2 (page 78).

(D) Remedies for breach of contract

(i) Introduction

By far the more important remedy for either party to a building contract in the event of the other party's breach of that contract is a claim for damages. While it is possible in unusual circumstances to obtain an order for specific performance, whereby one party is compelled to honour his contractual obligations, or alternatively to obtain an injunction prohibiting a party from acting in breach of contract, damages remain the prime remedy.

(ii) Damages for breach of contract

The essential purpose of damages is to compensate the innocent party for loss or injury suffered through the other party's breach. The innocent party is, so far as money can achieve it, to be placed in the same position as if the contract had been performed. If no damage has been suffered or no damage can be proven, nominal damages will be awarded.

However, it might in certain circumstances be possible for a contracting party to claim substantial damages even though no loss has been suffered. The House of Lords' cases of *St Martins Property Corporation Limited* v. *Sir Robert McAlpine Limited* and *Linden Gardens Trust Limited* v. *Lenesta Sludge Disposals Limited* (1994) have established that it might be possible for an employer under a building contract who sold the building which was the subject of the contract for full value and therefore in that sense would suffer no loss, to nevertheless claim substantial damages for breach of contract in respect of defective work. Further, the Court of Appeal has held that this can also be the case even where the employer under the building contract at no time actually owned the land on which the property was to be built: see *Darlington Borough Council* v. *Wiltshier Northern Limited* (1995). This entitlement to make a claim was an extension of the rule in *Dunlop* v. *Lambert* (1839) which had created an exception to the general rule.

However, the House of Lords appeared to hold that there was an acknowledged limitation to exception where the original parties to the contract contemplated that a separate contract would come into existence between the contractor and the actual owner regulating the liabilities between them (*The Albazero*) (1977) AC 774. It was thought that *The Albazero* exception (which related to contracts of carriage) might extend to the duty of care or collateral warranty type arrangements under construction projects. However the Court of Appeal has now held that this is not necessarily so.

In *Alfred McAlpine Construction Limited* v. *Panatown Limited* (1998) (Court of Appeal), the contractors had on the same day as the building contract with Panatown entered into an agreement with the building owners by virtue of a duty of care deed. Having considered the terms of both contracts, the court found that, in respect of the building contract, both parties contemplated that accounts would be settled between them, and that an anomaly would arise if the employer could not recover damages for defective work and the parties' expectations would be defeated. This was so even though the employer had suffered no loss, having sold the building on to the holder of the duty of care deed at a price based on a building which complied with the contract specification, whereas in truth it contained defects.

Equally clearly, the duty of care deed, a separate contract from the building contract, was intended to create a right of action in contract for the building owner against the contractor if the contractor was in breach of its terms. It was not intended to preclude the employer's right to receive substantial damages for the contractor's breach.

Any risk of double recovery arose from the fact that there were two contracts rather than from the fact that the employer was entitled to recover substantial damages in the circumstances. In the court's view, there would be no such risk if damages were recovered by the employer on behalf of the building owner, and such damages would have to be taken into account if the building owner made a separate claim.

The House of Lords' cases and this Court of Appeal case both leave in their wake difficulties in relation to how any damages obtained by the employer in these circumstances should be dealt with. The suggestion appears to be that the employer will hold such sums on account of the third party owner who has suffered the real loss. The legal basis for this is dubious and its practical application fraught with difficulties.

Difficulties in the assessment of damages do not prevent recovery. The fact that there are future losses involved might make assessment of damages very difficult but nevertheless an assessment must be made and damages in respect of losses incurred and future likely losses must be awarded at the same time. It is not possible to have a series of actions claiming damages for breach of contract as and when the amount of future losses is ascertained.

(iii) Causation and remoteness of damage

Although causation and remoteness are two different concepts, they are often closely related and are therefore discussed together here. It is a matter of causation if the question which requires an answer is, 'was the loss caused by the breach of contract?'. On the other hand, it is a matter of remoteness of damages if the question asked is, 'was the particular loss within the contemplation of the parties?'. The leading case on remoteness of damage is *Hadley* v. *Baxendale* (1854).

Facts:

The plaintiff's mill was brought to a standstill by the breakage of their only crank shaft. The defendant carriers failed to deliver the broken shaft to the manufacturer at the time when they had promised to do so. They were sued by the plaintiff who sought recovery of the profits which they would have made had the mill been started up again without the consequent delay due to the late delivery of the broken shaft.

Held:

The facts known to the defendant carriers were insufficient to show that they were reasonably aware that the profits of the mill would have been affected by an unreasonable delay in the delivery of the broken shaft. The loss was therefore too remote. Alderson B said as follows:

> 'Where two parties have made a contract which one of them has broken, the damages which the other party ought to receive in respect of such breach of contract should be such as may fairly and reasonably be considered either arising naturally, i.e. according to the usual course of things, from such breach of contract itself, or such as may reasonably be supposed to have been in the contemplation of both parties, at the time they made the contract, as the probable result of the breach of it. Now, if the special circumstances under which the contract was actually made were communicated by the plaintiffs to the defendants, and thus known to both parties, the damages resulting from the breach of such contract, which they would reasonably contemplate, would be the amount of injury which would ordinarily follow from a breach of contract under these special circumstances so known and communicated. But, on the other hand, if these special circumstances were wholly unknown to the party breaking the contract, he, at the most, could only be supposed to have had in his contemplation the amount of injury which would arise generally, and in the great multitude of cases not affected by any special circumstances, from such breach of contract.'

The principles laid down in the passage quoted above have been interpreted and re-stated in more recent times in *Victoria Laundry (Windsor) Limited* v. *Newman Industries Limited* (1949) and in *Koufos* v. *C. Czarnikow Limited (The Heron II)* (1969), and (in a construction context) in *Balfour Beatty Construction (Scotland) Limited* v. *Scottish Power plc* (1994).

In the *Victoria Laundry* case it was stated that firstly, the aggrieved party is only entitled to recover such part of the loss actually resulting as was at the time of the contract reasonably foreseeable as liable to result from the breach. Secondly, what was reasonably foreseeable depends on the knowledge possessed by the parties at the time of entering into the contract. Thirdly, that for this purpose, knowledge 'possessed' is of two kinds: one imputed, the other actual. Everyone is taken to know the 'ordinary course of things' and consequently what loss is liable to result from a breach of contract in that ordinary course, but to this knowledge which the party in breach of contract is *assumed* to possess (whether he actually possesses it or not) there may have to be added in a particular case knowledge which he *actually* possessed of special circumstances outside the 'ordinary course of things', of such a kind that a breach in those special circumstances would be liable to cause a particular loss.

In the *Heron II* case Lord Upjohn stated the broad rule as:

'What was in the assumed contemplation of both parties acting as reasonable men in the light of the general or special facts (as the case may be) known to both parties in regard to damages as a result of a breach of contract.'

These two cases were applied by the House of Lords in the *Balfour Beatty* case, which is of interest on the question of the extent to which a party to a contract is presumed to know about the business activities of the other party.

Facts:

The plaintiffs were main contractors for the construction of the roadway and associated structures forming part of the Edinburgh City bypass. They installed a concrete batching plant near the site and entered into an agreement with the defendants for a temporary supply of electricity. Part of the plaintiffs' work was the construction of a concrete aqueduct to carry the Union Canal over the road. During this part of the works the batching plant ceased to work in consequence of the rupturing of fuses provided by the defendants in their supply system. As a result of the failure and the resulting shutdown of the batching plant the construction of the aqueduct could not be completed by continuous pouring of concrete as required by the specification. The plaintiffs were obliged to demolish the partly constructed aqueduct and start again. The plaintiffs sued for damages.

Held:

The plaintiffs were not entitled to recover, the losses not being of a type which were within the defendants' contemplation. As Lord Jauncey put it:

'It must always be a question of circumstances what one contracting party is presumed to know about the business activities of the other. No doubt the simpler the activity of the one, the more readily can it be inferred that the other would have reasonable knowledge thereof. However, when the activity of A involves complicated construction or manufacturing techniques, I see no reason why B who supplies a commodity that A intends to use in the course of those techniques should be assumed, merely because of the order for the commodity, to be aware of the details of all the techniques undertaken by A and the effect thereupon of any failure of or deficiency in that commodity.'

(iv) Measure of damages and time of assessment

As a general rule, the assessment of damages will be based on the difference between the value of the subject matter of the contract in its defective or incomplete state, as compared to the value it would have had, had the contract been properly completed and fulfilled. However, in certain circumstances this may cause injustice or prove particularly difficult in terms of assessment, e.g. a contractor's failure to complete a building. In such a case the measure of damages is likely to be the difference between the contract price and the amount it actually costs the employer to complete the contract works substantially as it was originally intended.

The general rule is that the time at which the assessment of damages will take place will be the date when the cause of action accrues, i.e. in the case of breach of contract at the date of the breach, irrespective of when damage is suffered or known to be suffered as a result of the breach. However, in cases where the appropriate measure of damages is the cost of repair or completion of the outstanding work, damages are likely to be assessed at the time when the repairs or completion work ought reasonably to have been undertaken, and this may in appropriate circumstances be as late as the date of the hearing if the plaintiff was acting reasonably in not having the repairs carried out or the outstanding work completed before then.

The New Zealand case of *Bevan Investments* v. *Blackhall & Struthers* (1978) is an example of the flexible approach to the question of the most appropriate method for measuring damages in building contract cases and also as to the time at which such an assessment should be made. The learned authors of Building Law Reports summarised the court's findings in this case as follows (11 BLR at page 79):

'1. That the company was entitled to be put into the position it would have been in had the contract been performed;

2. That in building cases the loss should, *prima facie*, be measured by ascertaining the amount required to rectify the defects complained of and so give to the building owner the equivalent of a building on the land substantially in accordance with the contract;

3. That this rule was adopted unless the court was satisfied that some lesser basis of compensation could in all the circumstances be fairly employed;

4. That since the only practicable action in the present case was to complete according to the modified design, the cost of so doing was the reasonable measure of damages;

5. That in calculating those damages

(a) to avoid an element of betterment credit should be given for the hypothetical additional cost of a proper initial design;

(b) the assessment should be computed at the date of trial either because of the principle that damages were to be assessed by reference to the date when the reinstatement works could be reasonably carried out and, on the facts, it was reasonable to postpone the work until the issues of liability and damages were settled, or because such damage was not too remote in that it was foreseeable that the company might be unable to complete the works until a trial if the appellant failed to exercise the skill required of him.'

It can be seen therefore that the general principle of English law which provides that, where the proper measure of damages is the cost of repairs etc, the cost is to be calculated on the basis of prices prevailing at or within a reasonable time after the discovery of the defective work, is unlikely to be applied when in all the circumstances it is reasonable for the innocent party to postpone the repairs etc. Indeed, in appropriate circumstances the time of assessment may be as late as the date of the hearing itself. There may be good commercial reasons for delaying repairs and the court is entitled to take these into account.

There are two important potential qualifications to the general rule referred to at the beginning of this section, one of which was referred to in the *Bevan Investments* case:

(1) The general rule will apply unless the court is satisfied that some lesser basis of compensation can fairly be employed; and
(2) In the context of defects, although a plaintiff can only recover as damages the costs which the defendant ought reasonably to have foreseen that the plaintiff would incur, reasonable costs do not mean the minimum amount which, with hindsight, it could be held would have sufficed. When the nature of the repairs is such that the plaintiff can only make them with the assistance of expert advice, the defendant ought, arguably, to have foreseen that the plaintiff would take such advice and be influenced by it.

On the first of these qualifications, reference must be made to the decision of the House of Lords in *Ruxley Electronics & Construction Limited* v. *Forsyth* (1995).

Facts:

The appellants entered into a contract whereby they agreed to construct for the respondent a swimming pool having a maximum depth at its deepest point of 7ft 6in. As constructed, the pool had a maximum depth of only 6ft 9in, and that depth was not achieved at the intended deepest point. At first instance, the judge made certain findings of fact: that the shortfall in depth did not decrease the value of the pool; that the cost of reconstruction would be £21,560; that the respondent had no intention of building a new pool. The judge found that the expenditure on reconstruction would be unreasonable since the cost would be wholly disproportionate to the advantage, if any, to be gained. The judge awarded a sum of £2,500 as general damages for loss of amenity.

In the Court of Appeal the respondent's appeal was allowed, and an award was made of £21,500 being the estimated cost of the rebuilding of the pool. The appellants appealed to the House of Lords.

Held:

The proper application of the general principle that where a party sustains loss by virtue of breach of contract he is so far as money can do it to be placed in the same situation in respect of damages as if the contract had been performed was not the monetary equivalent of specific performance but required the court to ascertain the loss the plaintiff had in fact suffered by reason of the breach. The cost of reinstatement was not the only possible measure of damage for defective performance under a building contract, and is not the appropriate measure where the expenditure would be out of all proportion to the benefit to be obtained – even if the alternative measure of value, diminution in value, would lead to only nominal damages.

The second qualification flows from the decision in the case of *The Board of Governors of the Hospital for Sick Children and Another* v. *McLaughlin & Harvey plc and Others* (1987).

Facts:

Between 1977 and 1980 a new wing was constructed for the plaintiffs at Great Ormond Street Hospital. Soon after practical completion a walkway beam col- lapsed. As a result the plaintiffs commenced investigations which led to a detailed examination of the design and construction of the new wing, and, in due course, proceedings were commenced against the contractor, the architects and the structural engineers. The action was directed to be tried in a series of sub-trials, the first of which related to the question of the remedial work which the plaintiffs were entitled to carry out.

Held:

On the basis that the original design was negligent, the plaintiffs could recover the cost of remedial work which, by the date of the trial, they had carried out to the foundations of the building on the advice of their expert.

It is of course clear from this brief description of the Great Ormond Street case that there are limits to the building owners right of recovery. Although it appears not to be open to the contract breaker to criticise the course honestly taken by the injured person on the advice of his experts, even though it appears by the light of after events that another course might have saved loss, it is still open to the contract breaker to argue that the scheme employed involved an element of betterment, i.e. the work carried out was more than was necessary properly to address the defects.

If the reason for building work being abandoned before completion is due to the contractor accepting a repudiation by the employer, this gives the contractor the right to sue for damages including his lost profit, if any. There has been some debate as to whether or not the contractor might, instead of suing for damages, bring an action in *quantum meruit* (as much as it is worth) for the work carried out by him; in other words to ignore the original contract sum and claim a reasonable sum instead: see *Lodder* v. *Slowey* (1904). This could be of considerable advantage to the contractor where his tender price was uneconomic and could be demon- strated to be less than the value of the work actually carried out. The difficulty with this argument is that the employer would in such a case be penalised by having to pay the contractor a sum in excess of the true damages suffered by the contractor as a result of the breach of contract.

The decision in *Lodder* v. *Slowey* was considered in the case of *ERDC Construction Limited* v. *H. M. Love & Company* (1994), and although it was not criticised, the court resisted the temptation to extend the principle apparently accepted in *Lodder* v. *Slowey* to a situation where, although breaches of contract had occurred, the contract remained in existence. In short, the conclusion of the court was that although a claim on the basis of *quantum meruit* might succeed when a contract had been rescinded by repudiation and acceptance, in the present case the clai- mant had elected to affirm the contract and had necessarily waived any right to claim on the basis of *quantum meruit*.

(v) Mitigation

There is what is generally described as a duty on the part of the innocent party to take reasonable steps to mitigate the loss suffered as a result of the guilty party's breach of contract. It may be said very briefly that there are three rules. Firstly, a plaintiff cannot recover for losses resulting from a defendant's breach of contract where the plaintiff could have avoided the loss by taking reasonable steps. Secondly, if the plaintiff actually avoids or mitigates his loss, even by taking steps which went beyond what was reasonably necessary in all the circumstances, he cannot then recover such avoided loss. Thirdly, if the plaintiff incurs loss or expense by taking reasonable steps to mitigate the loss, he can claim such loss and expense incurred in taking the mitigating steps from the defendant, even if the result is that the losses flowing from the breach of contract have been exacerbated.

Strictly, the requirement to mitigate does not amount to a duty on the part of the innocent party, because there is no rule of law requiring such steps to be taken. However, there will be no right of recovery to the extent that it can be shown that any loss claimed could have been avoided by taking reasonable steps in mitigation.

(E) Limitation of actions

It is a matter of public policy that there should be an end to litigation and that a time should arrive when stale demands can no longer be pursued. Accordingly, the Limitation Act 1980 prescribes the time limits after the expiry of which a claim will become 'statute-barred'. However, to obtain the benefit of a defence of limitation, the defendant must plead it specifically in his defence.

So far as contracts not under seal or executed as a deed are concerned, by virtue of section 5 of the Act, no action founded on contract can be brought after the expiry of six years from the date on which the cause of action accrued. The cause of action accrues at the date of the breach of contract irrespective of when and if damage is suffered as a result of that breach. Typically, in building contracts, the last date therefore on which a cause of action for breach of contract will accrue will be the date of practical completion or substantial completion. If the contract is under seal or executed as a deed, then the period is twelve years instead of six.

For actions arising out of tortious acts, the limitation period is six years from the date on which the cause of action accrued, except in relation to actions for personal injuries when that period is three years. In the tort of negligence, the cause of action does not accrue until damage is suffered. However, in relation to actions for damages for negligence where there are latent defects or damage other than in respect of personal injuries, the limitation period is now subject to the changes introduced by the Latent Damage Act 1986 which amends the 1980 Act.

As a result of the 1986 Act there is a special time limit for commencing pro-ceedings where facts relevant to the cause of action were not known at the date that the cause of action accrued. It extends the present period of limitation to three years from the date on which the plaintiff knew or ought to have known the facts about the damage where these facts became apparent later than the usual six years from the date on which the cause of action accrued, namely, in negligence actions, the date when the damage was suffered whether discovered or not.

However, there is an overriding long stop which operates to bar all negligence claims involving latent defects or damage which are brought more than 15 years from the date of the defendant's breach of duty.

In summary, the time limit runs out in respect of such actions at whichever is the later of the following:

- Six years from the occurrence of damage
- Three years from the discovery of the damage following the expiry of the basic six year period, but subject to
- A final time limit of 15 years from the breach of duty.

A right of action is given to anyone who acquires property which is already damaged where the fact of such damage is not known and could not be known to him at the time that he acquired the interest.

The three year period commences on the earliest date on which the plaintiff had both the knowledge required for bringing an action and the right to bring such an action. However, in relation to successive owners of buildings etc. if a predecessor in title knew or ought to have known of the damage then the three year period begins at that time and does not start afresh when the new owner acquires his interest.

There are provisions dealing with what knowledge a plaintiff must have or ought to have had before the three year period begins to run. These cover such matters as the knowledge that the damage was attributable to negligence and, if a third party is involved, the identity of that party.

Any deliberate concealment of defective work will prevent time from running until the defect is or ought to have been discovered. Section 32 of the Limitation Act 1980 provides that in certain cases such as fraud, concealment or mistake, time will not begin to run until the plaintiff has discovered or could with reasonable diligence have discovered the fraud, concealment or mistake. Fraud in this sense means that the defendant has behaved or acted in an unconscionable manner, for example, if a building contractor knowingly takes a risk by allowing bad workmanship or materials to be covered up: see *Applegate* v. *Moss* (1971).

The Latent Damage Act has no application to contractual negligence, i.e. breach of a contractually imposed duty of care: see *Iron Trades Mutual Insurance Co Limited* v. *Buckenham* (1989).

(F) Effect of determination on the contractual provision for liquidated damages

Subject to what the express terms of the contract may say, the liability of the contractor to pay liquidated damages ends on the determination of the contractor's employment, whether under an express provision or under the general law. However, liquidated damages which have accrued to that date will probably still be deductible from or payable by the contractor.

CONSIDERATION OF THE RELEVANT CLAUSES OF IFC 98

Clause 7.1

Notices under section 7
7.1 Any notice or further notice to which clauses 7.2.1, 7.2.2, 7.2.3, 7.3.4, 7.9.1, 7.9.2, 7.9.3, 7.9.4, 7.10.3 and 7.13.1 refer shall be in writing and given by actual delivery or by special delivery or recorded delivery. If sent by special delivery or recorded delivery[1] the notice or further notice shall, subject to proof to the contrary, be deemed to have been received 48 hours after the date of posting (excluding[2] Saturday and Sunday and Public Holidays).

COMMENTARY ON CLAUSE 7.1

The provisions of clause 7.1 are of general application to section 7, i.e. irrespective of which party exercises the right of determination and regardless of the reason for that exercise. Having said this, the provisions of clause 7.1 are not exhaustive in that they deal only with questions of form and service, not timing and substance. These further aspects are dealt with below, in the commentary on clauses 7.2, 7.3, 7.9, 7.10 and 7.13.

Any notice referred to in clause 7.1 must be issued in written form, and the permitted manner of service is:

- Actual delivery
- Special delivery or recorded delivery.

These procedural requirements should be followed precisely, although there is some authority in English law for a common sense business construction to be given to commercial contracts (*Goodwin & Sons* v. *Fawcett* (1965), in which a notice required by the contract to be served by registered post which was in fact served by recorded delivery was held to be valid on this basis). Other jurisdictions have not been so benign: in the case of *Central Provident Fund Board* v. *Ho Bock Kee* (1981) before the Singapore Court of Appeal, a purported determination was held to be invalid when delivery was effected by hand rather than (as required by the contract) by registered post. In the case of *J. M. Hill & Sons Limited* v. *London Borough of Camden* (1980) Lord Justice Ormrod resisted an unduly strict construction of the requirements of the timing of a notice of determination, in the following terms:

> 'Nothing is more distasteful to me than to construe a business contract in this formalistic sense. I think it makes a mockery of the law. To construe this contract and to treat what happened in this case between the contractors and the local authority in the way which (counsel submits it should be treated) is to make, I think, a farce of such contractual provisions...
>
> There can be no doubt here that it never occurred to any sensible person that there was any defect whatever in the operation of (the notice provision). Everybody concerned knew perfectly well what was happening. No one was in the very slightest degree prejudiced by it or pretends that they were prejudiced by it or would be believed if they did. So that it is the most purely formal point.'

Having said this, it would be open to a party seeking to challenge the validity of a purported determination to point to a failure to comply with a clearly expressed

contractual requirement as to notice, particularly when the grounds for the purported determination would not amount to repudiatory conduct at common law.
Strict compliance is, therefore, advisable.

The right to give notice by actual delivery was introduced into IFC 84 in 1994 by
Amendment 7.

If sent by special delivery or recorded delivery, service is deemed to have been
effected 48 hours after the date of posting; this presumption is, however, rebuttable if there is sufficient evidence to demonstrate a different (later or earlier) date
of receipt. Saturdays, Sundays and public holidays are excluded when calculating
the 48 hour period.

NOTE TO CLAUSE 7.1

[1] '...actual delivery...'
Quite what is meant by these words is not made clear: it is possible to speculate
that what is envisaged is personal service, but the words in isolation do not
require so narrow a construction. Taking clause 7.1 as a whole, however, there
would be little point in specifying special delivery or recorded delivery as
acceptable means of service if actual delivery could be effected by ordinary post or
electronic transmission. In any event, the provision relating to deemed receipt is
only applicable to notice sent by special delivery or recorded delivery.

[2] '...(excluding...'
It seems that this exclusion does not apply where service by special delivery or
recorded delivery is the subject of proof, rather than presumption (and it does not
have any application in cases of actual delivery). It follows that care must be
exercised in the timing of actions taken in pursuance of any such notice.

Clauses 7.2 to 7.8

DETERMINATION BY EMPLOYER
Default by Contractor
7.2.1 If, before the date of Practical Completion[1], the Contractor shall make a default in any
one or more of the following respects:
(a) without reasonable cause he wholly or substantially[2] suspends the carrying out of
the Works, or
(b) he fails to proceed regularly and diligently with the Works, or
(c) he refuses or neglects to comply with a written notice or instruction from the
Architect/the Contract Administrator requiring him to remove any work, materials or
goods not in accordance with this Contract and by such refusal or neglect the Works are
materially affected, or
(d) he fails to comply with the provisions of either clause
3.1 (Assignment) or
3.2 (Sub-contracting) or
3.3 (Named sub-contractors), or
(e) he fails pursuant to the Conditions to comply with the requirements of the CDM
Regulations
the Architect/the Contract Administrator may give to the Contractor a notice[3] specifying
the default or defaults (the 'specified default or defaults').

7.2.2 If the Contractor continues a specified default for 14 days from receipt of the notice
under clause 7.2.1 then the Employer may on, or within 10 days from, the expiry of that

14 days by a further notice to the Contractor determine the employment of the Contractor under this Contract. Such determination shall take effect on the date of receipt of such further notice.

7.2.3   If

the Contractor ends the specified default or defaults, or

the Employer does not give the further notice referred to in clause 7.2.2

and the Contractor repeats a specified default (whether previously repeated or not) then, upon or within a reasonable time after such repetition, the Employer may by notice to the Contractor determine the employment of the Contractor under this Contract. Such determination shall take effect on the date of receipt of such notice.

7.2.4   A notice of determination under clause 7.2.2 or clause 7.2.3 shall not be given unreasonably or vexatiously.

Insolvency of Contractor

7.3.1   If the Contractor [Z]

makes a composition or arrangement with his creditors, or becomes bankrupt, or being a company,

makes a proposal for a voluntary arrangement for a composition of debts or scheme of arrangement to be approved in accordance with the Companies Act 1985 or the Insolvency Act 1986 as the case may be or any amendment or re-enactment thereof, or

has a provisional liquidator appointed, or

has a winding-up order made, or

passes a resolution for voluntary winding-up (except for the purposes of amalgamation or reconstruction), or

under the Insolvency Act 1986 or any amendment or re-enactment thereof has an administrator or an administrative receiver appointed

then

7.3.2   the Contractor shall immediately inform the Employer in writing if he has made a composition or arrangement with his creditors, or, being a company, has made a proposal for a voluntary arrangement for a composition of debts or scheme of arrangement to be approved in accordance with the Companies Act 1985 or the Insolvency Act 1986 as the case may be or any amendment or re-enactment thereof;

7.3.3   where a provisional liquidator or trustee in bankruptcy is appointed or a winding-up order is made or the Contractor passes a resolution for voluntary winding-up (except for the purposes of amalgamation or reconstruction) the employment of the Contractor under this Contract shall be forthwith automatically determined but the said employment may be reinstated if the Employer and the Contractor shall so agree;

7.3.4   where clause 7.3.3 does not apply the Employer may at any time[4], unless an agreement to which clause 7.5.2.1 refers has been made, by notice to the Contractor determine the employment of the Contractor under this Contract and such determination shall take effect on the date of receipt of such notice.

[Z] See Practice Note 24: after certain events an Insolvency Practitioner acts for the Employer

Corruption

7.4   The Employer shall be entitled to determine the employment of the Contractor under this or any other contract, if the Contractor shall have offered or given or agreed to give to any person any gift or consideration of any kind as an inducement or reward for doing or forbearing to do or for having done or forborne to do any action in relation to the obtaining or execution of this or any other contract with the Employer, or for showing or forbearing to show favour or disfavour to any person in relation to this or any other contract with the Employer, or if the like acts shall have been done by any person employed by the Contractor or acting on his behalf (whether with or without the knowledge of the Contractor), or if in relation to this or any other contract with the Employer the Contractor or any person employed by him or acting on his behalf shall have committed an offence under the Prevention of Corruption Acts 1889 to 1916, or, where the Employer is a local authority, shall have given any fee or reward the receipt of which is an offence under sub-section (2) of S.117 of the Local Government Act 1972 or any amendment or re-enactment thereof.

Insolvency of Contractor – option to Employer

7.5   Clauses 7.5.1 to 7.5.4 are only applicable where clause 7.3.4 applies.

7.5.1   From the date when, under clause 7.3.4, the Employer could first give notice to deter-mine the employment of the Contractor, the Employer, subject to clause 7.5.3, shall not be bound by any provisions of this Contract to make any further payment thereunder and the Contractor shall not be bound to continue to proceed with and complete the Works in compliance with clause 2.1.

7.5.2   Clause 7.5.1 shall apply until
         either
         .1   the Employer makes an agreement (a '7.5.2.1 agreement') with the Contractor on the continuation or novation or conditional novation of this Contract, in which case this Contract shall be subject to the terms set out in the 7.5.2.1 agreement
         or
         .2   the Employer determines the employment of the Contractor under this Contract in accordance with clause 7.3.4, in which case the provisions of clause 7.6 or clause 7.7 shall apply.

7.5.3   Notwithstanding clause 7.5.1, in the period before either a 7.5.2.1 agreement is made or the Employer under clause 7.3.4 determines the employment of the Contractor, the Employer and the Contractor may make an interim arrangement for work to be carried out. Subject to clause 7.5.4 any right of set-off which the Employer may have shall not be exercisable in respect of any payment due from the Employer to the Contractor under such interim arrangement.

7.5.4   From the date when, under clause 7.3.4, the Employer may first determine the employment of the Contractor (but subject to any agreement made pursuant to clause 7.5.2.1 or arrangement made pursuant to clause 7.5.3) the Employer may take reasonable measures to ensure that Site Materials, the site and the Works are adequately protected and that Site Materials are retained in, on the site of, or adjacent to the Works as the case may be. The Contractor shall allow and shall in no way hinder or delay the taking of the aforesaid measures. The Employer may deduct the reasonable cost of taking such measures from any monies due or to become due to the Contractor under this Contract (including any amount due under an agreement to which clause 7.5.2.1, or under an interim arrangement to which clause 7.5.3, refers) or may recover the same from the Contractor as a debt.

Consequences of determination under clauses 7.2 to 7.4

7.6   Without prejudice[5] to any proceedings including adjudication in which the validity of the determination is in issue, in the event of the determination of the employment of the Contractor under clause 7.2.2, 7.2.3, 7.3.3, 7.3.4 or 7.4 and so long as that employment has not been reinstated then:
        (a) the Contractor shall give up possession of the site of the Works subject to the orderly compliance of the Contractor with any instruction of the Architect/the Contract Administrator under clause 7.6(e);
        (b) the Employer may employ and pay other persons to carry out and complete the Works and to make good defects of the kind referred to in clause 2.10 and he or they may enter upon the site and the Works and use all temporary buildings, plant, tools, equipment and Site Materials, and may purchase all materials and goods necessary for the carrying out and completion of the Works and for the making good of defects as aforesaid; provided that where the aforesaid temporary buildings, plant, tools, equip-ment and Site Materials are not owned by the Contractor the consent of the owner thereof to such use is obtained by the Employer;
        (c) except where an insolvency event listed in clause 7.3.1 (other than the Contractor being a company making a proposal for a voluntary arrangement for a composition of debts or scheme of arrangement to be approved in accordance with the Companies Act 1985 or the Insolvency Act 1986 as the case may be or any amendment or re-enactment

thereof) has occurred the Contractor shall, if so required by the Employer or by the Architect/the Contract Administrator on behalf of the Employer within 14 days of the date of determination, assign to the Employer without payment the benefit of any agreement for the supply of materials or goods and/or for the execution of any work for the purposes of this Contract to the extent that the same is assignable;

(d) except where the Contractor has a trustee in bankruptcy appointed or being a company has a provisional liquidator appointed or has a petition alleging insolvency filed against it which is subsisting or passes a resolution for voluntary winding-up (other than for the purposes of amalgamation or reconstruction) which takes effect as a creditors voluntary liquidation, the Employer may pay any supplier or sub-contractor for any materials or goods delivered or works executed for the purposes of this Contract before the date of determination in so far as the price thereof has not already been paid by the Contractor. Payments made under this clause 7.6(d) may be deducted from any sum due or to become due to the Contractor or may be recoverable from the Contractor by the Employer as a debt;

(e) the Contractor shall when required in writing by the Architect/the Contract Administrator so to do (but not before) remove from the Works any temporary buildings, plant, tools, equipment, goods and materials belonging to him and the Contractor shall have removed by their owner any temporary buildings, plant, tools, equipment, goods and materials not owned by him. If within a reasonable time after any such requirement has been made the Contractor has not complied therewith in respect of temporary buildings, plant, tools, equipment, goods and materials belonging to him, then the Employer may (but without being responsible for any loss or damage) remove and sell any such property of the Contractor, holding the proceeds less all costs incurred to the credit of the Contractor;

(f) Subject to clauses 7.5.3 and 7.6(g) the provisions of this Contract which require any further payment or any release or further release to the Contractor of amounts withheld under clause 4.2.1 and, if applicable, under this clause 4.3 shall not apply; provided that clause 7.6(f) shall not be construed so as to prevent the enforcement by the Contractor of any rights under this Contract in respect of amounts properly due to be paid by the Employer to the Contractor which the Employer has unreasonably not paid and which, where clause 7.3.4 applies, have accrued 28 days or more before the date when under clause 7.3.4 the Employer could first give notice to determine the employment of the Contractor or, where clause 7.3.4 does not apply, which have accrued 28 days or more before the date of determination of the employment of the Contractor;

(g) Upon the completion of the Works and the making good of defects as referred to in clause 7.6(b) (but subject, where relevant, to the exercise of the right under clause 2.10 of the Architect/the Contract Administrator, with the consent of the Employer, not to require defects of the kind referred to in clause 2.10 to be made good) then within a reasonable time thereafter an account in respect of the matters referred to in this clause 7.6(g) shall be set out either in a statement prepared by the Employer or in a certificate issued by the Architect/the Contract Administrator:
- the amount of expenses properly incurred by the Employer including those incurred pursuant to clause 7.6(b) and of any direct loss and/or damage caused to the Employer as a result of the determination;
- the amount of any payment made to the Contractor;
- the total amount which would have been payable for the Works in accordance with this Contract.

If the sum of the first two amounts stated exceeds or is less than the third amount stated the difference shall be a debt payable by the Contractor to the Employer or by the Employer to the Contractor as the case may be.

Employer decides not to complete the Works

7.7.1 If the Employer decides after the determination of the employment of the Contractor not to have the Works carried out and completed, he shall so notify the Contractor in writing within 6 months from the date of such determination. Within a reasonable time from the date of such written notification the Employer shall send to the Contractor a statement of account setting out;

(a) the total value of work properly executed at the date of determination of the employment of the Contractor, such value to be ascertained in accordance with the Conditions as if the employment of the Contractor had not been determined, together with any amounts due to the Contractor under the Conditions not included in such total value;

(b) the amount of any expenses properly incurred by the Employer and of any direct loss and/or damage caused to the Employer as a result of the determination.

After taking into account amounts previously paid to the Contractor under this Contract, if the amount stated under clause 7.7.1(b) exceeds or is less than the amount stated under clause 7.7.1(a) the difference shall be a debt payable by the Contractor to the Employer or by the Employer to the Contractor as the case may be.

7.7.2   If after the expiry of the 6 month period referred to in clause 7.7.1 the Employer has not begun to operate the provisions of clause 7.6(b) and has not given a written notification pursuant to clause 7.7.1 the Contractor may require by notice in writing to the Employer that he states whether clauses 7.6(b) to 7.6(g) are to apply and, if not to apply, require that a statement of account pursuant to clause 7.7.1 be prepared by the Employer for submission to the Contractor.

Other rights and remedies

7.8   The provisions of clauses 7.2 to 7.7 are without prejudice to any other rights and remedies which the Employer may possess.

COMMENTARY ON CLAUSES 7.2 TO 7.8

Clauses 7.2 to 7.8 deal with the determination of the contractor's employment by the employer or by automatic operation of the contract and the consequences flowing from such determination. These provisions are similar to those found in JCT 80 following the review and revision to the determination provisions introduced by Amendment 11 to JCT 80 in 1992. The corresponding revision to IFC 84 was made by Amendment 7, dated April 1994.

Grounds for determination of contractor's employment by the employer in cases of default

The following are the grounds of default on the basis of which the employer may determine the contractor's employment under the contract.

(1)   If the contractor without reasonable cause wholly or substantially suspends the carrying out of the works before Practical Completion (clause 7.2.1(a)).

This ground for determination does not apply where the contractor has reasonable cause for suspending the carrying out of the works. A suspension brought about by delays outside the contractor's control will generally amount to a reasonable cause. Indeed, the wording may well implicitly require some form of culpability on the part of the contractor. It is not thought that what is reasonable is in any way linked to the grounds recognised by the contract as excusing delay.

The proviso in clause 7.2.4, that the notice of determination shall not be given unreasonably or vexatiously, must presumably mean that there are situations in which the contractor has without reasonable cause suspended, but where nevertheless it would be unreasonable or vexatious for the employer to determine the employment of the contractor.

This ground for determining the contractor's employment currently appears in JCT 98 (clause 27.2.1.1).

(2) If the contractor fails to proceed regularly and diligently with the works (clause 7.2.1(b)).

By clause 2.1 the contractor is under an express duty to proceed regularly and diligently with the works and failure to do so is a ground for determination. This ground for determining the contractor's employment currently appears in JCT 98 (clause 27.2.1.2).

The meaning of the words 'regularly and diligently' in the determination provisions of JCT 63 was considered by Mr Justice Megarry in the case of *London Borough of Hounslow* v. *Twickenham Garden Developments Limited* (1970) where he said (see 7 BLR page 120):

> 'These are elusive words, on which the dictionaries help little. The words convey a sense of activity, of orderly progress, and of industry and perseverance: but such words provide little help on the question of how much activity, progress and so on is to be expected. They are words used in a standard form of building contract in relation to functions to be discharged by the architect, and in those circumstances it may be that there is evidence that could be given, whether of usage among architects, builders and building owners or otherwise, that would be helpful in construing the words.'

The meaning of the words 'regularly and diligently' was more recently considered by the Court of Appeal in the case of *West Faulkner Associates* v. *London Borough of Newham* (1994). In that case, Lord Justice Brown said:

> 'My approach to the proper construction and application of the clause would be this. Although the contractor must proceed both regularly and diligently with the works, and although each word imports into that obligation certain discrete concepts which would not otherwise inform it, there is a measure of overlap between them and it is thus unhelpful to seek to define two quite separate and distinct obligations.
>
> What particularly is supplied by the word "regularly" is not least a requirement to attend for work on a regular daily basis with sufficient in the way of men, materials and plant to have the physical capacity to progress the works substantially in accordance with the contractual obligations.
>
> What in particular the word "diligently" contributes to the concept is the need to apply that physical capacity industriously and efficiently towards that same end.
>
> Taken together the obligation upon the contractor is essentially to proceed continuously, industriously and efficiently with appropriate physical resources so as to progress the works steadily towards completion substantially in accordance with the contractual requirements as to time, sequence and quality of work.'

Certainly the fact that the work is done defectively does not prevent it being carried out regularly: see *Lintest Builders* v. *Roberts* (1978) (10 BLR at page 129). The case of *Greater London Council* v. *Cleveland Bridge & Engineering Co Limited* (1984), particularly at first instance, may be relevant here (see page 78).

(3) If the contractor refuses or neglects to comply with a written notice or instruction from the architect to remove non-conforming, defective or improper work, materials or goods and such refusal or neglect has a material effect on the works (clause 7.2.1(c)).

The architect has an express power (clause 3.14) to issue instructions in regard to the removal of any work, materials or goods which are not in accordance with the contract. The instruction, to be valid, must expressly require the removal. Simply to condemn the work or material is insufficient: see *Holland Hannen & Cubitts* v. *Welsh Health Technical Services Organisation* (1981).

The wording of this provision suggests that it applies before practical completion. It does not therefore apply to a failure to make good (as opposed to removal etc.) during the defects liability period – see clause 2.10. This ground for determination currently appears in JCT 98 (clause 27.2.1.3).

(4) If the contractor fails to comply with the provisions of clauses 3.1 (assignment), 3.2 (sub-contracting), or 3.3 (named sub-contractors) (clause 7.2.1(d)).

The failure may be substantial, e.g. sub-contracting the whole of the labour element of the works without consent, or it may be much less substantial. In the latter event the proviso relating to unreasonable or vexatious notices of determination must be borne in mind.

There is a similar though not identical provision in JCT 98 (clause 27.2.1.4).

Whereas under JCT 98 failure to comply with the clauses on nomination is not a ground for the determination of the contractor's employment, in IFC 98 a failure by the contractor to comply with clause 3.3 dealing with named persons does qualify as a ground for determination. Bearing in mind that the mischief aimed at in JCT 98 was that of sub-contracting without consent, it is not easy to see the reasoning behind this additional ground in IFC 98.

(5) If the contractor fails, pursuant to the conditions, to comply with the requirements of the CDM Regulations (clause 7.2.1(e)).

This ground for determination was added by Amendment 8 in March 1995. It is restricted to a failure to comply with the Regulations, and might not, therefore, be invoked in the event of incidental failures, e.g. to comply with the requirement of clause 5.7.4 to provide information to the Planning Supervisor.

Determination in the event of the contractor's insolvency

In addition to the right of determination afforded to the employer when the contractor is in default, by virtue of clause 7.3 the employment of the contractor will or may be determined if any of a number of specified events occur which are likely to have a bearing on the contractor's financial ability to complete the works.

The provisions of clause 7.3 were drafted in their present form following a review by the Tribunal of the principles underlying determination in cases of insolvency. The review considered particularly the extent to which it remained desirable or appropriate that the contractual provisions should continue to provide for automatic determination which – in the view of the Tribunal – did not facilitate continuity, and could therefore cause delay and cost to both parties.

The full objects of the Tribunal in relation to insolvency-related determination are set out in Practice Note 24 issued by the Tribunal, and the result of the review is that the provision for determination is no longer automatic in all cases where an insolvency procedure is entered into; rather, the employer is in certain circumstances given an option. The purpose behind this is to allow a period following the happening of the specified insolvency events during which the contractual arrangements would remain in force while the parties seek to agree the way forward. In cases of bankruptcy or insolvency events resulting in liquidation, determination of the contractor's employment remains automatic.

The insolvency events listed under clause 7.3 are where:

(1) The contractor makes a composition or arrangement with his creditors (option to determine)
(2) The contractor becomes bankrupt (automatic determination)
(3) The contractor makes a proposal for a voluntary arrangement for a composition of debts or scheme of arrangement to be approved in accordance with the Companies Act 1985 or Insolvency Act 1986 (as amended or re-enacted) (option to determine)
(4) A provisional liquidator is appointed (automatic determination)
(5) A winding-up order is made (automatic determination)
(6) The contractor passes a resolution for a voluntary winding-up (except for the purposes of amalgamation or reconstruction) (automatic determination)
(7) An administrator or administrative receiver is appointed under the Insolvency Act 1986 (as amended or re-enacted) (option to determine).

The provision for automatic determination may still be open to question in terms of its ability to achieve its apparent objective. The intended consequences of a determination under the contract are dealt with in detail below but it can be said here that the contract purports to give the employer certain rights, e.g. to withhold payment and to make use of materials, goods and plant belonging to the contractor, which may prejudice other creditors of the contractor and in particular run foul of the statutory provisions as to the mandatory *pari passu* discharge of liabilities which cannot be contracted out of: see *British Eagle International Airlines Limited* v. *Compagnie Nationale Air France* (1975). It may be therefore that, even if this contractual provision is effective to determine the contractor's employment, the consequential provisions set out in clause 7.6 are not fully effective. However, it should be remembered that in such a situation, the employer will still have some very important rights of set-off under the mutual dealing provisions which apply on bankruptcy and liquidation.

Secondly, on a bankruptcy or liquidation the trustee in bankruptcy or the liquidator, as the case may be, has a statutory right to disclaim unprofitable contracts – see for example section 178 of the Insolvency Act 1986. The trustee's or liquidator's obligation is only to carry on the business so far as may be necessary for its beneficial winding-up. To the extent that the automatic determination provisions remain effective, this decision is removed from the trustee or liquidator. The notice of intention to disclaim a contract must generally be served within 28 days following an application in writing made by a person interested in that decision. A failure to respond to such a request in writing will be treated as an adoption of the contract. Whether the contractual provisions for determination are entirely consistent with the bankruptcy laws, is still perhaps open to doubt.

The scheme of the clause is as follows: if any of events (2), (4), (5) or (6) occur, the contractor's employment is automatically determined, but may be reinstated if both parties agree; if none of those events apply, but any of the other events occur, the employer may at any time determine the contractor's employment unless an agreement to which clause 7.5.2.1 refers has been made. A '7.5.2.1 agreement' is an agreement with the contractor on the continuation or novation or conditional novation of the contract.

This is the 'employer's option'. From the date when the employer could first give notice to determine the contractor's employment, subject to any interim arrangement for the work to be carried out made by the parties, the employer is not bound by any provision of the contract to make any further payment and the contractor is relieved of his obligation to proceed with and complete the works in compliance with clause 2.1. This state of affairs persists until either:

(1) A '7.5.2.1 agreement' is reached on the continuation, novation or conditional novation of the contract, at which point the terms of the '7.5.2.1 agreement' will regulate the parties' rights; *or*
(2) The employer determines the contractor's employment.

If the parties do make an interim arrangement for work to be carried out pending either of the above events, any right of set-off which the employer may have is – by virtue of clause 7.5.3 – not exercisable in respect of any payment due from the employer under the interim arrangement.

From the date when the employer is entitled to determine the contractor's employment, subject to any interim arrangement made under clause 7.5.3 or any '7.5.2.1 agreement', the employer is entitled to take reasonable measures to ensure that site materials, the site and the works are adequately protected and that site materials are retained. The contractor is to allow and must not hinder or delay the taking of these measures. The employer is entitled to deduct the reasonable cost of taking these measures from any monies due or to become due to the contractor under the contract. The employer is also entitled to deduct any such amounts from sums falling due under any interim arrangement or '7.5.2.1 agreement'.

Where one of the events occurs giving rise to automatic determination, the possibility remains that the contractor's employment might be reinstated. Determination is stated to occur forthwith, but it has been suggested that if the sense of the contract is such as to confer a discretionary remedy on one or other of the parties, it will make no difference that the wording requires automatic determination, because some act invoking the provision by the party entitled to do so must take place: see *Hudson's Building and Engineering Contracts*, 11th edition, page 1265. In any event the sensible course for the employer would be to inform the contractor, his trustee or liquidator, as the case may be, that the contractor's employment is determined to avoid any argument that there might have been a reinstatement of that employment by the inactivity of the employer being regarded as evidence of his willingness to treat the contractor's employment as reinstated.

The reduction in the grounds for automatic determination of the contractor's employment, and its replacement with an option to determine addresses the major criticism of the provisions for determination on insolvency as originally drafted. The fact of an administrative receiver or an administrator being appointed does not of itself put the contractor in breach of contract, and indeed – if permitted to – the contractor might under the guidance of an insolvency practitioner complete the contract satisfactorily. This is the rationale for the introduction of the 'employer's option'.

Although clause 7.3.3 expressly envisages the possibility of there being a reinstatement of the contractor's employment, this is not available as of right to the employer. Depending on the financial position of the contractor at the time of the purported automatic and immediate determination, it is conceivable that reinstatement would be to the employer's advantage, e.g. where there are insufficient funds to withhold and the contractor is able notwithstanding his difficulties to complete the works for the agreed contract sum. Having said this, where the financial balance is in the contractor's favour, it is perhaps doubtful whether there is sufficient incentive for the contractor's liquidator or trustee to agree to reinstatement. The prospect of reinstatement was, of course, much greater under the provisions as originally drafted, when automatic determination could occur on the happening of a much wider range of events.

Determination of contractor's employment by employer on grounds of corruption (clause 7.4)

As originally drafted, this ground applied only when the employer was a local authority. That restriction no longer applies, but it might well have been the case that any employer would have been entitled to determine on the occurrence of certain of the events listed in the clause, in view of the serious nature of the matters referred to.

Consequences of the determination of the contractor's employment by the employer and of automatic determination (clauses 7.6 and 7.7)

Clause 7.6 provides for the consequences of a determination of the contractor's employment under the contract on any of the grounds referred to above (i.e. under clauses 7.2.2, 7.2.3, 7.3.3, 7.3.4 or 7.4), so long as the contractor's employment has not been reinstated. The consequences may be summarised as follows:

(a) Contractor to relinquish possession of the site

Unlike JCT 98, IFC 98 contains an express contractual obligation on the contractor to give up possession of the site of the works. It is submitted that the purported determination of the contractor's employment will therefore bring to an end any licence which the contractor may have to occupy the site. Such licence is given by the contract and can be taken away by it. This is so even if the validity of the purported determination is challenged by the contractor (see the opening lines of clause 7.6). The decision in the case of *London Borough of Hounslow* v. *Twickenham Garden Developments Ltd* (1970), even if a correct statement of the law, which has been doubted (see pages 68), can have no application under this contract.

(b) Employer may engage others to complete the works and use the temporary buildings, plant, site materials, etc.

The employer is given the right to employ and pay others to complete the unexecuted work. This brings into consideration the general duty of a contracting party to mitigate his loss following a breach of contract by the other party. While a determination of the contractor's employment will not necessarily be the result of a breach of contract by the contractor, nevertheless the steps taken by the employer must be reasonable in all the circumstances. This point is discussed further below.

The employer has the right to enter the site and so do those engaged by him, to complete the unexecuted work, whether or not the contractor has complied with the requirement to give up possession of the site of the works. Clearly, neither the employer nor the new contractor would thereby have the right physically to remove the contractor from the site nor use force to obtain control of the temporary buildings etc. The contractor will, however, be a trespasser and the employer should have little difficulty in obtaining an interim injunction to remove the contractor. Whether the employer could obtain an injunction preventing the contractor from removing his temporary buildings etc. is much more doubtful.

The employer's rights in connection with temporary buildings, plant, site materials, etc. are, however, subject to the important proviso at the end of clause 7.6(b): if they are not owned by the contractor, the employer's rights are subject to consent being obtained from the owner. This express proviso perhaps only states what is legally obvious, i.e. that the parties to a contract cannot by that contract affect the rights of third parties. If, for example, the site materials were as a matter of law the property of the supplier who had provided them to the contractor, the provisions of the contract would not bind the supplier and he could take action to recover the goods or seek damages in the event that the employer had converted the materials to his own use. Whether or not the proviso has unintentionally worsened the employer's position is perhaps open to debate: circumstances could arise whereby possession of goods by the contractor falling short of ownership might by virtue of section 25(1) of the Sale of Goods Act 1979 be sufficient to enable title to pass to the employer. The express words of the proviso, nevertheless, require the true owner's consent to be obtained prior to the employer making use of them. As between the employer and the true owner, the proviso will have no effect, but the employer could find himself being challenged by the contractor's trustee or liquidator if consent has not been obtained. The subject of ownership of materials is dealt with earlier in this book on page 39.

Finally, the employer is expressly empowered to purchase all materials and goods necessary for the carrying out and completion of the works and the cost of these will form part of the account required under clause 7.6(g).

(c) Contractor to assign the benefit of agreements for the supply of materials or goods and/ or the execution of any work

The right of the employer to call for the assignment of the benefit of any agreement does not apply – with one exception – in the case of determination due to insolvency. The one exception is where the event in question is the contractor, being a company, making a proposal for a voluntary arrangement for a composition of debts or scheme of arrangement to be approved in accordance with the Companies Act 1985 or the Insolvency Act 1986. In any other case, the benefit would be an asset to be dealt with by the contractor's trustee or liquidator, and to the extent that the assignment would place the employer in a better position than other creditors, it would offend against the *pari passu* rule. The requirement to assign may be by the employer or by the architect on his behalf within 14 days of the date of determination, and is to be without payment. The obligation can only be enforced, however, to the extent that the benefit of the contract in question is assignable, and there would appear to be no obligation on the contractor to secure a right of assignment in anticipation of any request being made.

(d) Employer may pay suppliers or sub-contractors for materials or goods delivered or works executed for the purposes of the contract

Again, this provision is restricted in its application: it does not apply where the contractor has a trustee in bankruptcy appointed or being a company has a provisional liquidator appointed or has a petition alleging insolvency filed against it which is subsisting or passes a resolution for voluntary winding up (other than for the purposes of amalgamation or reconstruction) which takes effect as a creditors' voluntary liquidation. In all other circumstances, the employer can pay suppliers or sub-contractors to the contractor for any materials or work delivered or executed prior to the date of determination insofar as payment has not been paid by the contractor. Any payment made by the employer may be deducted from any sum due or to become due to the contractor (subject, where appropriate, to notice in accordance with section 111 of the Housing Grants, Construction and Regeneration Act 1996), or may be recovered by the employer as a debt. There is, of course, no need to make express contractual provision for such direct payment; however, what is required is that if such payments are brought within the framework of the contract there should be an express right to deduct or recover the amount paid.

(e) Removal from the works of goods, materials and plant, etc.

As and when required in writing by the architect, the contractor must remove from the works any temporary buildings, plant, tools, equipment, goods and materials owned by him, and must have removed by their owner any not owned by him. If he fails within a reasonable time to comply with this written requirement, the employer may remove (without responsibility for any damage) and sell any property of the contractor, holding the net proceeds to the credit of the contractor. The express reference in clause 7.6(e) to property not owned by the contractor is arguably necessary in order to ensure that the architect has the power to instruct its removal, where the existence of third party rights precludes the possibility of removal and sale by the employer. Quite what is to happen to third party property, in the unlikely event that the true owner does not claim it, is unclear.

The contractor's failure to remove the temporary buildings etc. would, quite apart from the provisions of the contract, put the employer in the position of an involuntary bailee and as such he may, though the position is not free from doubt, be able to avail himself of the wide powers of sale conferred by sections 12 and 13 of the Torts (Interference with Goods) Act 1977. However, the rights available under the Act would appear not to apply where the employer is aware that the temporary buildings etc. are not owned by the contractor.

(f) Further payment

Subject to any payment to which the contractor is due by virtue of an interim arrangement made under clause 7.5.3, and to the contractor's ultimate entitlement under clause 7.6(g), the provisions of the contract requiring any further payment or release of amounts held under clauses 4.2.1 and 4.3 no longer apply. This is subject to the proviso that the contractor is not prevented from enforcing any rights under the contract in respect of amounts properly due to be paid and which the employer has unreasonably not paid. In order to qualify under this proviso, the amount in question must have accrued 28 days or more before the date (if determination is not automatic) when the employer could first have given notice to determine the contractor's employment or (where determination is automatic) of determination of the contractor's employment. Presumably the justification for the 28 day period is that it is sufficiently proximate to the right to determine arising for the employer to have legitimate reservations about making payment, without non-payment itself being the cause of the contractor's insolvency.

(g) Financial consequences (clause 7.6(g))

The financial consequences of the determination are to be reflected either in a statement prepared by the employer or in a certificate issued by the architect. The relevant document will set out an account prepared within a reasonable time after completion of the works and the making good of defects. The defects in question are those dealt with by the machinery of the contract, i.e. those notified by the architect in accordance with clause 2.10. What this means is that the employer may not delay preparation of the account indefinitely, on the basis of an allegation that latent defects might appear; similarly, the employer may not delay on the basis that defects referred to under clause 2.10 have not been dealt with, in circumstances where the employer has consented to such defects not being made good.

The account must set out the following.

(i) The amount of expenses properly incurred by the employer and of any direct loss and/or damage caused to the employer

This can include losses equivalent to those recoverable as damages for breach of contract at common law: see *Wraight Limited* v. *P H & T (Holdings) Limited* (1968). The nature and extent of common law damages have been dealt with briefly earlier in this chapter on page 322.

Included in the amount will be any payments made in purchasing materials and goods or made to other persons to carry out and complete the works. The amount paid by the employer in this connection will be recoverable only to the extent that such expenditure was reasonably incurred. The employer has a duty to take reasonable steps to mitigate his losses. However, the approach of the courts – and so also therefore that of an arbitrator – is unlikely to be a strict one and what is reasonable in all the circumstances is likely to give the employer a good degree of room for manoeuvre. It will not be sufficient for the contractor simply to establish that the outstanding work could have been completed at less cost. He must show that the employer's actions were positively unreasonable.

(ii) The amount of any payment made to the contractor

The amount of any payment made ought not to be controversial. However, on publication, IFC 84 referred to the contractor being 'paid'. This was subsequently replaced with a reference to the amount due being paid or 'otherwise discharged', no doubt to make it absolutely clear that any legal obligation to pay could be discharged by legally setting off, e.g. liquidated damages. The wording has now reverted to a reference to the amount 'paid' to the contractor. For a possible explanation of the reason for this and its effect see earlier under note [2] to clause 1.11 (page 59) and note [16] to clauses 4.2 to 4.8 (page 222). What can be said here is that if 'paid' were to mean physically paid rather than a discharging of the legal obligation to pay, it would have the unfortunate and surely unintended effect that when the calculation required by the last paragraph of clause 7.11.3 is made (see the last paragraph under the next heading), any amount previously set-off, even lawfully, would in effect have to be recredited to the contractor.

(iii) The total amount which would have been payable for the works in accordance with the contract

This is a notional final account, in which values will be included for the work not carried out by the contractor and in respect of items which the contractor might not even have included in the contract sum (e.g. for variations carried out in the course of completing the works). On this point, it might be open to argument that the first item to be particularised in the account – the amount of expenses properly incurred by the employer – should not include the cost of variations which would have given rise to an entitlement on the part of the original contractor. However, if it is the case (as is likely) that the employer will have had to pay the substitute contractor more – both for the original and the varied works – than he would have had to pay the original contractor, then there would be a claim for the additional cost of both. This begs the question: how would an assessment be made of loss and expense, if circumstances arise during the completion works that would have given the original contractor an entitlement to claim? There would in fact be no logical basis on which the original contractor's theoretical loss could be assessed, and the likelihood is that any reimbursement to the substitute contractor in respect of such a claim would be excluded from the comparison; this would not, however, comply strictly with the requirements of the account in clause 7.6(g). If a strict comparison was attempted, there could be difficulty in relation to loss and expense payable to the substitute contractor over and above that which would have been payable to the original contractor. It is unclear whether the original contractor would be required to meet this particular loss.

The potential for disagreement in carrying out this theoretical comparison is great, all the more so because there is no guarantee that the exercise will be carried out by an independent certifier. The account may be drawn up either by the employer or the architect.

If the sum of the expenses properly incurred etc. plus payments made (or possibly discharged – see the last paragraph under the previous heading) exceeds the total amount which would have been payable, the difference is stated to be a debt payable by the contractor to the employer. If less, then the balance is payable by the employer to the contractor.

Employer decides not to complete (clause 7.7)

This provision mirrors clause 27.7 of JCT 98, and deals with the situation where the employer decides not to complete the works. It seeks to overcome the danger of unlimited delay in resolving the question of the contractor's entitlement.

Clause 7.7.1 provides that if the employer decides after the determination not to have the works carried out and completed, he must notify the contractor in writing within six months from the date of determination. The timetable for production of the account (i.e. 'a reasonable time') runs from the date of the notice. What constitutes a reasonable time will depend on the circumstances of the particular case, but the fact that the account will not need to contain particulars of the cost of the completion contract will, in many cases, make the preparation of the account a rather more straightforward, and therefore quicker, process.

The matters to be set out in the account are:

- The total value of work properly executed at the date of determination, ascertained in accordance with the conditions as if the contractor's employment had not been determined, together with any other amounts due under the conditions which are not included in the total value of work properly executed;
- The amount of any expenses properly incurred by the employer and of any direct loss and/or damage caused to the employer as a result of the determination.

The balance due is calculated by adding the amount of any payments made to the contractor to the amount of the expenses properly incurred and any employer's direct loss and/or damage, and comparing it with the valuation of work; if the expenses incurred and amounts paid exceed the valuation, then the difference is a debt payable by the contractor to the employer; if the balance is in the contractor's favour, it is recoverable by the contractor from the employer.

There is a further distinction between the provisions of clauses 7.6 and 7.7: whereas under clause 7.6 either the employer or the architect could prepare the account, under clause 7.7 no reference is made to the architect. Although it might be thought that the scope for dispute is reduced, in view of the absence of certain of the complicating factors referred to in connection with the preparation of the notional account under clause 7.6, the question of the direct loss and/or damage caused to the employer as a result of the determination is potentially highly contentious. No guidance is given by the JCT in Practice Note 24 as to the extent of the recoverable losses; however, the words used are 'direct loss and/or damage', but these words are likely to be treated as entitling the employer to recover as if its entitlement was to damages at common law.

By clause 7.7.2, if the six month period within which the employer is entitled to give notice of his decision not to have the works completed expires without the employer having taken any steps to employ others or to enter the site to carry on with the works, the contractor may by notice in writing to the employer require the employer to state whether or not the provisions of clause 7.6 will apply, and if not, the contractor can require the employer to produce the account under clause 7.7.

This appears to undermine the opening paragraph of clause 7.7.1, which, in its terms, gives the employer six months within which to give notice. The effect of clause 7.7.2 is to allow the employer to prepare the statement of account without having given the written notification.

The procedure for determination – default

When the determination is to take place on grounds of default by the contractor, the procedure must be instigated by the architect who is to give to the contractor a notice specifying the default or defaults on the basis of which the action is being taken. In the words of Lord Justice Brown in the Court of Appeal in the case of *West Faulkner Associates* v. *London Borough of Newham* (1994):

'The architect's notice is thus a necessary pre-condition of determination.'

The architect's notice operates as from receipt by the contractor. The contractor then has 14 days within which to end the specified default. If, however, the specified default continues for 14 days from receipt of the notice, the employer may within 10 days from the expiry of that 14 days by a further notice to the contractor determine the contractor's employment. Again, the determination notice takes effect on the date of receipt.

Clause 7.2.3 deals with the possibility of repetition of the specified default. If the contractor ends his default or the employer does not give the second notice determining the contractor's employment, and the contractor repeats a specified default (whether this is the first or subsequent repetition) then upon or within a reasonable time after the repetition, the employer may by notice determine the contractor's employment. It should be noted that this provision is confined to cases of repetition and would not apply to further unspecified default. It is possible that the reasoning behind this is that the provisions are intended to warn the defaulting party by the requirement of notice, and the further default might be inadvertent. It could be argued, however, that if the defaulting party is already on notice, albeit in relation to a specified default, this reasoning is somewhat artificial.

Perhaps the most significant change in procedural terms made by the Amendment 7 to section 7 is the requirement that the default notice should be given by the architect. As originally drafted, IFC 84 did not specify by whom the notice should be given, and the specific identification of this first notice as being the architect's responsibility can only add to the certainty of operation of the procedure. The difficulty that this might give rise to, however, is disagreement between the employer and his architect as to whether or not notice should be given: see *West Faulkner Associates* v. *London Borough of Newham* (1994). The amendment is not, however, surprising in view of the comments of Lord Justice

Ormrod in *J. M. Hill & Sons Limited* v. *London Borough of Camden* (1980). In relation to the provision in JCT 63 clause 25(1)(d) requiring the architect to serve the notice, he said:

> 'The condition requires that such a notice should be given by the architect. A very important qualification in my view ... It is easy to understand why, in condition 25, it should be the architect: the person who is independent and expert in these matters should be the person to give the certificate and not the employer, who might be peevish or uninformed in one way or another.'

In the case of *London Borough of Hounslow* v. *Twickenham Garden Developments* (1970) the architect, in giving the notice required under clause 25(1) of JCT 63, said:

> 'I therefore hereby give notice under clause 25(1) of the contract dated ... that in my opinion you have failed to proceed regularly and diligently with the works.'

Of this notice Mr Justice Megarry said as follows (7 BLR at page 115):

> 'I do not read the condition as requiring the architect, at his peril, to spell out accurately in his notice further and better particulars, as it were, of the particular default in question. All that I think the notice need do is to direct the contractor's mind to what is said to be amiss: and this was I think done by this notice.'

Unreasonable or vexatious notice

The notice purporting to determine the contractor's employment must not be given unreasonably or vexatiously. In regard to this Lord Justice Ormrod in *J. M. Hill & Sons Limited* v. *London Borough of Camden* (1980) at 18 BLR page 49 said:

> '... What the word "unreasonably" means in this context, one does not know. I imagine that it is meant to protect the employer who is a day out of time in payment, or whose cheque is in the post, or perhaps because the bank has closed, or there has been a delay in clearing the cheque, or something – something accidental or purely incidental so that the court could see that the contractor was taking advantage of the other side in circumstances in which, from a business point of view, it would be totally unfair and almost smacking of sharp practice.'

In *Lubenham Fidelities & Investments Company Limited* v. *South Pembrokeshire District Council* (1983) Judge Newey QC was asked to consider the determination provisions under JCT 63; he commented:

> 'Construction contracts often extend over long periods of time and involve the use of considerable resources in land and materials and if they are not completed the financial and other consequences can be very serious. I think that the inclusion of provisos in standard forms of contract, one of which was used in this case, are intended to prevent parties from standing on their legal rights when the effect of their so doing will be quite disproportionate to their grounds of complaint. I think that "unreasonably" in a proviso relates principally to lack of proportion.

''Vexatiously'' must mean something different from ''unreasonably''. I think that it imports an intention to harass or distress.

I think that surrounding facts and circumstances are relevant for the purposes of deciding what is unreasonable and/or vexatious. The fact that parties are negotiating at the time when a notice is served might well make the service of it unreasonable, but in my view there can be many other circumstances in which service would be unreasonable.'

Both dicta were considered by the Court of Appeal in the case of *John Jarvis Limited* v. *Rockdale Housing Association Limited* (1986), and giving the unanimous judgment of the Court of Appeal, Lord Justice Bingham said:

'When used in a legal context, the adverb ''vexatiously'' connotes an ulterior motive to oppress, harass or annoy. It was not seriously argued that this was such a case, and the Judge's findings make plain that it was not.

''Unreasonably'' as used in sub-clause 28.1.3.4 (of JCT 80) is a general term which can include anything which can be objectively judged to be unreasonable...

There is not in my view very much difference between the tests propounded by the parties, although I prefer the contractor's, since the sub-clause provides that notice shall not be given unreasonably and not that it may only be given reasonably.'

The case considered the determination by the contractor following a relatively brief postponement which had been necessary following the withdrawal from site of a nominated sub-contractor. The contractor's determination was upheld, Lord Justice Bingham saying:

'Weighing it all up, even a reasonable contractor might very well conclude that, rather than risk joining the long list of builders driven to failure by contracts that went wrong, he should in his own interests and those of his creditors exercise the right which the contract gave him.'

The procedure for determination – insolvency

It goes without saying that notice is not required in the event of automatic determination even though, as suggested earlier (page 340), it would be advisable to confirm to the contractor that his employment is being treated as determined. Where, however, an event identified in clause 7.3.1 occurs and the automatic determination provisions do not apply, determination is effected by a single notice issued by the employer. Clause 7.3.4 states that the employer may issue the notice at any time (unless a '7.5.2.1 agreement' has been made) and determination is effective on the date of receipt of the notice.

The procedure for determination – corruption

There is no procedure. Clause 7.4 merely entitles the employer to determine the contractor's employment under the contract (or, indeed, any other contract) if an event of the kind identified in the clause occurs. This perhaps reflects the comment made in relation to clause 7.4 at page 341, to the effect that an employer could well be entitled to regard any conduct by the contractor of the kind identified in clause 7.4 as repudiatory in any event.

Other rights and remedies

Although the clauses commented on above identify the only grounds on the basis of which the employer may determine the contractor's employment under the contract, there is a potentially significant reservation made by clause 7.8, i.e. that the employer's rights under clauses 7.2 to 7.7 are without prejudice to any other rights and remedies which the employer may possess.

Whether or not these express words of reservation are necessary has been the subject of some debate: in the case of *Architectural Installation Services Limited* v. *James Gibbons Windows Limited* (1989) the question arose as to whether or not a party to the contract which provided a right to terminate subject to compliance with a comprehensive code retained its common law right to determine, in circumstances where that code was not prefaced by the words 'without prejudice to other rights and remedies'. Judge Bowsher said:

> 'I would be sorry if the draftsmen of contract felt it necessary to include such legal verbiage in order to avoid unintended results of their drafting. Construction contracts are already sufficiently complicated when the draftsmen seek to state what they do mean. They should not be burdened with the additional task of stating what they do not mean. When someone has obviously gone to a great deal of trouble to draft a contract, and two commercial parties have agreed to a contract in those terms, the court should be very reluctant to step in and suggest that those two parties also agreed something which was not written down in the agreement between them.'

It might, however, be argued that because 'two commercial parties' had agreed something which was written down, i.e. a right of determination subject to a comprehensive code, they would not expect the court to step in and conclude that the parties had not thereby intended to exclude any common law right to regard the contract as being terminated which was not subject to that same code. Indeed, in the case of *Lockland Builders Limited* v. *John Kim Rickwood* (1995), the Court of Appeal upheld the judge at first instance who had concluded that since the building contract in question had provided a comprehensive machinery for determination and this right had not been expressed to be without prejudice to the parties' common law rights, it followed that the machinery and the common law rights could co-exist only in circumstances where the contractor displayed the clear intention not to be bound by his contract. In the absence of such circumstances the machinery of the contract created the only effective way in which the agreement could be determined. Of Judge Bowsher's comments in the *Architectural Installation Services Limited* v. *James Gibbons Windows Limited* case, Lord Justice Russell in the Court of Appeal said:

> 'With all respect to the learned Judge, for my part I do attach significance to the absence of such words as "without prejudice to other rights and remedies", and

I do not think that to include them would, as Judge Bowsher thought, involve verbiage in drafting.'

The leading texts appear to favour the view that in the absence of express provision, contractual determinations are not intended as a substitute for, or to exclude, common law rights: see *Hudson's Building and Engineering Contracts*, 11th edition, page 1246 and *Keating on Building Contracts*, 6th edition, page 162. The latter suggests, however, the theoretical possibility that such an exclusion might arise by implication, and perhaps in the light of the *Lockland* case, that view could be expressed more strongly.

NOTES TO CLAUSES 7.2 TO 7.8

[1] '... Practical Completion ...'
As originally drafted, IFC 84 merely referred to 'completion'. This amendment perhaps avoids the possibility of determination occurring on the basis of an allegation that the contractor has defaulted in relation to his residual obligations in respect of defects (although there is a distinct possibility that a notice of determination in such circumstances would be regarded as having been given unreasonably or vexatiously).

[2] '... or substantially ...'
Again, these words have been added, IFC 84 as originally drafted requiring the suspension of the whole of the works. It is possible that the *de minimis* rule would have applied to the former wording, so that if the contractor's suspension was virtually complete, although not absolutely so, the employer could still rely on this ground. What the original wording may not have covered, however, was the situation in which the contractor did not wholly suspend the carrying out of the works but instead left merely a token presence on the site. It is thought that this possibility may be behind the introduction of these words.

[3] '... the Architect ... may give ... notice ...'
Although permissive in its terms, a failure by the architect to serve a notice might amount to a breach of the architect's contract with the employer: see *West Faulkner Associates* v. *London Borough of Newham* (1994).

[4] '... at any time ...'
These words in clause 7.3.4 suggest that the employer's rights are unlimited in terms of time; it is, however, possible that arguments of waiver or estoppel might be advanced on behalf of the contractor, in the event that the employer acted in such a way as to suggest that he intended not to rely on his strict legal rights, particularly where the contractor has acted to his detriment in reliance on the employer's apparent acceptance of any particular situation.

[5] 'Without prejudice ...'
The use of this phrase at the beginning of clause 7.6 should ensure that the contractor's compliance with any obligation placed on him in the sub-clauses of clause 7.6 is not to be treated as an admission in any arbitration or proceedings in which the validity of the determination is contested by the contractor.

Clauses 7.9 to 7.12

DETERMINATION BY CONTRACTOR
Default by Employer
7.9.1 If the Employer shall make default in any one or more of the following respects:
(a) he does not pay[1] by the final date for payment the amount properly due[2] to the Contractor in respect of any certificate for payment and/or any VAT due on that amount pursuant to the VAT Agreement, or
(b) he interferes with or obstructs the issue of any certificate due under this Contract, or
(c) he fails to comply with the provisions of clause 3.1 (*Assignment*), or
(d) he fails pursuant to the Conditions to comply with the requirements of the CDM Regulations
the Contractor may give to the Employer a notice specifying the default or defaults ('the specified default or defaults').

Suspension of uncompleted Works
7.9.2 If, before the date of Practical Completion, the carrying out of the whole or substantially the whole of the uncompleted Works[3] is suspended for the continuous period of one month[4] by reason of one or more of the following events:
(a) .1 where an Information Release Schedule has been provided, failure of the Architect/the Contract Administrator to comply with clause 1.7.1, or
.2 failure of the Architect/the Contract Administrator to comply with clause 1.7.2 , or
(b) the Architect's/the Contract Administrator's instructions issued under clauses
1.4   (*Inconsistencies*)
3.6   (*Variation*)
3.15 (*Postponemen*)
unless caused by reason of some negligence or default of the Contractor, his servants or agents or of any person employed or engaged upon or in connection with the Works or any part thereof, his servants or agents other than the Employer or any persons employed or engaged by the Employer, or
(c) delay in the execution of work not forming part of this Contract by the Employer himself or by persons employed or otherwise engaged by the Employer as referred to in clause 3.11 or the failure to execute such work or delay in the supply by the Employer of materials and goods which the Employer has agreed to supply for the Works or the failure so to supply, or
(d) failure of the Employer to give in due time ingress to or egress from the site of the Works or any part thereof through or over any land, buildings, way or passage adjoining or connected with the site and in the possession and control of the Employer, in accordance with the relevant Contract Documents, after receipt by the Architect/the Contract Administrator of such notice, if any, as the Contractor is required to give, or failure of the Employer to give such ingress or egress as otherwise agreed between the Architect/the Contractor Administrator and the Contractor,
the Contractor may give to the Employer a notice specifying the event or events ('the specified suspension event or events').

7.9.3 If
– the Employer continues a specified default, or
– a specified suspension event is continued
for 14 days from receipt of the notice under clause 7.9.1 or clause 7.9.2 then the Contractor may on, or within 10 days from, the expiry of that 14 days by a further notice to the Employer determine the employment of the Contractor under this Contract. Such determination shall take effect on the date of receipt of such further notice.

7.9.4 If
– the Employer ends the specified default or defaults, or
– the specified suspension event or events cease, or
– the Contractor does not give the further notice referred to in clause 7.9.3 and
– the Employer repeats (whether previously repeated or not) a specified default, or

    – a specified suspension event is repeated for whatever period (whether previously repeated or not) whereby the regular progress of the Works is or is likely to be materially affected

then, upon or within a reasonable time after such repetition, the Contractor may by notice to the Employer determine the employment of the Contractor under this Contract. Such determination shall take effect on the date of receipt of this notice.

7.9.5   A notice of determination under clause 7.9.3 or clause 7.9.4 shall not be given unreasonably or vexatiously.

Insolvency of Employer

7.10.1   If the Employer [aa]

makes a composition or arrangement with his creditors, or becomes bankrupt, or being a company,

makes a proposal for a voluntary arrangement for a composition of debts or scheme of arrangement to be approved in accordance with the Companies Act 1985 or the Insolvency Act 1986 or any amendment or re-enactment thereof as the case may be, or

has a provisional liquidator appointed, or

has a winding-up order made, or

passes a resolution for voluntary winding-up (except for the purposes of amalgamation or reconstruction), or

under the Insolvency Act 1986 or any amendment or re-enactment thereof has an administrator or an administrative receiver appointed

then

7.10.2   the Employer shall immediately inform the Contractor in writing if he has made a composition or arrangement with his creditors, or, being a company, has made a proposal for a voluntary arrangement to be approved in accordance with the Companies Act 1985 or the Insolvency Act 1986 as the case may be or any amendment or re-enactment thereof;

7.10.3   the Contractor may by notice to the Employer determine the employment of the Contractor under this Contract. Such determination shall take effect on the date of receipt of such notice. Provided that after the occurrence of any of the events set out in clause 7.10.1 and before the taking effect of any notice of determination of his employment issued by the Contractor pursuant to clause 7.10.3 the obligation of the Contractor to proceed with and complete the Works in compliance with clause 2.1 shall be suspended.

[aa] See Practice Note 24: after certain insolvency events an Insolvency Practitioner acts for the Employer

---

Consequences of determination under clause 7.9 or 7.10

7.11   In the event of the determination of the employment of the Contractor under clause 7.9.3, 7.9.4 or 7.10.3 and so long as that employment has not been reinstated the provisions of clauses 7.11.1, 7.11.2 and 7.11.3 shall apply; such application shall be without prejudice to the accrued rights or remedies of either party or to any liability of the classes mentioned in clause 6.1 which may accrue either before the Contractor or any sub-contractors, their servants or agents or others employed on or engaged upon or in connection with the Works or any part thereof other than the Employer or any person employed or engaged by the Employer shall have removed his or their temporary buildings, plant, tools, equipment, goods or materials (including Site Materials) or by reason of his or their so removing the same. Subject to clauses 7.11.2 and 7.11.3 the provisions of this Contract which require any further payment to the Contractor shall not apply.

7.11.1   The Contractor shall, with all reasonable dispatch and in such manner and with such precautions as will prevent injury, death or damage of the classes in respect of which before the date of determination he was liable to indemnify the Employer under clause 6.1, remove from the site all his temporary buildings, plant, tools, equipment, goods and materials (including Site Materials) and shall ensure that his sub-contractors do the same, but subject always to the provisions of clause 7.11.3(e).

7.11.2 Within 28 days of the determination of the employment of the Contractor the Employer shall pay to the Contractor the amount withheld under clause 4.2.1 and, if applicable, under clause 4.3 by the Employer prior to the determination of the employment of the Contractor but subject to any right of the Employer to continue to withhold such amounts which have accrued before the date of determination of the Contractor's employment.

7.11.3 The Contractor shall with reasonable dispatch prepare an account setting out the sum of the amounts referred to in clauses 7.11.3(a) to 7.11.3(e):
(a) the total value of work properly executed at the date of determination of the employment of the Contractor, such value to be ascertained in accordance with the Conditions as if the employment of the Contractor had not been determined, together with any amounts due to the Contractor under the Conditions not included in such total value; and
(b) any sum ascertained[5] in respect of direct loss and/or expense under clause 4.11 (whether ascertained before or after the date of determination); and
(c) the reasonable cost of removal pursuant to clause 7.11.1; and
(d) any direct loss and/or damage caused to the Contractor by the determination; and
(e) the cost of materials or goods (including Site Materials) properly ordered for the Works for which the Contractor shall have paid or for which the Contractor is legally bound to pay, and on such payment in full by the Employer such materials or goods shall become the property of the Employer.
After taking into account amounts previously paid to the Contractor under this Contract the Employer shall pay to the Contractor the amount properly due in respect of this account within 28 days of its submission by the Contractor to the Employer.

Other rights and remedies
7.12   The provisions of clauses 7.9 to 7.11 are without prejudice to any other rights and remedies which the Contractor may possess.

COMMENTARY ON CLAUSES 7.9 TO 7.12

Clauses 7.9 to 7.12 deal with the determination of the contractor's employment by the contractor and the consequences flowing from this. The grounds of determination are dealt with under three separate headings: default by employer; suspension of uncompleted works; and insolvency of employer. For practical purposes, however, a determination under either of the first two categories will be procedurally indistinguishable.

Default by employer

The specific grounds are as follows:

(1) If the employer does not pay in accordance with the contract the amount properly due in respect of any certificate and/or VAT (clause 7.9.1(a))
For an interim certificate the employer has 14 days from its date within which to pay under clauses 4.2 and 4.3. In the case of the final certificate sum, if any, payable to the contractor, this is not due until the 28 days from the date of the final certificate under clause 4.6.

If the failure to pay continues after the contractor has served notice specifying the default or is repeated, the contractor can by notice determine his own employment. As originally drafted, IFC 84 made reference to the employer not paying an amount properly due under each of the separate payment clauses, and this raised the question of whether or not it would count as a repetition, if the employer failed to make payment under clause 4.3, having previously failed to make a payment under clause 4.2, and similarly if he failed to make payment under clause 4.6 having previously failed to make payment under either of clauses 4.2 or 4.3. As presently worded, any subsequent non-payment would appear to be a repetition.

In the case of a failure to pay on the final certificate, the right of determination might be regarded as having limited value, in view of the fact that the remedies available under clause 7.11 would offer little practical assistance.

It is to be observed that the specified default is a failure to discharge the amount 'properly due', so that great care is required on the part of contractors when considering action under this clause. In *Lubenham* v. *South Pembrokeshire District Council* (1986) the Court of Appeal held that the amount due was the amount stated as due on the face of an interim certificate. This was so even if the interim certificate did not reflect the amount properly due. In that case deductions had been made in respect of retention, alleged defective work and delay, none of which ought to have appeared either in the manner or in the amount shown on the certificate. This was so, in the words of Lord Justice May:

> 'Although it ought to have been apparent to all concerned that, at least in respect of the deduction for alleged defective work, these certificates were not in accordance with the relevant contracts...'

In those circumstances, the contractor's remedy lay in arbitration.

It ought to be pointed out, however, that this aspect of the *Lubenham* case has been the subject of adverse comment – see *Hudson's Building and Engineering Contracts*, 11th edition, paragraph 6.193. Further, there is relatively recent authority suggesting that if the architect's failure to correctly certify payment is due to partiality, unfairness or even perhaps unreasonableness, this would put the employer in breach of an implied term that the employer will ensure that the architect will act lawfully, fairly and reasonably: see *John Barker Construction Limited* v. *Landon Portman Hotel Limited* (1996) and *Balfour Beatty Civil Engineering Limited* v. *Docklands Light Railway Limited* (1996). In some circumstances this could result in the employer falling foul of the next following ground. These authorities might now have to be reviewed, however, in the light of the decision of the House of Lords in *Beaufort Developments (NI) Ltd* v. *Gilbert Ash NI Ltd* (1998).

(2) If the employer interferes with or obstructs the issue of any certificate due under the contract (clause 7.9.1(b))

Interference or obstruction by the employer in relation to the issue of any certificate by the architect is a very serious matter. It is certainly a serious breach of contract and may well be seen as undermining the independence of the architect in certain circumstances. This is sometimes not fully appreciated by officers within an employer's organisation, particularly auditors within a local or other public authority. Also, lay councillors might not always understand the architect's independent role especially when he is an employee of the authority.

The procedure for determination is discussed below; however, on a point of information there was always a requirement under IFC 84 that the contractor must first give notice specifying the default, and it is only if the default continues or is subsequently repeated that the notice of determination may be issued. Under JCT

80 as originally drafted, the right of determination on this ground could be exercised forthwith upon notice. By amendment JCT 98 now requires the issue of a notice specifying the default prior to the right arising.

Interference with or obstruction of the issue of any certificate is not limited to matters concerning the document itself, but extends to the actions necessary for the preparation of the certificate, e.g. allowing the architect access to the site: see *Burden* v. *Swansea Corporation* (1957).

(3) If the employer fails to comply with the provisions of clause 3.1 (clause 7.9.1(c))
This ground did not appear in IFC 84 as originally drafted, being introduced as part of Amendment 7 dated April 1994, and mirrors the right afforded to the employer to determine under clause 7.2.1(d).

(4) If the employer fails pursuant to the conditions to comply with the requirements of the CDM Regulations (clause 7.9.1(d))
This ground was inserted by Amendment 8 dated March 1995, and mirrors the employer's right to determine in clause 7.2.1(e).

Suspension of uncompleted works

If the carrying out of the whole or substantially the whole of the uncompleted works is suspended (before the date of practical completion) for the continuous period of one month by reason of:

(a)   failure of the architect to provide information under clause 1.7.1 or drawings, details and instructions under clause 1.7.2, whichever is applicable (clause 7.9.2(a)) (see commentary on similar words used in clause 2.4.7, page 91);

(b)   instructions in relation to:
     (i)   inconsistencies (clause 1.4)
     (ii)   variations (clause 3.6)
     (iii) postponement (clause 3.15)
     unless caused by negligence or default on the part of the contractor, his servants or agents or any person employed or engaged upon or in connection with the works or any part thereof, his servants or agents other than the employer or any persons employed or engaged by the employer (clause 7.9.2(b));

(c)   delay or failure by the employer or others employed or engaged by him under clause 3.11 in relation to work not forming part of the contract or the supply by the employer of materials or goods which he has agreed to supply (clause 7.9.2(c)) (see commentary on similar words used in clause 2.4.8 and clause 2.4.9, page 91);

(d)   failure of the employer to give in due time agreed ingress or egress over adjoining property over which he exercises possession and control (clause 7.9.2(d)) (see commentary on the identical wording of clause 2.4.12, page 92).

The first point to note is that in relation to all these grounds the employer can be said to have some responsibility and in many cases the matter will or should be within his control. Provisions enabling the contractor to determine his own employment in respect of what might be called neutral events, namely force majeure, loss or damage caused by specified perils, civil commotion, instructions necessitated by default of those over whom the employer does not have control, hostilities and terrorist activity, are dealt with separately in IFC 98 in clause 7.13 (see below).

It is interesting to note that there is no provision at all in IFC 98 for determination following a prolonged suspension due to an instruction as to opening up and testing issued under clause 3.12 (unlike JCT 98 in this respect).

Clause 7.5.3(a) of IFC 84 as originally published, which was the equivalent of the current clause 7.9.2(b) (see (b) above) ended with the words 'unless caused by reason of some negligence or default of the contractor?'. By Amendment 3 of July 1988 this clause was amended by adding a reference to the contractor's servants, agents, or any person employed or engaged upon or in connection with the works or their servants or agents, other than the employer or those for whom the employer is responsible or any local authority or statutory undertaker executing works solely in pursuance of its statutory obligations. Clause 7.9.2(b) has been further amended, deleting the express reference to local authorities or statutory undertakers (transferred to clause 7.13 as a neutral event). The additional wording introduced by Amendment 3 was intended to make it clear that the negligence or default of the contractor did not have to be personal and could include that of his sub-contractors.

In the case of *John Jarvis Limited* v. *Rockdale Housing Association Limited* (1987) the Court of Appeal had held in relation to clause 28.1.3.4 of JCT 80 that identical words to those in IFC 84 clause 7.5.3, before Amendment 3, referred to negligence or default of the contractor himself, that is, the management or employees of the contractor, and did not extend to nominated sub-contractors (and presumably did not extend to domestic sub-contractors either). The contractor in that case was thus able to determine his own employment after a one month suspension where the cause of postponement was the need for redesign of the piling system following defective work on the part of a nominated sub-contractor. JCT 80 has since been amended by Amendment 4 of July 1987 in a similar way to the amendment to IFC 84, except that while under IFC 84 named sub-contractors would be regarded as being included in the additional wording, under JCT 80 nominated sub-contractors are expressly excluded (see clause 28.2.2.2) so that if the facts of the *Jarvis* case were to recur, the result under JCT 80, even as amended, would be the same.

It may be that in any event the different arrangements of the determination clauses in IFC 84 even as originally published (compare clauses 7.5.3 and 7.8.1 of the unamended IFC 84 with clause 28.1.3 of the unamended JCT 80) would have produced a different result. Part at least of the reasoning of the Court of Appeal in support of the decision was that there was a reference in clause 28.1.3.2 of JCT 80 to '... unless caused by the negligence of the contractor, his servants or agents or any sub-contractor, his servants or agents...'. This reasoning might not apply with equal force to IFC 84 as originally drafted, where the equivalent provision was to be found in an entirely separate clause – clause 7.8. Nevertheless, the additional words introduced by Amendment 3 put the matter beyond doubt.

The period of suspension is fixed at one month for the listed matters. Although this contract is not designed for contract periods beyond 12 months (see Practice Note 20), some employers may still regard this period as too short in many situations.

Insolvency of employer (clause 7.10)

This ground is similar to that in clause 7.3 in relation to the contractor's bank- ruptcy, liquidation, etc. except that the occurrence of any of the events listed in clause 7.10.1 will only ever give rise to the option of determination. There are no grounds on which the contractor's employment will determine automatically in the case of the employer's insolvency.

As it is the contractor's choice to determine his own employment, there is no provision here for reinstatement of the contractor's employment by agreement with the employer's trustee in bankruptcy, administrator, liquidator, adminis- trative receiver, etc. (or with the employer himself in appropriate circumstances) although there could of course be such an arrangement made between the parties outside the terms of the contract. This is expressly recognised in the opening sentence of clause 7.11.

Clause 7.10 does not discriminate, as did JCT 80 as originally published, between a private edition of the contract (where the equivalent of this provision is to be found) and the local authorities edition (where there was no such provision). The position under both JCT 98 and IFC 98 is that all employers are covered.

Any bankruptcy or liquidation must by definition have taken effect before a notice under clause 7.10.3 to determine can be issued on such grounds. Clearly there is a real chance in such an event that either the notice will be invalidated as an interference with the bankruptcy laws, or even if valid to determine the con- tractor's employment, it will be ineffective so far as many of the consequences set out in clause 7.11 are concerned, where such consequences would disturb the due and fair distribution of the employer's assets following the bankruptcy or liquidation.

Many of the comments made above in relation to the contractor's bankruptcy, liquidation, etc. are relevant here (see page 339).

The service of the notices

The notices to be given in the event of determination by the contractor are subject to the formalities discussed in relation to the commentary on clause 7.1 above (page 331). The notice provisions of IFC 98 and JCT 98 are (on the basis of current revisions) now in line, and in the case of both default and suspension, a warning notice specifying the default or the suspension event must be given prior to the further notice of determination.

In the case of both the notice of specified default and the notice of specified suspension, *receipt* of notice triggers the procedure; dispatching the notice is not sufficient. If either the specified default or specified suspension event continues for 14 days from receipt of the notice, then the contractor may on or within 10 days from expiry of that 14 days issue the further notice determining his employment.

Determination takes effect on the date of receipt of the further notice. Many of the general comments made earlier in relation to the service of notices and the proviso that the notice shall not be given unreasonably or vexatiously where the employer determines the contractor's employment are, where appropriate, relevant here also (see page 348).

The requirement for a first notice is of course a major benefit to the employer, particularly perhaps the inattentive or inefficient one, as it is less likely that he will be taken by surprise. For example, if an instruction results in a postponement which suspends the carrying out of the uncompleted work for at least one month, the contractor cannot simply determine his own employment. He must first serve a notice specifying on which ground he is entitled to determine and the employer has at least the opportunity of remedying the situation. A similar provision in JCT 80 (at the time) just might have enabled the employer in the *Jarvis* case to head off a determination by the contractor of his own employment.

There are provisions in clause 7.9.4 relating to repetition of the specified default or specified suspension event which correspond with the provisions contained in clause 7.2.

In the case of insolvency under clause 7.10, although no provision exists for automatic determination, the notice requirements differ from cases of default and suspension in that there is only a single notice. If the contractor opts to bring his employment to an end, he is required by clause 7.10.3 to give notice to the employer and determination takes effect on the date of receipt of the notice. In common with clause 7.5.1, upon the occurrence of any of the events set out in clause 7.10.1, the contractor's obligations to proceed with and complete the works are suspended.

The consequences of a determination by the contractor of his own employment (clause 7.11) are as follows.

Clause 7.11 provides for the consequences of the determination of the contractor's employment under clauses 7.9 and 7.10. The contractor's rights and entitlement are expressly given without prejudice to the accrued rights or remedies of either party. Quite why this express reservation is contained in clause 7.11, when it does not appear in the corresponding provisions of clause 7.6 in the event of the employer determining the contractor's employment, is not clear. To the extent necessary the employer will presumably rely on the provisions of clause 7.8, but since the provisions of clause 7.12 mirror – for the benefit of the contractor – the provisions of clause 7.8, there appears to be a duplication. Either way, care should be taken expressly to preserve the contractor's other rights and remedies where any action is taken which might appear to be inconsistent with those rights.

The effect of clause 7.11 is to substitute for the payment provisions under the contract, a new regime, as set out in clause 7.11, but this is expressed to be subject to clauses 7.11.2 and 7.11.3. Those provisions deal, respectively, with release of amounts withheld under clauses 4.2.1 and 4.3 (i.e. retention) and the preparation of the account (discussed in detail below).

The consequences of a determination under clauses 7.9 or 7.10 are as follows:

(1) The contractor shall with all reasonable dispatch and in a safe manner remove from the site all his temporary buildings, plant, tools, equipment, goods and materials and shall ensure that his sub-contractors do the same.

(2)   Within 28 days of the determination, the employer is to pay the contractor the amount withheld under clauses 4.2.1 and 4.3 prior to the determination; this is, however, subject to any right the employer may have to continue to withhold such amounts accruing before the date of determination.

(3)   The contractor is to prepare an account setting out the sum of amounts referred to in clauses 7.11.3(a) to (e) i.e.:

  –   The total value of work properly executed at the date of determination (ascertained in accordance with the conditions as if the employment of the contractor had not been determined) together with amounts due under the conditions not included in such total value
  –   Any sum ascertained in respect of direct loss and/or expense
  –   The reasonable cost of removal
  –   Any direct loss and/or damage caused to the contractor by the determination
  –   The cost of materials or goods properly ordered for which the contractor has paid or is legally bound to pay (such materials becoming the property of the employer on payment in full by the employer).

  After taking into account previous payments made to the contractor, the employer is to pay the contractor the amount properly due in respect of the account within 28 days of its submission.

The wording of the final paragraph of clause 7.11.3 differs from the corresponding provision of JCT 98 (clause 28.4) in that the latter provision states that payment is to be without deduction of retention. It is not thought that this marks a distinction of substance: rather, it is attributable to the different way in which retention is dealt with in JCT 98, i.e. by express deduction, whereas under IFC 98 the entitlement is only expressed in terms of 95% or $97\frac{1}{2}$% of the total.

The preparation of the account is, accordingly, not subject to an express timetable; however, the timing of payment is entirely within the contractor's control in that it is his obligation to prepare the account.

The main focus of attention is likely to be the claim by the contractor in respect of direct loss and/or damage. This is equivalent to that which could be recovered as damages for breach of contract at common law: see *Wraight Limited* v. *P. H. & T. (Holdings) Limited* (1968). This topic has been discussed in detail earlier on page 344.

As mentioned above, the loss and/or damage provision does not extend to what might be termed the 'neutral events' which can give rise to a determination and which are now separately treated in clauses 7.13 to 7.19.

Like JCT 98, IFC 98 contains no express provision entitling the contractor to take possession of unfixed goods or materials which may have become the property of the employer together with a lien over them in respect of monies due.

NOTES TO CLAUSES 7.9 TO 7.12

[1] '...pay...'
The wording of IFC 84 as originally drafted referred only to the employer failing to 'pay'; the word 'discharge' was introduced by amendment, which potentially widened the range of defences available to the employer. The original wording

has, however, been restored, and references throughout the contract to payment being 'otherwise discharged' have been deleted. For a possible explanation of this see under note [2] to clause 1.11 (page 59) and under note [16] to clauses 4.2 to 4.8 (see page 222).

[2] '...properly due...'
It is submitted that the amount properly due is the amount which should be contained in a payment certificate. Any deduction by way of lawful set-off, e.g. liquidated damages, takes effect as a deduction from the amount due; it does not form part of the calculation of how much is due, rather it is relevant to how much is to be paid. This appears to be the effect of the requirements of the Housing Grants, Construction and Regeneration Act 1996 sections 110 and 111 which require for their operation both a due date and a final date for payment, and see for example clause 4.2 of IFC 98. This analysis supports the contention that the obligation on the employer to 'pay ... the amount properly due...', see note [1] above, can be satisfied by the discharge, other than by physical payment, of the legal obligation to pay. If it were otherwise, it would mean that such other means of discharge would not be a payment under this clause and the contractor would have grounds to determine.

The employer may also have certain common law rights of set-off (discussed earlier in this book on page 195). If the employer serves the appropriate notice and purports to set-off his claim, thereby raising at least a prima facie argument of valid set-off, then it is suggested that the contractor needs to take great care before purporting to determine his own employment on this ground, since, if he is mistaken, his act may well amount to a repudiation, leaving the employer with little alternative but to accept it.

It is submitted that provided the employer pays the certified sum he has for the purposes of this clause paid the sum properly due even if the certificate contains a clear undervaluation: see *Lubenham Fidelities & Investments Company Limited* v. *South Pembrokeshire District Council* (1983) and earlier in this chapter (page 355).

[3] '...substantially the whole of the uncompleted Works...'
The distinction between this form of words can be contrasted with the words used in clause 7.2.1(a), in which reference is made to substantial suspension rather than suspension of substantially the whole of the uncompleted works. For a possible justification for this difference in wording see earlier under note [2] to clauses 7.2 to 7.8 (page 351).

[4] '...month...'
This means calendar month – see Law of Property Act 1925, section 61.

[5] '...any sum ascertained...'
Clause 7.11.3(b) refers to 'any sum ascertained' whether before or after the date of determination. This raises the question as to the duty of the architect or quantity surveyor after the contractor's employment has been determined to carry out and ascertain loss and expense on receipt (whether before or after the date of determination) of an application from the contractor. If there is no such duty, there is a gap in the contractual machinery and the contractor would have to rely on his

common law remedy of damages for breach of contract, if indeed there had been a breach. It is submitted that the words used are sufficiently wide to require an ascertainment to be carried out in respect of any loss and expense incurred for which the contract entitles the contractor to reimbursement provided that the appropriate application is made, particularly bearing in mind that the contractual provisions continue as necessary after a determination of the contractor's employment.

Clauses 7.13 to 7.19

DETERMINATION BY EMPLOYER OR CONTRACTOR
Grounds for determination of the employment of the Contractor

7.13.1   If, before the date of Practical Completion, the carrying out of the whole or substantially the whole of the uncompleted Works is suspended by reason of one or more of the events stated in clause 7.13.1(a), 7.13.1(b) and 7.13.1(c) for a period of three months or by reason of one or more of the events stated in clause 7.13.1(d), 7.13.1(e) and 7.13.1(f) for a period of one month:

(a) force majeure, or

(b) loss or damage to the Works occasioned by any one or more of the Specified Perils, or

(c) civil commotion, or

(d) the Architect's/the Contract Administrator's instructions issued under clauses

1.4   (*Inconsistencies*)

3.6   (*Variations*)

3.15   (*Postponement*)

which have been issued as a result of the negligence or default of any local authority or statutory undertaker executing works solely in pursuance of its statutory obligations, or

(e) hostilities involving the United Kingdom (whether war be declared or not), or

(f) terrorist activity,

then the Employer or the Contractor may upon the expiry of the aforesaid relevant period of suspension give notice to the other that unless the suspension is terminated within 7 days after the date of receipt of that notice the employment of the Contractor under this Contract will determine 7 days after the date of receipt of such notice; and the employment of the Contractor shall so determine 7 days after receipt of such notice.

7.13.2   The Contractor shall not be entitled to give notice under clause 7.13.1 in respect of the matter referred to in clause 7.13.1(b) where the loss or damage to the Works occasioned by any one or more of the Specified Perils was caused by some negligence or default of the Contractor, his servants or agents or of any person employed or engaged upon or in connection with the Works or any part thereof, his servants or agents other than the Employer or any person employed or engaged by the Employer or by any local authority or statutory undertaker executing work solely in pursuance of its statutory obligations.

7.13.3   A notice of determination under clause 7.13.1 shall not be given unreasonably or vexatiously.

Consequences of determination under clause 7.13.1

7.14   Upon determination of the employment of the Contractor under clause 7.13.1 the provisions of this Contract which require any further payment to the Contractor shall not apply; and the provisions of clauses 7.15 to 7.19 shall apply.

7.15   The Contractor shall give up possession of the site of the Works subject to the orderly compliance of the Contractor with any instruction of the Architect/the Contract Administrator under clause 7.16.

7.16   The Contractor shall, with all reasonable dispatch and in such manner and with such precautions as will prevent injury, death or damage of the classes in respect of which before the date of determination of his employment he was liable to indemnify the Employer under clause 6.1, remove from the site all his temporary buildings, plant, tools, equipment, goods and materials (including Site Materials) and shall ensure that his sub-contractors do the same, but subject always to the provisions of clause 7.18.4.

7.17   The Employer shall pay to the Contractor one half of the amount withheld under clause 4.2.1 and, if applicable, under clause 4.3 by the Employer prior to the determination of the employment of the Contractor within 28 days of the date of determination of the Contractor's employment and the other half as part of the account to which clause 7.18 refers but subject to any right to continue to withhold such amounts which have accrued before the date of such determination.

7.18   The Contractor shall, not later than 2 months after the date of the determination of the Contractor's employment, provide the Employer with all documents necessary for the preparation of the account to which this clause refers. Subject to due discharge by the Contractor of this obligation the Employer shall with reasonable dispatch prepare an account setting out the sum of the amounts referred to in clauses 7.18.1 to 7.18.4 and, if clause 7.19 applies, clause 7.18.5:

7.18.1 the total value of work properly executed at the date of determination of the employment of the Contractor, such value to be ascertained in accordance with the Conditions as if the employment of the Contractor had not been determined, together with any amounts due to the Contractor under the Conditions not included in such total value; and

7.18.2 any sum ascertained in respect of direct loss and/or expense under clause 4.11 (whether ascertained before or after the date of determination); and

7.18.3 the reasonable cost of removal pursuant to clause 7.16; and

7.18.4 the cost of materials or goods (including Site Materials) properly ordered for the Works for which the Contractor shall have paid or for which the Contractor is legally bound to pay, and on such payment in full by the Employer such materials or goods shall become the property of the Employer; and

7.18.5 any direct loss and/or damage caused to the Contractor by the determination. After taking into account amounts previously paid to the Contractor under this Contract the Employer shall pay to the Contractor the amount properly due in respect of this account within 28 days of its submission by the Employer to the Contractor.

7.19   Where determination of the employment of the Contractor has occurred in respect of the matter referred to in clause 7.13.1(b) and the loss or damage to the Works occasioned by any one or more of the Specified Perils was caused by some negligence or default of the Employer or of any person for whom the Employer is responsible, then upon such determination of the employment of the Contractor the account prepared under clause 7.18 shall include the amount, if any, to which clause 7.18.5 refers.

COMMENTARY ON CLAUSES 7.13 TO 7.19

Clauses 7.13 to 7.19 deal with the determination of the employment of the contractor by either the employer or the contractor where the carrying out of the whole, or substantially the whole, of the uncompleted works is suspended by reason of one or more of the events listed under clause 7.13 for the period of time specified in the clause (three months or one month depending on the event).

The common feature of the grounds giving rise to the right to determine under this clause (with the partial exception in respect of clause 7.13.1(b) – see clause 7.19) is that they do not involve fault on the part of either party to the contract.

Although a 'no fault' provision was in earlier versions of IFC 84, the range of grounds on which determination might occur was extended, principally by Amendment 7 of April 1994.

In addition to the extra grounds for determination that have been added by amendment, other changes have occurred which may or may not have significance: as originally drafted, the rights under clauses 7.8 and 7.9 (now replaced by clauses 7.13 to 7.19) were expressed to be without prejudice to any other rights or remedies which the employer or contractor may possess. The significance of the 'without prejudice' wording in other parts of section 7 has been dealt with earlier, and it might be thought that because of the 'no fault' nature of the grounds for determination under this part of section 7, there would be no other rights or remedies to be prejudiced by the express provisions. This is not necessarily the case. For example, instructions issued under clauses 1.4, 3.6 or 3.15 resulting from the negligence or default of a local authority or statutory undertaker might well – as a matter of contractual risk – be categorised as the default of one or other of the contracting parties, giving rise to a right to claim damages (and possibly the right to seek other remedies). This poses the question: does the absence of the 'without prejudice' wording make a difference, in view of the fact that those words appear expressly under the other categories of determination in section 7? It seems at least arguable that the elimination of these words and/or the comparison with clauses 7.8 and 7.12 is intended to reflect the intention that clauses 7.13 to 7.19 are the exhaustive rights and remedies of the parties in the circumstances envisaged.

Grounds for determination by employer or contractor

If there is a suspension of the carrying out of the whole or substantially the whole of the uncompleted works before practical completion, and that suspension endures:

(1)   for more than three months by reason of:
- force majeure;
- loss or damage to the works occasioned by a specified peril;
- civil commotion; or
(2)   for more than one month by reason of:
- architect's instructions under clauses 1.4, 3.6 or 3.15 issued as a result of the negligence or default of any local authority or statutory undertaker executing works solely in pursuance of its statutory obligations;
- hostilities involving the UK;
- terrorist activity;

then the employer or the contractor may upon the expiry of the relevant period give notice to the effect that unless the suspension is terminated within seven days after the date of receipt of the notice, the employment of the contractor will thereupon determine.

The first ground, namely force majeure, is of uncertain meaning. What is certain is that its interpretation in this contract will be affected by the context in which it appears, so that while in one contract it may extend to freak weather, national strikes, embargoes, etc., in another contract, where those other matters are separately listed or treated, the meaning of force majeure will not extend to such matters. In IFC 98 clause 2.4 (events for extensions of time) listed separately in addition to force majeure are 15 further grounds for extension of time. Its meaning in clause 2.4 does not therefore include any of those other events.

In clause 7.13.1 apart from force majeure there is also reference to a further five matters entitling either party to operate the clause. This raises the question: does force majeure when used in clause 7.13.1 include matters which, as a matter of construction, it would be taken not to include in the context of clause 2.4? Although the same words used in the same contract will generally be construed as carrying the same meaning, in which case the words will exclude the wider range of events identified in clause 2.4, this construction will not necessarily prevail where the context in which the words are used differs in such a way as to make it clear that another meaning is to be given to them. It is submitted, on balance, that the same meaning is likely to be given to the words 'force majeure' in both clauses. See the commentary to clause 2.4.1 (page 88) for a further discussion of the meaning of force majeure.

In relation to loss or damage to the works occasioned by specified perils, clause 7.13.2 provides that the contractor shall not be entitled to determine his employment under this head where the loss or damage was caused by the neg-ligence or default of the contractor, his servants or agents, etc., which for this purpose excludes the employer and, it is tentatively submitted, any local authority or statutory undertaker executing work solely in pursuance of its statutory obli-gations. This tentativeness is as a result of the difficulties of construction in this clause, particularly the effect of the word 'by' in front of 'any local authority' near the end of clause 7.13.2.

The contractor's negligence does not of course prevent the employer from determining the contractor's employment under clause 7.13.1. It may seem sur-prising that clause 7.13.2 does not purport to disentitle the employer from determining the contractor's employment where the employer or his servants or agents have negligently caused the loss or damage occasioned by the specified peril. This could of course happen, for example by negligence on the part of the employer's direct contractors. While generally to construe words in a contract so as to enable one party to exercise such a fundamental right as a result of its own negligence or the negligence of those for whom that party is contractually responsible would be unlikely without express clear wording to that effect, the fact that here the contractor's right to do so is expressly taken away, whereas the employer's is not, may well, it is submitted, allow the employer to operate clause 7.13.1 in such circumstances.

There is in fact a sound logic behind this. The employer owns or has control of the land. If damage is caused to the works by a specified peril, e.g. fire, it could fundamentally affect the employer's general position, e.g. where fire destroys not only the works but also the employer's existing buildings. The employer should be free to decide not to carry on with the works even if the fire was due to the employer's negligence. This is no doubt why clause 7.19 has been included pro-viding, exceptionally, for the contractor to be able to recover direct loss or damage in such circumstances pursuant to clause 7.18.5.

The suspension must last for the period of time specified, i.e. three months or one month as the case may be. However, unlike the one month's suspension entitling the contractor to determine his own employment under clause 7.9.2, the suspension under clause 7.13.1 is not expressly stated to be continuous. It is certainly arguable therefore that intermittent periods of suspension can be added together, and if they exceed the three or one month period as the case may be then the notice of determination can be given.

It is unlikely that individual periods of less than three months or one month as the case may be arising from different heads which when added together exceed the required period can be used as a basis for determination of the contractor's employment. If this had been intended it would, presumably, have been stated expressly.

Service of the notice

Only one notice is required, but unlike the single notice referred to in cases of insolvency (and unlike the single notice provision in the predecessor to clause 7.13 of IFC 98, i.e. the old clause 7.8) the single notice is conditional: the contractor's employment will determine seven days after receipt of the notice unless the suspension is terminated within that period. The available time is short, but it does at least afford the employer an opportunity to take action.

A notice of determination under clause 7.13 is not to be given unreasonably or vexatiously – see earlier commentary under clause 7.2.4 and 7.9.5.

Consequences of determination under clause 7.13 (clauses 7.14 to 7.19)

The consequences that flow from a determination under clause 7.13 are similar to those applying where the contractor determines his own employment under clauses 7.9 or 7.10. Under the provisions of IFC 84 as originally drafted, in the event of a 'no fault' determination the contractor had the same rights as if he had operated the determination provisions as a result of default by the employer, except that he was not entitled to recover direct loss and/or damage caused by the determination.

By virtue of Amendment 7, 'no fault' determination now has its own detailed provisions. The differences from the scheme of clause 7.11 are:

- The timetable for release of retention is slightly longer, the contractor being entitled only to half the retention within 28 days of the determination, the balance being payable as part of the account;
- Responsibility for preparation of the account does not lie with the contractor, as under clause 7.11; the contractor has an obligation not later than two months after the date of the determination, to provide the employer with all documents necessary for the preparation of the account, and thereafter the employer must with reasonable dispatch prepare the account;
- The constituent elements of the account are identical to those listed under clause 7.11.3(a) to (e), with the exception of the entitlement to direct loss and/or damage; this is only recoverable where clause 7.19 applies, i.e. where determination of the contractor's employment has occurred as a result of suspension due to loss or damage occasioned by a specified peril caused by negligence or default of the employer or a person for whom the employer is responsible.

Determination under clause 6.3C.4.4

It is possible for either party to determine the employment of the contractor under clause 6.3C.4.4 where, following loss or damage to the works, it is just and equitable so to do – see earlier commentary on clause 6.3C on page 307 for the probable reasoning behind this provision. However, the notice must be sent within 28 days of the occurrence of such loss or damage. Within seven days of receipt the contractor or employer, as the case may be, may in effect challenge the determination and require the question of whether it is just and equitable to determine the contractor's employment to be determined by adjudication, arbitration or litigation.

There is a further, optional, right of determination where Amendment TC/94/IFC is incorporated. This deals with the non-availability of terrorism cover under the All Risks Insurance (whether taken out by the contractor under clause 6.3A or the employer under clause 6.3B or clause 6.3C) and affords the employer the right under clause 6.3A.5.2.2, 6.3B.4.2.2 or 6.3C.1A.1.2 as appropriate to notify the contractor (upon notification by insurers that terrorism cover under the Joint Names Policy is to cease and will no longer be available) that his employment is determined.

Chapter 11
Interpretation of the contract

CONTENT

This chapter briefly considers section 8, which deals with matters of contract interpretation, construction and definition.

SUMMARY OF GENERAL LAW

Where a contract contains clauses dealing with the interpretation or definition of phrases or words, then clearly reference will have to be made to that clause in deciding the meaning of words or phrases to be found in the contract. However, general principles of law relating to the construction of contracts will also be highly relevant. It is not possible in this book to do more than mention this topic and, if interested, the reader should refer to textbooks which deal with it in more detail, e.g. *Chitty on Contracts*, 27th edition, at paragraphs 12.039 to 12.079; *Keating on Building Contracts*, 6th edition, Chapter 3.

CONSIDERATION OF THE RELEVANT CLAUSES OF IFC 98

Clause 8.1

Interpretation etc.
References to clauses, etc.
8.1 Unless otherwise specifically stated a reference in the Articles of Agreement, the Conditions, the Supplemental Conditions or the Appendix to any clause means that clause of the Conditions or the Supplemental Conditions.

COMMENTARY ON CLAUSE 8.1

This clause calls for no comment.

Clause 8.2

Articles etc. to be read as a whole
8.2 The Articles of Agreement, the Conditions, the Supplemental Conditions and the Appendix are to be read as a whole and the effect or operation of any article or clause in the Conditions or the Supplemental Conditions or item in or entry in the Appendix must therefore unless otherwise specifically stated be read subject to any relevant qualification or modification in any other article or any of the other clauses in the Conditions or the Supplemental Conditions or item in or entry in the Appendix.

COMMENTARY ON CLAUSE 8.2

This clause makes it quite clear that the articles, clauses and items in the appendix which make up the contract conditions are to be read together. Any particular article, clause etc. must not therefore be considered in isolation but must be read in the light of other articles or clauses etc. and, unless otherwise expressly stated, must be read subject to any qualification or modification in any other article or clause etc.

Clause 8.3

Definitions

8.3 Unless the context otherwise requires or the Articles of Agreement or the Conditions or the Supplemental Conditions or an item or entry in the Appendix specifically otherwise provides, the following words and phrases in the Articles of Agreement, the Conditions, the Supplemental Conditions and the Appendix shall have the meanings given below:
Activity Schedule:
means the schedule of activities as attached to the Appendix with each activity priced and with the sum of those prices being the Contract Sum excluding provisional sums, prime cost sums and Contractor's profit thereon and the value of work for which Approximate Quantities are included in the Contract Documents: see clause 4.2.1(a).
Adjudication Agreement:
see clause 9A.2.1.
Adjudicator:
means any individual appointed pursuant to clause 9A as the Adjudicator.
All Risks Insurance:
see clause 6.3.2.
Appendix:
means the Appendix to the Conditions as completed by the parties.
Approximate Quantity:
means a quantity in the Contract Documents identified therein as an approximate quantity [bb].
Articles or Articles of Agreement:
means the Articles of Agreement to which the Conditions are annexed, and references to any recital are to the recitals set out before the Articles.
Base Date:
means the date stated in the Appendix.
CDM Regulations:
means the Construction (Design and Management) Regulations 1994 or any remaking thereof or any amendment to a regulation therein.
Contract Sum Analysis (see Second recital):
means an analysis of the Contract Sum provided by the Contractor in accordance with the stated requirements of the Employer.
Excepted Risks:
means ionising radiations or contamination by radioactivity from any nuclear fuel or from any nuclear waste from the combustion of nuclear fuel, radioactive toxic explosive or other hazardous properties of any explosive nuclear assembly or nuclear component thereof, pressure waves caused by aircraft or other aerial devices travelling at sonic or supersonic speeds.
Form of Tender and Agreement NAM/T:
means the Form issued under that name by the JCT for use where a person is to be named as a sub-contractor under clause 3.3.
Health and Safety Plan:
means where it is stated in the Appendix that all the CDM Regulations apply, the plan provided to the Principal Contractor and developed by him to comply with regulation 15(4) of the CDM Regulations and, for the purpose of regulation 10 of the CDM Regulations, received by the Employer before any construction work under this Contract has started;

and any further development of that plan by the Principal Contractor during the progress of the Works.

Information Release Schedule:
means the Schedule referred to in the Fourth recital or as varied pursuant to clause 1.7.1.

Joint Fire Code:
means the Joint Code of Practice on the Protection from Fire of Construction Sites and Buildings Undergoing Renovation which is published by the Building Employers Confederation (now Construction Confederation), the Loss Prevention Council and the National Contractors' Group with the support of the Association of British Insurers, the Chief and Assistant Chief Fire Officers Association and the London Fire Brigade which is current at the Base Date.

Parties:
means the Employer and the Contractor named as the Employer and the Contractor in the Articles of Agreement.

Party:
means the Employer or the Contractor named as the Employer or the Contractor in the Articles of Agreement.

person:
means an individual, firm (partnership) or body corporate.

Planning Supervisor:
means the person named in article 5 or any successor duly appointed by the Employer as the Planning Supervisor pursuant to regulation 6(5) of the CDM Regulations.

Practical Completion:
See clause 2.9.

Price Statement:
means the Price Statement referred to in clause 3.7.1.2 Option A.

Principal Contractor:
means the Contractor or any other contractor duly appointed by the Employer as the Principal Contractor pursuant to regulation 6(5) of the CDM Regulations.

provisional sum:
where the Contract Documents include bills of quantities, includes a sum provided in such bills for work whether or not identified as being for defined or undefined work. [bb]

Public Holiday:
means Christmas Day, Good Friday or a day which under the Banking and Financial Dealings Act 1971 is a bank holiday. [cc]

[cc] Amend as necessary if different Public Holidays are applicable.

Schedules of Work:
means an unpriced schedule referring to the Works which has been provided by the Employer and which if priced by the Contractor (as mentioned in the Second recital) for the computation of the Contract Sum is included in the Contract Documents.

Site Materials:
means all unfixed materials and goods delivered to, placed on or adjacent to the Works and intended for incorporation therein.

Specified Perils:
means fire, lightning, explosion, storm, tempest, flood, bursting or overflowing of water tanks, apparatus or pipes, earthquake, aircraft and other aerial devices or articles dropped therefrom, riot and civil commotion, but excluding Excepted Risks.

Sub-Contract Conditions NAM/SC:
means the Sub-Contract Conditions NAM/SC incorporated by reference in article 1.2 of Section III of the Tender and Agreement NAM/T.

Supplemental Conditions:
means the clauses set out or referred to after the Appendix and referred to in clauses
4.9(a)   (*Tax etc. fluctuations*),
4.9(b)   (*Formula fluctuations*),
5.5   (*VAT*), *and*
5.6   (*Statutory tax deduction*).

[bb] General Rules 10.1 to 10.6 of the Standard Method of Measurement 7th Edition provide:

10.1

Where work can be described and given in items in accordance with these rules but the quantity of work required cannot be accurately determined, an estimate of the quantity shall be given and identified as an approximate quantity.

10.2

Where work cannot be described and given in items in accordance with these rules it shall be given as a Provisional Sum and identified as for either defined or undefined work as appropriate.

10.3

A Provisional Sum for defined work is a sum provided for work which is not completely designed but for which the following information shall be provided:
(a) The nature and construction of the work.
(b) A statement of how and where the work is fixed to the building and what other work is to be fixed thereto.
(c) A quantity or quantities which indicate the scope and extent of the work.
(d) Any specific limitations and the like identified in Section A35.

10.4

Where Provisional Sums are given for defined work the Contractor will be deemed to have made due allowance in programming, planning and pricing Preliminaries. Any such allowance will only be subject to adjustment in those circumstances where a variation in respect of other work measured in detail in accordance with the rules would give rise to adjustment.

10.5

A Provisional Sum for undefined work is a sum provided for work where the information required in accordance with rule 10.3 cannot be given.

10.6

Where Provisional Sums are given for undefined work the Contractor will be deemed not to have made any allowance in programming, planning and pricing Preliminaries.

COMMENTARY ON CLAUSE 8.3

Certain words and phrases are given specific meanings for the purpose of construing the contract. Many of them are self-explanatory and need little comment. Where relevant, reference is made to the meaning of these words or phrases in the discussion of the various articles or clauses in which they appear throughout this book.

Clause 8.4

'The Architect'/'The Contract Administrator'
8.4 Where the person named in article 3 is entitled to the name 'Architect' under and in accordance with the Architects Act 1997 the term 'the Contract Administrator' shall be deemed to have been deleted throughout the contract but where the person named is not so entitled, the term 'the Architect' shall be deemed to have been deleted.

COMMENTARY ON CLAUSE 8.4

The right to use the title 'Architect' is governed by the Architects Act 1997 (refer to page 18).

Clause 8.5

Priced Specification or priced Schedules of Work

8.5   Where in the Conditions there is a reference to the 'Specification' or the 'Schedules of Work' then, where the Second recital alternative A applies, such reference is to the Specification or the Schedules of Work as priced by the Contractor unless the context otherwise requires.

COMMENTARY ON CLAUSE 8.5

There is no comment to be made on this clause.

Chapter 12

Settlement of disputes: adjudication – arbitration – legal proceedings

CONTENT

This chapter looks at section 9 of the contract together with article 8 and article 9 of IFC 98 dealing with adjudication, arbitration and legal proceedings. The adjudication provisions are intended to satisfy the requirements of section 108 of the Housing Grants, Construction and Regeneration Act 1996. The arbitration provisions are intended to be an 'arbitration agreement' under section 6 of the Arbitration Act 1996, which defines 'arbitration agreement' as an agreement to submit to arbitration present or future disputes. The option is available in IFC 98 to choose either arbitration or litigation as the appropriate tribunal for the final determination of disputes and differences.

SUMMARY OF GENERAL LAW

Adjudication and arbitration are two methods of resolving disputes or differences. Others include conciliation and mediation. Whilst neither conciliation nor mediation feature expressly within IFC 98, there is of course nothing to prevent the parties by agreement seeking to resolve their differences in this way. For those interested in mediation, the Tribunal has produced Practice Note 28 *Mediation on a building contract or sub-contract dispute*, which contains a brief description of mediation and its purpose and also contains examples of a mediation agreement, an agreement appointing a mediator and an agreement following the resolution of a dispute after mediation.

Adjudication and arbitration compared

According to the *Shorter Oxford English Dictionary* to 'adjudicate' means to determine judicially. An 'adjudicator' is someone who settles a question. Naturally enough the definition refers to a judicial function. However, historically in the construction industry, adjudication may not bear all of the hallmarks of a judicial exercise. This is particularly so where the contractual imposition of very tight time scales can deprive a party of a reasonable opportunity of presenting its case to its best advantage.

In seeking to understand the nature of the adjudication process in the construction industry, it is instructive to consider the attributes of adjudication alongside the attributes of arbitration. Mustill & Boyd *Commercial Arbitration*, 2nd edition, at pages 41–42, sets out the essential attributes of arbitration:

'(i)   The agreement pursuant to which the process is, or is to be, carried on ('the procedural agreement') must contemplate that the tribunal which carries on the process will make a decision which is binding on the parties to the procedural agreement.

(ii)   The procedural agreement must contemplate that the process will be carried on between those persons whose substantive rights are determined by the tribunal.

(iii)   The jurisdiction of the tribunal to carry on the process and to decide the rights of the parties must derive either from the consent of the parties, or from an order of the court or from a statute the terms of which make it clear that the process is to be an arbitration.

(iv)   The tribunal must be chosen, either by the parties, or by a method to which they have consented.

(v)   The procedural agreement must contemplate that the tribunal will determine the rights of the parties in an impartial manner, with the tribunal owing an equal obligation of fairness towards both sides.

(vi)   The agreement of the parties to refer their disputes to the decision of the tribunal must be intended to be enforceable in law.

(vii)   The procedural agreement must contemplate a process whereby the tribunal will make a decision upon a dispute which is already formulated at the time when the tribunal is appointed.'

If a provision for dispute resolution in a contract covers these matters then even if it is described as an adjudication clause it will be an arbitration agreement and therefore subject to the Arbitration Act 1996. This issue was considered in the case of *Cape Durasteel Ltd* v. *Rosser and Russell Building Services Ltd* (1995). In this case a plaintiff sub-sub-contractor claimed against the defendant sub-contractor for various sums in connection with refurbishment works. The sub-contractor's own standard terms were relevant and clauses 23.1 and 23.2 provided as follows:

'23   Settlement of Disputes

23.1   In the event of any dispute arising out of or in connection with the Sub-Contract the parties agree to refer such dispute to adjudication to a person agreed upon or failing agreement to some person appointed by the President for the time being of the Chartered Institute of Building Services Engineers.

23.2   Such reference to adjudication shall not (unless the Contractor decides and notifies otherwise) be opened until after Practical Completion or alleged completion of the Principal Contract Works.'

The sub-contractor sought to have the sub-sub-contractor's claim through the courts stayed under section 4 of the Arbitration Act 1950 on the grounds that there was an arbitration agreement rather than only an adjudication clause.

Held

–   The use of the word '*adjudication*' was not decisive as to whether or not there was a binding arbitration agreement;

- The question to be answered was whether or not the agreement to refer disputes to another person for decision had the essential features of an arbitration agreement as stated in Mustill & Boyd (referred to above);
- In this case clause 23.1 did involve an arbitration agreement even though it referred to adjudication.

It is important to point out that there was no separate arbitration clause in this contract. As this was the only dispute resolution clause it was not surprising that the court held that it was in truth an arbitration clause.

It is clear from this decision and others that the courts will look at the overall nature of the clause and its context in the contract as a whole in order to determine its true status as an arbitration clause or something different to arbitration. In the *Cape Durasteel* case Judge Humphrey Lloyd QC said:

'It is clear on the authorities that, ..., the test to be applied is the customary one of ascertaining the presumed intention of the parties from their contract and its circumstances. It is plain that "adjudication" taken by itself means a process by which a dispute is resolved in a judicial manner. It is equally clear that "adjudication" has as yet no settled special meaning in the construction industry (which is not surprising since it is a creature of contract and contractual procedures utilising an "adjudicator" vary as do forms of contract). Even if it were to have the special meaning accorded to it in some sections of the construction industry where it describes the initial determination of certain classes of dispute in a summary manner, the force of which is tempered by its ephemeral status as there are concomitant provisions for the decision to be reviewed and if necessary reversed by an arbitrator, I would see no reason why it should have that meaning in this contract...'

Construction industry adjudication

The two essential features of adjudication in the construction industry can be seen to be firstly a tight time scale within which an adjudicator must make a decision; and secondly, while the decision itself must be complied with, nevertheless an aggrieved party is entitled to refer the underlying dispute or difference to arbitration or litigation as the case may be. These two attributes are closely connected. If tight time-scales are in operation, particularly where the dispute or difference raises complex issues and a considerable amount of relevant material, it is important that if the adjudicator's decision is unacceptable to one of the parties, that party should be able to refer the dispute for resolution under a fuller and more considered method such as arbitration or litigation.

Rapid adjudication can, on occasions, prejudice a party in a number of ways. For example, not all relevant points may be able to be made; not all relevant documentation may be available within the time-table allowed; it may not be possible in the time permitted to be represented by someone of a party's choice, whether a lawyer or other suitably trained representative; further, it may be that in the time permitted, it is not possible, even though it may be desirable, to have an oral hearing, for example, where there is a dispute as to a factual matter which might best be resolved by having the benefit of oral examination and cross-examination of witnesses.

Historically, adjudication was introduced into the construction industry at the sub-contract level in order to control the main contractor's common law right of set-off against sub-contractors. It was introduced in the 1970s to regulate the position between main contractor and sub-contractor where the main contractor was withholding money from the sub-contractor alleging delay or disruption on the part of the sub-contractor. Its purpose was to obtain a quick decision based on a written form of procedure in connection with the vital question of cash flow as between main contractor and sub-contractor. It was a question of who should hold the money while the dispute was sorted out, generally in arbitration. The adjudicator could either uphold the deduction made by the main contractor; or order the contractor to pay some or all of the amount to the sub-contractor; or order the sum or any part of it to be paid to a stakeholder. The issue was at the same time referred to arbitration and the arbitration would continue if either party was unhappy with the adjudicator's decision and the dispute could not be settled between them.

Subsequently, e.g. in supplementary provisions incorporated into the JCT 81 With Contractor's Design form of contract, adjudication was introduced covering a wider area of disputes, and more recently, in some standard forms of contract, e.g. see the New Engineering and Construction Contract, adjudication has been introduced to cover all disputes.

We now have statutory adjudication. This has been considered generally earlier in this book (see page 25).

Arbitration

Arbitration is a very common and long-standing means of resolving disputes in the construction industry. Almost always the parties arbitrate because they have entered into a contract containing an arbitration clause which amounts to an arbitration agreement within section 6 of the Arbitration Act 1996 (formerly within section 32 of the Arbitration Act 1950). Arbitration generally is now governed by the Arbitration Act 1996 (the Act) replacing, in particular, the Arbitration Acts 1950, 1975 and 1979.

The Act is in part a consolidation of earlier legislation and common law and in part a modification of it. Its essential thrust is to help facilitate speed and cost effectiveness in the arbitral process, and to increase the scope for the parties to control the proceedings, with however, fall-back provisions to apply where the parties have not agreed what should happen. The provisions of the Act have been significantly influenced by the Model Law on Arbitration, drafted by UNCITRAL, the International Trade Law Committee of the United Nations.

An interesting and unusual feature of the Act is that it contains a statement of overriding objectives, a kind of mission statement. Section 1 provides:

'The provisions of this Part are founded on the following principles, and shall be construed accordingly:
(a)   the object of arbitration is to obtain the fair resolution of disputes by an impartial tribunal without unnecessary delay or expense;
(b)   the parties should be free to agree how their disputes are resolved, subject only to such safeguards as are necessary in the public interest;

(c)   in matters governed by this Part the court should not intervene except as provided by this Part.'

The general duty of the arbitrator is set out in clause 33, which provides

'(1)   The Tribunal shall
      (a)   act fairly and impartially as between the parties, giving each party a reasonable opportunity of putting his case and dealing with that of his opponent, and
      (b)   adopt procedures suitable to the circumstances of the particular case, avoiding unnecessary delay or expense, so as to provide a fair means for the resolution of the matters falling to be determined.
  (2)   The Tribunal shall comply with that general duty in conducting the arbitral proceedings, in its decisions on matters of procedure and evidence and in the exercise of all other powers conferred on it.'

In passing it is worth noting that the duty of the arbitrator is to act 'fairly' as well as impartially. The duty of the adjudicator – see section 108(2)(e) of the Housing Grants, Construction and Regeneration Act 1996 – is to act impartially. There is no reference to the adjudicator acting fairly. This difference may be significant. Having regard to the very tight time-scales which govern adjudication proceedings, even for the most complex cases, it may be that the adjudicator will not be able to act fairly although he will still be able to act impartially.

The adjudicator has a duty to reach a decision within a stated time. Within that framework he must treat the parties equally in the sense that he must not show partiality. However, supposing a sub-contractor is brought into the adjudication process by the main contractor who is claiming that the sub-contractor carried out no work on site between, say, 15 and 20 July. The sub-contractor says that he had one operative on site during that time carrying out work. There is accordingly a dispute as to the facts. The adjudicator requires the parties to produce statements from relevant witnesses. There is also the possibility of a very brief oral hearing. Each party is given the same period of time (10 days) in which to produce their statements which will just enable the adjudicator to consider them and call for an oral hearing if he wishes. The relevant operative of the sub-contractor is, however, in hospital in Australia and in a coma. It is anticipated that he will surface from the coma within about three weeks and should then make a full recovery.

In these circumstances it is submitted that all that the adjudicator can do is to decide the issue on the evidence he actually receives. He cannot extend the time-scales without the consent of the main contractor and, it is submitted, he cannot opt out of the process by making no decision at all. This is unfair to the sub-contractor though the adjudicator has acted impartially. Having said all this, if the adjudicator simply refuses to make a decision, even though this may be a breach of his duties, while the adjudicator may put at risk his fees and expenses, there is little that the main contractor can really do about it, except seek the appointment of another adjudicator.

The essential attributes of arbitration as taken from Mustill & Boyd *Commercial Arbitration*, 2nd edition, have been set out above. It is not possible in a book of this kind dealing with IFC 98 generally, to deal in detail with the subject of arbitration. Reference should be made therefore to standard works on the subject such as that of Mustill & Boyd. However, two points are worth particular mention. Firstly, should one party seek to litigate the dispute rather than to arbitrate in accordance with an arbitration agreement entered into, the courts will upon application stay such court proceedings. The court has no discretion to allow the proceedings to continue: see section 9 of the Act; *Halki Shipping Corporation* v. *Sopex Oils Ltd* (1997); *Davies and Middleton & Davies Ltd* v. *Toyo Engineering Corporation* (1997); and *Ahmad Al-Naimi* v. *Islamic Press Agency Incorporated* (1998). In this last case, Judge Bowsher QC sitting in the Technology and Construction Court, held that a section 9 stay should be granted even where there was a dispute between the parties as to whether their contractual dispute was covered by an arbitration agreement at all.*

The matter can only continue in the courts if the defendant has taken an appropriate step in the proceedings to answer the substantive claim; or where the court is satisfied that the arbitration agreement is null and void, inoperative or incapable of being performed.

Secondly, it should be noted that sections 89 to 91 of the Act lay down an alternative scheme for regulation of consumer arbitration agreements. They achieve this by extending the application of the Unfair Terms in Consumer Contracts Regulations (1994) (the Regulations) to arbitration agreements. The Regulations, which implemented the European Unfair Terms in Consumer Contracts Directive 1993, apply to contracts between consumers (i.e. a natural person making a contract for purposes other than business) and a commercial concern supplying goods or services relating to a business (including a profession, government department or local authority).

Under the Regulations, a contract term is unfair, and is rendered unenforceable in certain circumstances (referred to earlier in this book – see page 43). The purpose of section 89 is to apply the Regulations to consumer arbitration agreements and to repeal the Consumer Arbitration Agreements Act 1988 so that consumer arbitration agreements are governed by a single set of rules.

Schedule 3 to the Regulations sets out a list of items which are presumed to be unfair, although this is rebuttable. Item (q) refers to any term which excludes or hinders the consumer's right to legal redress, in particular the imposition of an obligation to go to arbitration. It will be far from easy for this presumption of unfairness to be rebutted in relation to arbitration clauses in contracts with consumers. However, as the arbitration agreement in IFC 98 is optional only, and as the choice of arbitration or litigation will generally be that of the employer (the consumer), it could be said that as the obligation to go to arbitration is inserted by the consumer, the optional arbitration provision is not unfair.

Section 89 itself is a curious provision in that the Regulations apply independently of it to consumer arbitration agreements so that section 89 simply declares the existing state of the law. However, sections 89 and 90 do have a purpose in that they extend the scope of the Regulations in some respects when compared with the 1988 Act.

A particular point to note is that the 1988 Act did not apply to consumer arbitration agreements which related to a sum in excess of £3000. Section 91 of the 1996 Act contains power to likewise limit the application of the Regulations by excluding

---

* In *Birse Construction Ltd* v. *St. David Ltd* (1999) Judge Humphrey Lloyd QC regarded the Ahmad Al-Naimi case as one on its own particular facts. If on a section 9 application the question is raised as to whether there is a valid arbitration agreement, this is likely to be determined by the court itself.

arbitration agreements relating to sums in excess of any specified limit. Under section 89, the Unfair Arbitration Agreements (Specified Amount) Order 1996 – SI 1996/3211 has been issued specifying the amount of £3000 as being the maximum sum at stake if an arbitration agreement is to be unfair for the purpose of the Regulations. However, bearing in mind that the consumer appears to still have a right to rely directly on the Regulations themselves, the fact that the sum at stake is greater than the specified limit and so is outside section 89 does not mean that the arbitration clause is automatically valid as it still has to pass the tests in the Regulations themselves. It has been suggested that the result is that this section is largely pointless, and that any order made under it is essentially meaningless; see *Arbitration Act 1996 – an Annotated Guide* by Robert Merkin, published by LLP Ltd, London.

CONSIDERATION OF THE RELEVANT CLAUSES OF IFC 98

9   **Settlement of disputes – Adjudication – Arbitration – Legal Proceedings [dd]**

[dd] It is open to the Employer and the Contractor to resolve disputes by the process of Mediation: see Practice note 28 'Mediation on a Building Contract or Sub-Contract Dispute'.

9A   **Adjudication**
Application of clause 9A

9A.1   Clause 9A applies where, pursuant to article 8, either Party refers any dispute or difference arising under this Contract to adjudication.

Identity of Adjudicator

9A.2   The Adjudicator to decide the dispute or difference shall be either an individual agreed by the Parties or, on the application of either Party, an individual to be nominated as the Adjudicator by the person named in the Appendix ('the nominator'). [ee] Provided that

9A.2.1   no Adjudicator shall be agreed or nominated under clause 9A.2 or clause 9A.3 who will not execute the Standard Agreement for the appointment of an Adjudicator issued by the JCT (the 'JCT Adjudication Agreement' [ff]) with the Parties [ee], and

9A.2.2   where either Party has given notice of his intention to refer a dispute to adjudication then
  - any agreement by the Parties on the appointment of an Adjudicator must be reached with the object of securing the appointment of and the referral of the dispute or difference to the Adjudicator within 7 days of the date of the notice of intention to refer (see clause 9A.4.1);
  - any application to the nominator must be made with the object of securing the appointment of and the referral of the dispute or difference to the Adjudicator within 7 days of the date of the notice of intention to refer.

Upon agreement by the Parties on the appointment of the Adjudicator or upon receipt by the Parties from the nominator of the name of the nominated Adjudicator the Parties shall thereupon execute with the Adjudicator the JCT Adjudication Agreement.

[ee] The nominators named in the Appendix have agreed with the JCT that they will comply with the requirements of clause 9A on the nomination of an adjudicator including the requirement in clause 9A.2.2 for the nomination to be made with the object of securing the appointment of, and the referral of the dispute or difference to, the Adjudicator within 7 days of the date of the notice of intention to refer; and will only nominate adjudicators who will enter in the 'JCT Adjudication Agreement'.

[ff] The JCT Adjudication Agreement is available from the retailers of JCT Forms.
A version of this agreement is also available for use if the Parties have named an Adjudicator in their contract[*].

[*] Author's note: in which case amendments are required to clauses 9A.2; 9A.3; 8.3 and the Appendix; these amendments should be based on those provided in the Amendment 12 and Guidance Notes to IFC 84 issued April 1998 (pages 43 and 44) and which are reproduced on page 381 of this book.

Death of Adjudicator – inability to adjudicate

9A.3   If the Adjudicator dies or becomes ill or is unavailable for some other cause and is thus unable to adjudicate on a dispute or difference referred to him, then either the Parties may agree upon an individual to replace the Adjudicator or either Party may apply to the nominator for the nomination of an adjudicator to adjudicate that dispute or difference; and the Parties shall execute the JCT Adjudication Agreement with the agreed or nominated Adjudicator.

Dispute or difference – notice of intention to refer to Adjudication – referral

9A.4.1   When pursuant to article 8 a Party requires a dispute or difference to be referred to adjudication then that Party shall give notice to the other Party of his intention to refer the dispute or difference, briefly identified in the notice, to adjudication. If an Adjudicator is agreed or appointed within 7 days of the notice then the Party giving the notice shall refer the dispute or difference to the Adjudicator ('the referral') within 7 days of the notice. If an Adjudicator is not agreed or appointed within 7 days of the notice the referral shall be made immediately on such agreement or appointment. The said party shall include with that referral particulars of the dispute or difference together with a summary of the contentions on which he relies, a statement of the relief or remedy which is sought and any material he wishes the Adjudicator to consider. The referral and its accompanying documentation shall be copied simultaneously to the other Party.

9A.4.2   The referral by a Party with its accompanying documentation to the Adjudicator and the copies thereof to be provided to the other Party shall be given by actual delivery or by FAX or by special delivery or recorded delivery. If given by FAX then, for record purposes, the referral and its accompanying documentation must forthwith be sent by first class post or given by actual delivery. If sent by special delivery or recorded delivery the referral and its accompanying documentation shall, subject to proof to the contrary, be deemed to have been received 48 hours after the date of posting subject to the exclusion of Sundays and any Public Holiday.

Conduct of the Adjudication

9A.5.1   The Adjudicator shall immediately upon receipt of the referral and its accompanying documentation confirm the date of that receipt to the Parties.

9A.5.2   The Party not making the referral may, by the same means stated in clause 9A.4.2, send to the Adjudicator within 7 days of the date of the referral, with a copy to the other Party, a written statement of the contentions on which he relies and any material he wishes the Adjudicator to consider.

9A.5.3   The Adjudicator shall within 28 days of the referral under clause 9A.4.1 and acting as an Adjudicator for the purposes of S.108 of the Housing Grants, Construction and Regeneration Act 1996 and not as an expert or an arbitrator reach his decision and forthwith send that decision in writing to the Parties. Provided that the Party who has made the referral may consent to allowing the Adjudicator to extend the period of 28 days by up to 14 days; and that by agreement between the Parties after the referral has been made a longer period than 28 days may be notified jointly by the Parties to the Adjudicator within which to reach his decision.

9A.5.4   The Adjudicator shall not be obliged to give reasons for his decision.

9A.5.5   In reaching his decision the Adjudicator shall act impartially and set his own procedure; and at his absolute discretion may take the initiative in ascertaining the facts and the law as he considers necessary in respect of the referral which may include the following:
    .1 using his own knowledge and/or experience;
    .2 opening up, reviewing and revising any certificate, opinion, decision, requirement or notice issued, given or made under the Contract as if no such certificate, opinion, decision, requirement or notice had been issued, given or made;
    .3 requiring from the Parties further information than that contained in the notice of referral and its accompanying documentation or in any written statement provided by the Parties including the results of any tests that have been made or of any opening up;

.4 requiring the Parties to carry out tests or additional tests or to open up work or further open up work;

.5 visiting the site of the Works or any workshop where work is being or has been prepared for this Contract;

.6 obtaining such information as he considers necessary from any employee or representative of the Parties provided that before obtaining information from an employee of a Party he has given prior notice to that Party;

.7 obtaining from others such information and advice as he considers necessary on technical and on legal matters subject to giving prior notice to the Parties together with a statement or estimate of the cost involved;

.8 having regard to any term of this Contract relating to the payment of interest, deciding the circumstances in which and the period for which a simple rate of interest shall be paid.

9A.5.6 Any failure by either Party to enter into the JCT Adjudication Agreement or to comply with any requirement of the Adjudicator under clause 9A.5.5 or with any provision in or requirement under clause 9A shall not invalidate the decision of the Adjudicator.

9A.5.7 The Parties shall meet their own costs of the Adjudication except that the Adjudicator may direct as to who should pay the cost of any test or opening up if required pursuant to clause 9A.5.5.4.

Adjudicator's fee and reasonable expenses – payment
9A.6.1 The Adjudicator in his decision shall state how payment of his fee and reasonable expenses is to be apportioned as between the Parties. In default of such statement the Parties shall bear the cost of the Adjudicator's fee and reasonable expenses in equal proportions.

9A.6.2 The Parties shall be jointly and severally liable to the Adjudicator for his fee and for all expenses reasonably incurred by the Adjudicator pursuant to the Adjudication.

Effect of Adjudicator's decision
9A.7.1 The decision of the Adjudicator shall be binding on the Parties until the dispute or difference is finally determined by arbitration or by legal proceedings [gg] or by an agreement in writing between the Parties made after the decision of the Adjudicator has been given.

9A.7.2 The Parties shall, without prejudice to their other rights under this Contract, comply with the decision of the Adjudicator; and the Employer and the Contractor shall ensure that the decision of the Adjudicator is given effect.

9A.7.3 If either Party does not comply with the decision of the Adjudicator the other Party shall be entitled to take legal proceedings to secure such compliance pending any final determination of the referred dispute or difference pursuant to clause 9A.7.1.

Immunity
9A.8 The Adjudicator shall not be liable for anything done or omitted in the discharge or purported discharge of his functions as Adjudicator unless the act or omission is in bad faith and this protection from liability shall similarly extend to any employee or agent of the Adjudicator.

[gg] The arbitration or legal proceedings are not an appeal against the decision of the Adjudicator but are a consideration of the dispute or difference as if no decision had been made by an Adjudicator.

Amendments to clause 9A where the Parties wish to name an Adjudicator in the Contract [see footnote [ff] to clause 9A.2.1]
Clause 9A.2 Delete the text and insert:
9A.2 The Adjudicator to decide the dispute or difference shall be the individual named as the Adjudicator in the Appendix with whom the Parties have executed the Standard Agreement for the appointment of an Adjudicator named in a contract issued by the Joint Contracts Tribunal (the 'JCT Named Adjudicator Agreement')*; provided that,

unless the Parties have otherwise agreed, the individual is not an employee of, or otherwise engaged by, either Party.
Delete footnote [m.4] [now [ee]]

---

* The JCT Named Adjudication Agreement, whose text is set out in the Guidance Notes to this Amendment, is available from the retailers of JCT Forms (see Appendix 3 to this book).

Clause 9A.3 Delete the text and insert:

9A.3   If the Adjudicator dies or becomes ill or is unavailable for some other cause and is thus unable to adjudicate on a dispute or difference referred to him, the Parties may either agree an individual to replace the Adjudicator or either Party may apply to the person named in the Appendix as the appointor of the individual to be appointed as Adjudicator, or, if no such appointor is so named, to the President or a Vice-President of the Royal Institute of British Architects. Provided that if the Adjudicator named in the Appendix is unable by reason of illness or other cause to adjudicate on a dispute or difference referred to him any appointment under clause 9A.3 shall not terminate the Adjudication Agreement of that individual with the Parties.

Clause 8.3 Definition of Adjudicator
Delete the text and insert:
Adjudicator:
the individual named in the Appendix as the Adjudicator or the individual appointed as the Adjudicator pursuant to clause 9A.3.
Appendix: delete the entry with reference to clause 9A.2
Insert an additional entry:

9A.2   Adjudication   Name and address of Adjudicator

_____

_____

_____

Fee of Adjudicator   * Lump sum or

£_____

    * Hourly rate

£_____

    * Complete as applicable

COMMENTARY ON CLAUSE 9A – ADJUDICATION

Introduction

Clause 9A sets out the adjudication procedure for dealing with disputes or differences. It is clearly intended to satisfy the requirements of section 108 of the Housing Grants, Construction and Regeneration Act 1996 (the 1996 Act), for which see earlier in this book – page 25.

Article 8 to IFC 98 provides as follows:

'If any dispute or difference arises under this Contract either Party may refer it to adjudication in accordance with clause 9A.'

The first point to note is that it is only disputes or differences which *arise under* the contract which are covered by the clause. This is to be contrasted with article 9A dealing with arbitration which refers to disputes or differences '... as to any matter or thing of whatsoever nature arising under this Contract or in connection therewith . . .'. The latter words have been held to give the arbitrator jurisdiction to deal with matters such as misrepresentation or negligent misstatement: see *Ashville Investments* v. *Elmer Contractors* (1989). Such matters as these may therefore not be covered by the adjudication clause. Furthermore, it is submitted that, unlike an arbitrator, the adjudicator has no power to decide on a dispute or difference which relates to his jurisdiction, e.g. whether he has been properly appointed.

While the adjudication clause may have its limitations in scope, it does not, like article 9A dealing with arbitration, have any express exclusions in relation to value added tax or the statutory tax deduction scheme (see earlier under the commentary to the articles – page 21). These were omitted from arbitration because in each case there was a statutory method of dealing with disputes. Accordingly it would have been logical to have likewise excluded them from the adjudication process. However, as section 108 appears to grant parties to a construction contract a right to refer *any* dispute or difference arising under the contract to an adjudicator, it was probably thought that to except any particular type of dispute might leave the IFC 98 adjudication provisions open to attack on the basis that they did not comply with the section 108 requirements.

Although clause 9A is clearly intended to satisfy the requirements of section 108 of the Act, IFC 98 is not without its possible problems in this respect; see earlier page 181 dealing with clause 3.13 and page 311 dealing with clause 6.3(C).

Procedures for the adjudication process have deliberately been kept to an absolute minimum, giving the appointed adjudicator as much discretion as possible in determining how the adjudication process should operate. The requirement that the adjudicator's decision must generally be reached within 28 days of the referral of the dispute or difference to him means that detailed and rigid procedures are best avoided. It is hoped that the adjudication process will be fast and flexible in practice.

Apart from clause 9A, there is also a standard form of adjudication agreement setting out basic terms between the adjudicator and the parties, dealing with the adjudicator's appointment, fees, expenses and the termination of the adjudicator's appointment.

The adjudicator's decision is not permanently binding unless the parties wish it to be. However, if the dispute or difference is referred to arbitration or litigation, the adjudicator's decision must nevertheless be complied with in the meantime.

The adjudicator can be appointed in a number of ways. It is anticipated that the most common method will be for the parties either to agree the name of an adjudicator once the dispute or difference has arisen or, failing this, for either party to seek the nomination of an adjudicator (dealt with in more detail below). Clause 9A proceeds on the assumption that one or other of these two means of appointment will take place. However, it may be that the parties wish or are prepared to name an adjudicator when entering into the contract. To facilitate this, the Tribunal in the Guidance Note attached to Amendment 12 to IFC 84 issued April 1998, provided alternative clauses to replace the existing 9A.2 and 9A.3, together with an adjudication agreement specifically designed for this situation (see pages 43 to 48 of the Guidance Note). It also includes for this purpose an insertion into clause 8.3 of a slightly different definition of the adjudicator and an amendment to the appendix to IFC 84 (and which is suitable for use with IFC 98) with a place to insert the name and address of the adjudicator together with his lump sum fee or agreed hourly rate (see earlier page 381 for these amendments).

Naming an adjudicator at the outset has the advantage that as he is already appointed, he should be in a position to quickly deal with disputes or differences referred to him as no time will be lost in going through the appointment process. As against this however, many parties will prefer to wait until a dispute has arisen before seeking an adjudicator in order to make sure that the appointee is someone appropriately qualified and experienced to deal with the particular area of dispute which has arisen. It is worth pointing out that the adjudicator is entitled to obtain such technical or legal advice as he considers necessary.

Enforcement of adjudicators' decisions is to be through the courts and not through the arbitration process. More is said about this below.

SUMMARY OF ADJUDICATION PROCEDURE

Commencement of the adjudication

The adjudication procedure commences with one party giving written notice to the other of his intention to refer a dispute or difference to adjudication. The notice must briefly identify the dispute or difference.

Identity and appointment of adjudicator

As mentioned above the adjudicator may be named in the contract or alternatively agreed by the parties following the dispute or difference arising, failing which the adjudicator is appointed by either party applying to a nominating body identified in the appendix to IFC 98. The nominating body is to be selected in the appendix from one of the following:

President or a Vice-President or Chairman or a Vice-Chairman of:
• Royal Institute of British Architects
• Royal Institution of Chartered Surveyors
• Construction Confederation
• National Specialist Contractors Council.

The selection is by way of deletion of unwanted nominating bodies from the appendix entry. In the event that no selection has been made in the appendix, the nominator will be the President or a Vice-President of the Royal Institute of British Architects.

If the adjudicator is named in the contract and the amended clauses (referred to above) have been inserted into clauses 9A, 8.3 and the appendix, then the appendix item will provide the name of the adjudicator and his fee.

The parties to IFC 98 must execute the adjudication agreement. If the adjudicator is named in the contract, the parties should have executed the JCT Named Adjudicator Agreement (a copy of which is set out in Appendix 3 of this book – see page 413) before or at the time of entering into the IFC 98 Contract itself. If the adjudicator is appointed after the dispute or difference has arisen then the parties must execute the JCT Adjudication Agreement (a copy of which is set out in Appendix 4 of this book – see page 418) as soon as an adjudicator has been agreed or nominated as the case may be.

If the adjudicator is named in the contract, he must not, unless the parties have otherwise agreed, be an employee of, or otherwise engaged by, either party. Accordingly the architect cannot, unless the parties agree, act as the adjudicator.

In all cases the adjudicator must be an individual and not a corporate body or a partnership.

Upon his appointment, the adjudicator will execute the adjudication agreement.

Timing of appointment

Where the adjudicator has not been named in the contract, the procedure for appointment must be such as to have the object of securing the appointment of the adjudicator and the referral of the dispute or difference to him within seven days of the notice of intention to refer. To this end, there is a footnote [ee] indicating that the nominators named in the appendix have agreed with the Tribunal that they will comply with the requirements of clause 9A and in particular clause 9A.2.2 for the nomination to be made with the object of meeting this time limit and that they will only nominate adjudicators who will be willing to enter into the JCT Adjudication Agreement. This drafting clearly reflects the requirements of section 108(2)(b) of the 1996 Act which states that any contractual adjudication procedure must provide a timetable with the object of securing the appointment of the adjudicator and the referral of the dispute to him within seven days of the notice of adjudication

Replacement of adjudicator

(i) Pre-named

Where an adjudicator who has already been named in the contract becomes ill or unavailable for some other cause and so is unable to act as an adjudicator, the parties can agree another individual, failing which either party can apply to the President or a Vice-President of the Royal Institute of British Architects to make a replacement appointment. The parties must then enter into the JCT Adjudication Agreement with the replacement adjudicator which will however be specific to the particular dispute or difference. The adjudication agreement already entered into with the named adjudicator will remain in place so that the named adjudicator can act in relation to other disputes or differences which may occur and for which he may be fit and available to act.

(ii) Adjudicator selected after dispute or difference has arisen

If the adjudicator who has been selected either by agreement or by a nominator after the dispute or difference has arisen, dies or becomes ill or unavailable and so is unable to adjudicate, the parties can agree another individual or go back to the nominator selected in the appendix for a further nomination. The parties must then enter into the JCT Adjudication Agreement with the replacement adjudicator.

The referral

The party serving the notice of intention to adjudicate must within seven days of the notice refer the dispute or difference to the adjudicator by sending to the adjudicator particulars of the dispute or difference with a summary of his contentions, the relief being sought and any material which he wishes the adjudicator to consider. This must be simultaneously copied to the other party. This can be sent by actual delivery, by special delivery or recorded delivery, in which case, subject to proof otherwise, it will be deemed to have been received 48 hours after the date of posting excluding Sundays and public holidays; or it can be sent by facsimile transmission, in which case, for record purposes, it is necessary for it to also be sent by first class post or actual delivery. The adjudicator is to immediately acknowledge receipt to the parties. If for some reason the adjudicator has not been agreed or appointed within this seven day period then the referral must be made immediately such agreement or appointment has been made.

Non-referring party's response

The non-referring party must respond within seven days of the date of referral by sending to the adjudicator a written statement containing the contentions on which he relies and any material which he wishes the adjudicator to consider. At the same time a copy must be sent to the referring party.

The decision

The adjudicator must reach his decision within 28 days of the date of referral and forthwith send that decision in writing to the parties. However, the referring party is entitled to extend that 28 days, if the adjudicator wishes it, by a further 14 days. It is therefore the referring party who controls the timetable. There is no power for the non-referring party to extend time. In addition there is no power for the adjudicator of his own volition to do so. Having said this, if the adjudicator fails to make his decision within the time allowed, there is not a great deal that the parties can do about it except to terminate his appointment, which disentitles the adjudicator from recovering his fees and expenses (see clause 5 of the JCT Adjudication Agreement).

The time-scales are such that it is imperative, particularly for the non-referring party, to react quickly once the adjudication process is set in motion. It will very often require the input of considerable resources on a concentrated short-term basis if the non-referring party is not to be prejudiced. The referring party does of course have the period prior to serving notice of intention to adjudicate in which he can prepare for adjudication.

The adjudicator is not obliged to give reasons for his decision. Indeed bearing in mind the speed at which the process is meant to operate, there may be little or no time for a reasoned decision to be given. In most cases an adjudicator would be ill advised to provide his reasoning process, which may show flaws, even if the result is right. It will unnecessarily expose the adjudicator and his decision to attack by an aggrieved party, probably by application to the courts.

The adjudicator is to act impartially. This requirement is not only obviously sensible but is a reflection of section 108(2)(e) which requires any contractual adjudication scheme to impose a duty on the adjudicator to so act.

Again reflecting section 108 (this time sub-clause (2)(f)), the adjudicator is enabled to take the initiative in ascertaining the facts and the law. This is clearly an important power for the adjudicator to have if he is named in the contract, as when the dispute or difference arises, it may not be a matter with which he is sufficiently experienced or in which he is adequately qualified. In such a case he can take the appropriate advice.

Effect of decision

The adjudicator's decision is binding unless and until the underlying dispute or difference is finally determined by arbitration or legal proceedings or by some written settlement of the dispute or difference between the parties made after the adjudicator's decision.

The parties must comply with the decision and both employer and contractor are to ensure that the adjudicator's decision is given effect. The employer must therefore, it is suggested, make sure that the architect takes account of any adjudicator's decision in administering the contract. For example, if there is a dispute or difference as to whether certain work carried out by the contractor amounts to a variation, and the adjudicator decides that it is, then it must be valued and added to the contract sum under the contract in the usual way.

The requirement for the parties to honour the adjudicator's decision pending a resolution of the underlying dispute or difference by arbitration or litigation puts the parties into a kind of twilight zone. During this time the parties must abide by the decision. Clause 9A.7.2 expressly so states. What is to happen if they take irrevocable steps. For example, an adjudicator may decide that the employer has been guilty of a breach of contract which is sufficiently serious to amount to a repudiation of the contract by the employer, entitling the contractor to bring the contract to an end. The contractor may therefore do so. Some time later an arbitrator or a court of law may hold that the employer's conduct did not amount to a breach of contract at all, or that if it did, it was not a repudiation of the contract. What is now the contractor's position? Is he required to return to the site? If he is, what is the position in relation to liquidated and ascertained damages and extensions of time, and indeed loss and expense involved in re-establishing the site? What about all the sub-contracts which will have been brought to an end and which will have involved the contractor in possibly paying large sums for which he will have sought recovery from the employer? Will the contractor be treated as having been in breach of contract once the arbitrator has given his decision? Presumably not as the contractor will, in treating the contract as at an end, have done no more than given effect to the adjudicator's decision.

All of this and much more is left completely in the air. To have a situation where decisions can be made during the progress of the works, particularly legal decisions which alter the course of the contract, but which are binding only in a temporary fashion is both worrying and indeed astonishing. It is the result of the provisions of section 108 of the 1996 Act and will undoubtedly lead to injustice and on occasions to irreparable financial damage and possibly insolvency following an adjudicator's wrong decision.

If either party fails to comply with the adjudicator's decision, the other party can take legal proceedings to enforce it. The provision in clause 41A.7.3, entitling the parties to enforce the adjudicator's decision by taking legal proceedings, is likely to defeat any application for a stay of legal proceedings to arbitration under section 9 of the Arbitration Act 1966. The provisions of this clause are to be contrasted with the statutory Scheme for Construction Contracts (Part 1) which makes no such express provision. A decision of the courts on the effect of section 9 on an attempt to enforce a statutory Scheme adjudication award has been eagerly awaited; so too has a decision on what would be the court's approach where an allegation was made that the adjudicator's 'decision' was no decision at all on the grounds that it was in some way invalid.

The first case concerning adjudication pursuant to the Housing Grants, Construction and Regeneration Act 1996 provisions has come before the courts. In *Macob Civil Engineering Ltd* v. *Morrison Construction Ltd* (1999) (Technology and Construction Court) the adjudicator appointed had acted under the statutory Scheme as the parties' own adjudication procedure failed to comply with the requirements of section 108 of the Act (see earlier page 25). He decided that the main contractor should pay the sub-contractor the sum of £302,000 plus interest. When payment was not made the sub-contractor applied to the court.

The main contractor contended that the decision was invalid on the ground that the adjudicator had failed to comply with the principles of natural justice, *inter alia*, in not giving the parties a proper opportunity of making representations. Mr Justice Dyson said:

> 'The intention of Parliament ... was plain. It was to introduce a speedy mechanism for settling disputes ... on an interim basis, and requiring the decisions ... to be enforced pending ... final determination ... by arbitration, litigation or agreement.'

If the mere contention that a decision was invalid meant that there was no decision at all, this would substantially undermine the effectiveness of the scheme for adjudication. The judge went on:

> 'The timetable for adjudication is very tight... Many would say unreasonably tight, and likely to result in injustice. Parliament must be taken to have been aware of this.'

He held that by analogy with the position in public law, the question of the interim status of a decision which is challenged and subsequently found to be invalid was a matter of construction of the statutory Scheme, the governing Act, and the background against which it was passed. Adopting that purposive approach he was in no doubt that the decision whose validity was challenged, was nevertheless capable of enforcement pending a final outcome both as to the validity of the challenge and on the merits of the underlying dispute itself.

On the question of the main contractor's application for a stay of the enforcement proceedings pursuant to section 9 of the Arbitration Act 1996, the judge refused the application. The main contractor had served a notice of arbitration which both challenged the validity of the decision – in effect claiming that it was not a 'decision' at all – and had also sought to arbitrate as to the merits of the underlying dispute.

The judge held that it was inconsistent to argue on both counts. If the merits were referred to arbitration then the effect of section 108(3) of the Act was to render the adjudicator's decision binding and enforceable in the interim. The main contractor could not have it both ways.

This last aspect of the decision is not easy to fathom and it might be that on a future occasion a court might stay the enforcement of a decision under statutory Scheme adjudication procedure to arbitration. A further point to note is that this case concerned a statutory Scheme adjudication in connection with which a statutory instrument provides for it to be binding and enforceable pending a final outcome. The analogy with the position under public law may not therefore be as appropriate where the decision is made pursuant to a contractual scheme. Even here, however, the decision is still one which statute requires the contractual provisions to state as binding.

One thing is clear: the Court has made a clear statement of policy that it will wherever possible support this statutory provision for adjudication, whether under the contract or the statutory Scheme.*

Under clause 9A.7.3 it is not open to the party seeking to enforce the decision to unilaterally take the matter to arbitration for enforcement. Even if the underlying dispute or difference is to go to arbitration, disputes or differences in connection with the enforcement of any decision of an adjudicator are excluded from arbitration.

It is submitted that non-compliance by a party with the decision of an adjudicator will, *ipso facto*, be a sufficient dispute or difference to exclude it from the arbitration agreement which is contained in article 9A and clause 9B. Even if this were not so, it is submitted that the terms of clause 9A.7.3 entitling a party to take legal proceedings to secure compliance with an adjudicator's decision will be an adequate basis on which to proceed to the courts without the fear of a successful application by the other party under section 9 of the Arbitration Act 1996 to stay the proceedings on the basis that there is a valid agreement to arbitrate covering the dispute or difference.

Adjudicator's powers

The adjudicator has absolute discretion to set his own procedures. This should encourage the maximum flexibility and optimal use of time. His powers include, but are not limited to, the following:

(1) Using his own knowledge and/or experience. This is clearly a very desirable power to have. He need not just rely therefore on the representations made to

---

* In *Outwing Construction Ltd v. Randell and Son Ltd* (1999) the Technology and Construction Court has made it clear that it will generally support applications to shorten the usual time-scales so that the enforcement can be dealt with promptly.

him but can make use of his own knowledge and experience in reaching his decision.

(2) Opening up, reviewing and revising certificates, opinions, decisions, requirements or notices under the contract as if such certificate etc. had not been issued etc. This gives the adjudicator the same sort of powers as the arbitrator (see clause 9B.2). As the arbitrator and the courts have such powers, it is clearly appropriate that the adjudicator should expressly be given them to avoid any possible argument that otherwise he would not have them. This power will, for instance, enable the adjudicator to stand in the shoes of the architect or quantity surveyor and make a decision without being fettered by any previous actions of the architect or quantity surveyor.

(3) Requiring from the parties further information, including the results of any tests that have been made or of any opening up.

(4) Requiring the parties to carry out tests or to open up work. This is clearly a useful power to have. The cost of any such testing or opening up can be the subject of an order from the adjudicator as to which party shall bear the cost.

(5) Visiting the site or any workshop where work is being carried out or prepared.

(6) Obtaining necessary information from employees or representatives of a party.

(7) Obtaining information and advice on technical and legal matters subject to giving prior notice to the parties together with a statement or estimate of the cost involved. This is clearly an important power to have and enables an adjudicator to seek technical or legal advice in situations where he considers this to be necessary. While adjudication is intended to be quick and inexpensive, it could on occasions involve significant expenditure where expert help is enlisted. Although the parties are entitled to receive a statement or estimate of the cost involved, this is for information purposes. They have no easy means of controlling such costs.*

(8) Having regard to any term of the contract relating to payment of interest, deciding the circumstances in which and the period for which simple interest shall be paid. The adjudicator therefore has express power to award simple interest. Quite apart from this power, any failure by the employer to pay money due by the final date for payment will attract interest under the contract in any event.

Effect of non-compliance by a party

If either party either fails to execute the adjudication agreement or fails to comply with any requirement of the adjudicator or with any provision or requirement in clause 9A generally, this will not invalidate the adjudicator's decision. This is clearly a necessary provision intended to thwart any attempts by an unwilling

---

* At the time of writing the Tribunal is in the process of inserting additional procedures requiring the adjudicator, at the request of either party, to obtain the opinion of an expert in relation to the reasonableness of an architect's instruction to open up and test pursuant to clause 3.13.

party to the adjudication who may seek to delay or nullify the process by failing to respond as required by clause 9A or the adjudicator.

Costs and fees

(i) Costs

The parties must meet their own costs in participating in the adjudication process. The only partial exception to this is in relation to the power of the adjudicator to require a party to carry out tests or to open up work and then to direct who should pay for this. If no direction is given such costs will lie where they fall.

(ii) Adjudicator's fees and expenses

The adjudication agreement has a schedule in which will be inserted either a lump sum figure or an hourly rate for the adjudicator's fees. Expenses will be on top of this.

Both parties are liable for all of the adjudicator's fees and expenses. He can therefore go to either party to recover these in full. As between the parties themselves, the adjudicator has power to apportion liability for fees and expenses. If the adjudicator makes no apportionment then the parties, as between themselves, are each responsible for one half of the fees and expenses. So, if one person pays all of the fees and expenses of the adjudicator and no apportionment is made, there will be a right to recover one half from the other party. In the event of fees and expenses being of a significant amount, there is clearly a risk involved should the party who has paid the fees and expenses be unable to recover them from the other party who may have financial problems. It is difficult to see how this problem can be avoided.

Immunity for adjudicator

The adjudicator is not liable for anything done or omitted in the discharge of his functions unless the act or omission is in bad faith. This protection from liability is extended to any employee or agent of the adjudicator. This provision mirrors section 108(4) of the 1996 Act. The parties will therefore be unable to sue the adjudicator if his decision has been negligently made. It would be necessary to establish bad faith on the part of the adjudicator and this would require the clearest evidence. In practice it is likely to prove very difficult.

It is just possible that someone who is not a party to the contract but who is foreseeably likely to be affected by it may be able to sue the adjudicator in negligence if a duty of care can be shown to exist between the adjudicator and such third party. If the negligent decision leads to personal injury or damage to property of someone likely to be affected as a consequence of such a decision then an action would lie. However, it is submitted that the decision of the adjudicator would not relieve either party or the architect or any other professional advisor from their duty of care to avoid physical harm to third parties. If an adjudicator were brave (and perhaps foolish) enough to decide for example that the balconies on a hotel were safe this should in no way influence the actions of the employer, architect or any structural engineer involved in determining what should be done in the circumstances. If the view is formed, apart from the adjudicator's decision, that the balconies are unsafe then steps should be taken accordingly. It is unlikely therefore that an adjudicator would be the cause of the injury or damage in circumstances such as these.

The immunity is extended to employees and agents of the adjudicator. Depending on what is meant by 'agent' in these circumstances, it might be argued that this immunity extends to any expert engaged by the adjudicator to provide technical or legal advice.

Clause 9B – Arbitration

9B   Arbitration
A reference in clause 9B to a Rule or Rules is a reference to the JCT 1998 edition of the Construction Industry Model Arbitration Rules (CIMAR) current at the Base Date.

9B.1.1   Where pursuant to article 9A either Party requires a dispute or difference to be referred to arbitration then that Party shall serve on the other Party a notice of arbitration to such effect in accordance with Rule 2.1 which states:

'Arbitral proceedings are begun in respect of a dispute when one party serves on the other a written notice of arbitration identifying the dispute and requiring him to agree to the appointment of an arbitrator';

and an arbitrator shall be an individual agreed by the parties or appointed by the person named in the Appendix in accordance with Rule 2.3 which states:

'If the parties fail to agree on the name of an arbitrator within 14 days (or any agreed extension) after:
(i)   the notice of arbitration is served, or
(ii) a previously appointed arbitrator ceases to hold office for any reason,
either party may apply for the appointment of an arbitrator to the person so empowered.'

By Rule 2.5:

'the arbitrator's appointment takes effect upon his agreement to act or his appointment under Rule 2.3, whether or not his terms have been accepted.'

9B.1.2   (a) Where two or more related arbitral proceedings in respect of the Works fall under separate arbitration agreements, Rules 2.6, 2.7 and 2.8 shall apply thereto.
(b) After an arbitrator has been appointed either Party may give a further notice of arbitration to the other Party and to the Arbitrator referring any other dispute which falls under article 9A to be decided in the arbitral proceedings and Rule 3.3 shall apply thereto.

9B.2   Subject to the provisions of article 9A and clause 4.7 (*Effect of final certificate*) the Arbitrator shall, without prejudice to the generality of his powers, have power to rectify this Contract so that it accurately reflects the true agreement made by the Parties, to direct such measurements and/or valuations as may in his opinion be desirable in order to determine the rights of the parties and to ascertain and award any sum which ought to have been the subject of or included in any certificate and to open up, review and revise any certificate, opinion, decision, requirement or notice and to determine all matters in dispute which shall be submitted to him in the same manner as if no such certificate, opinion, decision, requirement or notice had been given.

9B.3   Subject to clause 9B.4 the award of such Arbitrator shall be final and binding on the Parties.

9B.4   The Parties hereby agree pursuant to S.45(2)(a) and S.69(2)(a) of the Arbitration Act 1996 that either Party may (upon notice to the other Party and to the Arbitrator):

9B.4.1   apply to the courts to determine any question of law arising in the course of the reference; and

9B.4.2   appeal to the courts on any question of law arising out of an award made in an arbitration under this Arbitration Agreement.

9B.5   The provisions of the Arbitration Act 1996 or any amendment thereof shall apply to any arbitration under this Contract wherever the same, or any part of it, shall be conducted. [hh]

9B.6   The arbitration shall be conducted in accordance with the JCT 1998 edition of the Construction Industry Model Arbitration Rules (CIMAR) current at the Base Date. Provided that if any amendments to the Rules so current have been issued by the JCT after the Base Date the Parties may, by a joint notice in writing to the Arbitrator, state that they wish the arbitration to be conducted in accordance with the Rules as so amended.

[hh] It should be noted that the provisions of the Arbitration Act 1996 do not extend to Scotland. Where the site of the Works is situated in Scotland then the forms issued by the Scottish Building Contract Committee which contain Scots proper law adjudication and arbitration provisions are the appropriate documents. The SBCC issues guidance in this respect.

COMMENTARY ON CLAUSE 9B – ARBITRATION

Introduction

By clause 1.15 of IFC 98, whatever the nationality, residence or domicile of the Employer, the Contractor or any sub-contractor and wherever the Works are situated, the law of England is the applicable law for the IFC 98 contract. The Arbitration Act 1996 (the Act) applies to any arbitration under IFC 98, in whatever country the arbitration proceedings are conducted. If this is not intended to be the case then clause 9B.5 should be amended. It should be noted that the Act has no application to Scotland.

Article 9A refers to '... any dispute or difference as to any matter or thing of whatsoever nature arising under this Contract or in connection therewith'. A similar phrase also appeared in clause 35 of JCT 63, and was given a liberal interpretation by the Court of Appeal in the case of *Ashville Investments Ltd* v. *Elmer Contractors Ltd* (1987) so as to bestow on the arbitrator a jurisdiction which includes the power to grant rectification of the contract arising out of alleged innocent or negligent misrepresentation.

It should be remembered that certain matters are expressly excluded from arbitration by article 9A itself (see the commentary on Article 9A earlier page 21). Further, it should be remembered that many (though not all – see earlier page 382) disputes or differences arising under the arbitration agreement, may be referred by either party to an adjudicator for a decision, although this will not prevent the dispute or difference also being referred to an arbitrator. There seems nothing to prevent one party adjudicating a dispute or difference at the same time as he or the other party is referring it to arbitration. This is an unlikely event but is possible.

Finally, clearly some issues arising under the contract are not covered as they will not amount to a dispute or difference. Examples are where the contract provides for the employer and contractor to agree on a matter, e.g. clause 3.7.1.1, allowing the parties to agree a price for a variation without applying the contract's valuation rules; or where the employer has an unfettered discretion, e.g. clause 2.10 (under which the employer can decide to have an appropriate deduction to the contract sum rather than require the contractor to make good defects), and the similar provision in clause 3.9 (in relation to inaccurate setting out); and note also clause 3.1 where assignment of the contract by one party depends on obtaining the consent of the other party. The arbitration agreement would on the other hand include an issue as to whether consent had been unreasonably delayed or withheld where consent under the contract is expressly qualified in this way, e.g. clauses 1.7, 3.2, 3.3 and 3.11.

SUMMARY OF IFC 98 ARBITRATION PROCEDURE

The parties elect through appendix entries whether they will refer disputes or differences to arbitration under clause 9B or litigation under clause 9C. If arbitration is the selected tribunal the procedure is as follows.

The JCT 1998 edition of the Construction Industry Model Arbitration Rules (CIMAR)

Arbitration under IFC 98 will be conducted in accordance with the CIMAR Rules (the Rules) current at the base date stated in the appendix to IFC 98. If the Rules are subsequently amended, the parties can agree by a joint notice in writing to the arbitrator that they wish the arbitration to be conducted under the amended Rules.

As the Rules apply to all JCT contracts, they are published separately from the IFC 98 Contract. It is not intended to comment in detail on the Rules themselves although they will be mentioned as necessary when dealing with the various sub-clauses of 9B which make specific reference to the Rules. The JCT edition of the CIMAR Rules provides firstly for some supplementary procedures and secondly for some advisory procedures, the latter only applying if both parties expressly agree to them after the commencement of the arbitration. These advisory procedures provide, *inter alia*, timetables for the submission of statements of case and deal with the consequences of a failure to comply with the timetable and as such can be particularly prejudicial in effect. It is unlikely that a respondent in arbitration proceedings will agree very often to their application.

The Rules provide three main procedures for dealing with disputes: a short hearing procedure (Rule 7); a documents only procedure (Rule 8); and a full procedure (Rule 9).

The notice and commencement of the arbitration

Under IFC 98, arbitration is commenced when one party serves on the other a notice of arbitration identifying the dispute and requiring the other party to agree to the appointment of an arbitrator. The service of this notice marks the commencement of the arbitration for the purposes of the Limitation Act 1980 section 34 (and see sections 13 and 14 Arbitration Act 1996) and for the purposes of the 28 day time limit contained in clause 4.7. This is reflected both in clause 9B.1.1 and in Rule 2.1.

A notice which merely indicates an intention to commence arbitration proceedings will not suffice: see *Cruden Construction Ltd* v. *Commission for the New Towns* (1994). Similarly the service of a notice purporting to refer a dispute to arbitration but which fails to require the other party to agree to the appointment of an arbitrator will be insufficient to be a commencement of the arbitration: see *Vosnoc Ltd* v. *Transglobal Projects Ltd* (1997)

As arbitration depends on an agreement to arbitrate between the parties and also because the notice referring the dispute or difference to arbitration will generally stop time from running for limitation purposes, it is important that the notice describes the dispute or difference in the broadest possible terms, otherwise the recipient of the notice may argue at a later date that a claim actually being made in the arbitration as identified in the statement of case or claim is outside the scope of the written notice. Indeed, the objecting party may even be able to challenge the arbitrator's award (section 67 of the Act) or its enforcement (section 66(3) of the Act) where it deals with matters outside the scope of the written notice, provided that the objecting party has properly reserved his position (section 31 of the Act) as to the arbitrator's lack of authority, despite having gone ahead with the arbitration including the claim objected to.

Some specimen notices referring disputes or differences to arbitration opt for the broadest possible wording such as 'I require you to concur with me in the appointment of an arbitrator for the settlement of the disputes which have arisen between us...' (see *Russel on Arbitration*, 20th edition, published by Stephens & Sons). While this may be the safest course it may not be the most helpful. A possible answer is to be as specific as possible in defining the disputes or differences which have arisen and to conclude with some such phrase as 'and without prejudice to the particularity of the foregoing any other disputes or differences which have arisen between us'.

A notice of arbitration cannot cover future disputes, although by clause 9B.1.2(b), after an arbitrator has been appointed either party may give further notice of arbitration referring any other dispute to be decided in the same arbitral proceedings, in which case Rule 3.3 will apply. Rule 3.3 provides that if in such circumstances the other party does not consent to the other dispute being referred to the arbitrator, the arbitrator himself will decide whether or not it should be consolidated within the same arbitral proceedings or whether it should not be consolidated in which case a separate arbitration will have to be commenced. Even if subsequent disputes are consolidated in this way, it will not result in commencement of the arbitration of the other dispute being related back to the original notice. Therefore for limitation purposes, and it is suggested for the purposes of the 28 day time limit in clause 4.7 (referred to above) the appropriate date will be the date of the subsequent notice. This is reinforced by Rule 3.6 which provides that arbitral proceedings in respect of any other dispute are begun when the notice of arbitration for that dispute is served (and see section 13 of the Act).

By Rule 3.1, a notice of arbitration may include two or more disputes.

Appointment of arbitrator

The arbitrator is to be an individual. He is to be appointed either by agreement by the parties within 14 days of the notice of arbitration (or any extended period agreed by the parties) or, failing such an agreement, by an appointing body selected through the appendix to IFC 98. The appointing body will be whichever of the following is selected in the appendix: the President or Vice-President of the Royal Institute of British Architects, the Royal Institution of Chartered Surveyors; or the Chartered Institute of Arbitrators. Like the position with JCT 98, which provides for the President or a Vice-President of the Royal Institute of British Architects to make the appointment where no selection is made in the appendix, IFC 98 (unlike IFC 84) has a similar fall-back position.

The appointment takes effect either on the arbitrator's agreement to act (where the parties have agreed on his appointment); or upon his appointment by the appointing body selected in the appendix, whether or not his terms of appointment have been accepted. Presumably it is therefore an appointment subject to a condition subsequent so that if there is no agreement as to the arbitrator's terms the appointment will subsequently fail. While unlikely, if this does occur it could cause considerable inconvenience both in time and cost, especially if the arbitration runs while terms are being negotiated. However, if the arbitration does get under way while terms are still being negotiated, it could well be that either there will have been an acceptance of terms by conduct at some stage or alternatively, if the outstanding issue relates to the arbitrator's fees, the arbitrator may have to continue and claim such reasonable fees and expenses as are appropriate in the circumstances – see section 28 of the Act.

It is clearly sensible for the parties to tie up the terms of appointment as soon as possible. However, the arbitrator's terms of appointment appear to be a good area for spoiling tactics on the part of unco-operative respondents.

By clause 9B.1.2(b), after an arbitrator has been appointed, either party may give a further notice of arbitration to which Rule 3.3 will apply (see above page 395).

Arbitrator's powers

The arbitrator's powers stem from the Act and also from clause 9B itself together with the rules. As already noted, it is not proposed in this book to comment in any detail on the Act although it is worth noting that section 30 makes it clear that unless the parties agree otherwise (and there is no such agreement otherwise in clause 9B) the arbitrator can rule on an issue relating to his own substantive jurisdiction as to:

'(a) whether there is a valid arbitration agreement,
(b) whether the tribunal is properly constituted, and
(c) what matters have been submitted to arbitration in accordance with the arbitration agreement.'

This confirms but significantly extends the decision of the Court of Appeal in *Harbour Assurance Co. Ltd* v. *Kansa General International Insurance Co. Ltd* (1993).

Clause 9B.2 confers significant powers on the arbitrator including the following:

(1) Rectification

The arbitrator has power to rectify the contract so that it accurately reflects the true agreement made between the parties. It is probable that even without this express power to rectify, the scope of the wording of the arbitration clause '. . . any dispute or difference as to any matter or thing of whatsoever nature arising under this contract *or in connection therewith* . . .' (Article 9A) is sufficient to enable the arbitrator to rectify the contract in many instances: see *Ashville Investments Ltd* v. *Elmer Contractors Ltd* (1987) discussed earlier at page 393.

(2) Directing measurements or valuations

The arbitrator can direct such measurements or valuations as may in his opinion be desirable in order to determine the rights of the parties.

(3) Open up, review and revise certificates, opinions, decisions, requirements or notices

The express reference in clause 9B.2 to the opening up etc. of certificates, opinions, decisions, requirements or notices must relate to those given in circumstances where the architect uses his professional judgment or discretion under the contract. In other words, the arbitrator's power to open up would not extend to what was in truth a decision of the employer through the architect, e.g. to vary the works. In practice it is the role of the architect which will be reviewed here; for example, decisions as to extensions of time, certificates relating to interim valuations, opinions as to whether defects have been made good so as to justify a certificate of making good defects, and so on.

So far as architect's instructions are concerned, while the arbitrator clearly has power to determine whether they are validly issued in the sense of being empowered by the conditions, he has no power to open and review etc. where the architect not only has the power but also has a discretion as to whether to issue the instruction at all, e.g. as to the expenditure of a provisional sum.

Should the parties choose to litigate, the courts possess similar powers to open up and review etc.: see *Beaufort Developments (NI) Ltd* v. *Gilbert-Ash NI Ltd and Others* (1998) overriding *Northern Regional Health Authority* v. *Derek Crouch Construction Company Ltd and Others* (1984).

All of the express powers given to the arbitrator in clause 9B.2 are, however, made subject to clause 4.7 (final certificate conclusive evidence as to certain matters).

Applications to determine and appeals on questions of law to the courts

The arbitrator's award is final and binding on the parties, subject however to the reference in clause 9B.4 to the effect that the parties agree pursuant to section 45(2)(a) and section 69(2)(a) of the Act, that either party may upon notice to the other party and to the arbitrator:

- Apply to the courts to determine any question of law arising in the course of the reference; and
- Appeal to the courts on any question of law arising out of an award made in the arbitration.

To take a point of law to the courts either for a determination during the course of the arbitration (section 45) or by way of appeal arising out of the award (section 69) requires either the agreement of all other parties to the proceedings or (in the case of a determination) with the permission of the arbitrator, the court being satisfied that the determination of the question is likely to produce substantial savings in costs and that the application was made without delay; or (in the case of an appeal) with leave of the court – which is only given if the court is satisfied that the determination of the question will substantially affect the rights of one or more of the parties; that the question is one which the arbitrator was asked to determine; and that on the basis of the findings of fact in the award, the decision of the arbitrator on the question is obviously wrong or is of general public importance and is open to serious doubt and that it is just and proper in all the circumstances for the court to determine the question.

   It can be seen therefore that in the absence of agreement there are considerable hurdles to overcome. Given the background to section 45(2)(a) and section 69(2)(a) (originally contained in sections 1(3)(a) and 2(1)(b) of the Arbitration Act 1979 although referring to the consent rather than agreement of the parties) and the clear intention to severely restrict the ability of a dissatisfied party taking a point of law to the courts, the obtaining of agreement in advance of any dispute or difference arising by means of a provision in the arbitration agreement is highly significant. Some might suggest that to embody such agreement in the arbitration agreement before any dispute or difference has arisen could be regarded as an attempt to circumvent the spirit, if not the letter, of sections 45 and 69 of the Act.

   In the case of *Finelvet A.G.* v. *Vinava Shipping Co. Ltd (The Chrysalis)* (1983) Mr Justice Mustill (as he then was) said:

> 'Counsel for the charterers has, however, contended that the position is different in the present case, because the parties had agreed in advance that there should be a right of appeal on any question of law. This shows, so it is maintained, that the parties wanted an authoritative ruling on the question of frustration and this they would not get from a mode of appeal which precluded the judge from substituting his own opinion for that of the arbitrator on the "judgmental" stage of the reasoning process. I am afraid that I cannot read the agreement as showing any such intention. Its obvious purpose was to save the time and expense involved in a contested application under section 1(3)(b) of the 1979 Act...'

This appears to recognise implicitly that the agreement in advance is adequate and effective.

   In *Wates Construction (South) Ltd* v. *Bredero Fleet Ltd* (1993) it was accepted as common ground, without argument, that similar provisions in clause 41.6 of JCT 80 were effective. In the case of *Vascroft (Contractors) Ltd* v. *Seeboard Plc* (1996), a case involving similar provisions in the form of sub-contract known as DOM/2, similar provisions were held to be effective. Finally, in the case of *Amec Building*

Ltd v. *Cadmus Investment Co. Ltd* (1996) it was held that although the consent provisions in clause 41.6 of JCT 80 were effective, nevertheless the courts still had some discretion to refuse to allow a point of law to be taken, e.g. if it had not been argued before the arbitrator and could and should have been.

Further, in *Taylor Woodrow Civil Engineering Ltd* v. *Hutchinson IDH Developments Limited* (1998), it was held that the earlier provisions of IFC 84 which referred to sections 1(3)(a) and 2(1)(b) of the Arbitration Act 1979 were effective even where the arbitration had commenced after 31 January 1997 (when the Arbitration Act 1996 came into force) and was therefore governed by the 1996 Act.

Related disputes consolidation and joinder

Arbitration is consensual. Therefore a person who has not agreed to arbitrate cannot be compelled to. All of the JCT arbitration agreements are contained in contracts to which there are two parties. It is not unusual, however, for a dispute between the parties under one contract to affect a party under a related contract with one of the disputing parties. For example, an employer may be in dispute with a contractor regarding allegedly defective work. The contractor may likewise be in dispute with a named sub-contractor in respect of the same work. The contractor is a party to both the main and named sub-contract but of course the named sub-contractor is not a party to the main contract and the employer is not a party to the named sub-contract. Both contracts have arbitration agreements but without more if a dispute were to be the subject of arbitration, it would necessarily result in two separate sets of arbitration proceedings in respect of the same issues – one between the employer and the contractor and the other between the contractor and the sub-contractor. There may be different arbitrators and different decisions may be made even though based on the same evidence. If this potential problem arises in litigation the court has ample powers to bring all of the affected parties within the same proceedings so that the risk of inconsistent findings as well as the duplication of costs is avoided. The Act and clause 9B.1.2, together with the Rules, go some way to achieving a similar result in arbitration, hopefully with resultant saving in time and cost.

Any number of potential difficulties can arise in the operation of consolidation or joinder provisions – see for example the attempt, fortunately unsuccessful, in the case of *A. Monk & Co. Ltd* v. *Devon County Council* (1978) (a tripartite arbitration between employer, main contractor and sub-contractor) to tie the hands of the main contractor in his dispute with the employer following a settlement between the main contractor and the sub-contractor. If co-operation and common sense prevail, the problems are not insurmountable. Even so, these potential problems are likely to be a potent factor in deciding whether to choose litigation as the dispute resolution procedure for the contract.

If there is already an arbitration running in respect of a dispute or difference which gives rise to related issues in a dispute in respect of the works which falls within an arbitration agreement under a different contract (e.g. an existing dispute between employer and contractor under IFC 98 regarding workmanship which is likewise a dispute between the main contractor and a sub-contractor under a sub-contract), then provided the other contract also includes the Rules, whichever appointing body has the responsibility to appoint an arbitrator in the subsequent related dispute, must give due consideration as to whether the same or a different arbitrator should be appointed in respect of it. The same arbitrator is to be appointed unless there are sufficient grounds shown for not doing so (Rule 2.6). This applies whether the same or a different appointing body is involved in relation to the two different arbitration agreements – Rule 2.7. This obligation may be discharged by one appointing body arranging for another appointing body to make the appointment – Rule 2.8.

Rule 3 deals with consolidation and joinder and may be summarised as follows.

To avoid any doubt, it is provided in Rule 3.1 that a notice of arbitration may include two or more disputes if they fall under the same arbitration agreement.

If one party serves a notice of arbitration on another party, that other party is entitled, before an arbitrator has been appointed, to give notice of arbitration in respect of any other disputes under the same arbitration agreement, in which case the disputes are consolidated. Once the arbitrator has been appointed, either party can seek to bring further disputes between them within the same arbitration. If one party does not consent to this, the arbitrator decides whether or not to consolidate the new dispute. If it is not consolidated the party seeking to consolidate will have to commence a separate arbitration in the usual way – rule 3.3.

Where the same arbitrator is appointed in related arbitration proceedings, including therefore disputes between employer and main contractor on the one hand and main contractor and sub-contractor on the other, where common issues arise, the arbitrator may order concurrent hearings. This falls short of joinder or full consolidation of the two separate arbitrations. For instance, separate awards will be delivered.

If all the parties agree, the related arbitrations can be fully consolidated which will amount to a joinder of the separate arbitrations rather than just concurrent hearings. Unless the parties agree otherwise, a single award will then be issued which will bind all parties.

The arbitrator has power to revoke previous orders as to concurrent hearings or consolidation – see Rules 3.7 to 3.12 and also generally section 35 of the 1996 Act.

Clause 9C – Legal proceedings

Where article 9B applies, any dispute or difference shall be determined by legal proceedings pursuant to article 9B.

COMMENTARY ON CLAUSE 9C – LEGAL PROCEEDINGS

Where pursuant to article 9B, the entry in the appendix makes it clear that clause 9B is not to apply, then clause 9C applies so that the final determination of disputes and differences goes to the courts. If no deletion is made in the appendix to the conditions then by default clause 9B will apply and disputes or differences will be referred to arbitration.

The courts will have the same powers as an arbitrator would have (see clause 9B.2) to open up, review and revise any certificate, opinion, decision, requirement or notice in order to determine all matters in dispute. This has now been confirmed by the House of Lords in *Beaufort Developments (NI) Ltd* v. *Gilbert-Ash NI Ltd and Others* (1997) overruling the Court of Appeal's decision in *Northern Regional Health Authority* v. *Derek Crouch Construction Co. Ltd* (1984) to the effect that such powers were reserved to the arbitrator and were not available to the courts.

It should be appreciated that although generally IFC 98 is a consolidation of IFC 84 together with Amendments 1 to 12 inclusive, clause 9C has appeared in IFC for the first time in its 98 edition and was not included in IFC 84 or any of its amendments.

Chapter 13

The Appendix to IFC 98

IFC 98 is a standard form of contract. For it to work in respect of any individual project, certain variables have to be provided, for example, date for possession, date for completion, level of liquidated and ascertained damages etc. In addition, it is often desirable if options are to be provided, for the appropriate option to be selected in a systematic and obvious way. This is achieved by means of an appendix to the IFC 98 conditions. It is a way of tailoring the contract to the individual project.

The appendix to IFC 98 is set out here. The various items referred to in it are discussed as necessary in commenting on the clauses to which such entries relate (the relevant clauses are identified on the left hand side of the appendix against each relevant item). Accordingly it is not intended here to discuss them further.

Appendix

Clause etc.	Subject	
Third recital	CDM Regulations	* All the CDM Regulations apply/ Regulations 7 and 13 only of the CDM Regulations apply
Articles 9A and 9B Clauses 9B and 9C	Dispute or difference – settlement of disputes	* Clause 9B applies * Delete if disputes are to be decided by legal proceedings and clause 9C is thus to apply *See the Guidance Note to JCT 80 Amendment 18 on factors to be taken into account by the Parties considering whether disputes are to be decided by arbitration or by legal proceedings*
1·16	Electronic data interchange	The JCT Supplemental Provisions for EDI *apply/do not apply If applicable: the EDI Agreement to which the Supplemental Provisions refer is: * the EDI Association Standard EDI Agreement * the European Model EDI Agreement
2·1	Date of Possession	_____
2·1	Date for Completion	_____
2·2 and 2·4·14 and 4·11(a)	Deferment of the Date of Possession	Clause 2·2 *applies/does not apply Where clauses 2·2 applies: period of deferment _____weeks (period not to exceed 6 weeks)
2·4·10 and 2·4·11	Extension of time for inability to secure essential labour or goods or materials	Clause 2·4·10 *(labour)* *applies/does not apply Clause 2·4·11 *(goods or materials)* *applies/does not apply
2·7	Liquidated damages	at the rate of £ _____ per_____.
2·10	Defects liability period (if none stated is 6 months from the day named in the certificate of Practical Completion of the Works)	_____
4·2	Period of interim payments if interval is not one month	_____

* Delete as applicable.

Appendix *continued*

Clause etc.	Subject	
4·2(b)	Advance payment	Clause 4·2(b) *applies/does not apply If applicable: the advance payment will be **£ _____ / _____ % of the Contract Sum and will be paid to the Contractor on _____ and will be reimbursed to the Employer in the following amount(s) and at the following time(s) _____ _____ An advance payment bond *is/is not required
4·2·1(a)	Valuation	A priced Activity Schedule *is/is not attached to this Appendix
4·2·1(c)·1	Listed items – uniquely identified	*For uniquely identified listed items a bond as referred to in clause 4·2·1(c)·1 in respect of payment for such items is required for £ _____ *Delete if no bond is required
4·2·1(c)·2	Listed items – not uniquely identified	*For listed items that are not uniquely identified a bond as referred to in clause 4·2·1(c)·2 in respect of payment for such items is required for £ _____ *Delete if clause 4·2·1(c)·2 does not apply
4·9(a) and C7	Supplemental Condition C: Tax etc. fluctuations	Percentage addition _____% [ii]
4·9(b)	Formula fluctuations (not applicable unless Bills of Quantities are a Contract Document)	Supplemental Condition D [ii] *applies/does not apply

---

*Delete as applicable.

**Insert either a money amount or a percentage figure and delete the other alternative.

[ii] In accordance with clause 4·9, if Supplemental Condition D is not stated to apply then Supplemental Condition C applies.

Appendix *continued*

Clause etc.	Subject	
D1	Formula Rules (only where Supplemental Condition D applies)	rule 3: Base Month
		rule 3: Non-Adjustable Element [jj]
		_____ (not to exceed 10%)
		rules 10 and 30(i): *Part I/Part II of Section 2 of the Formula Rules is to apply
5·5	Value added tax: Supplemental Condition A	Clause A1·1 of Supplemental Condition A *applies/does not apply [o]
6·2·1	Insurance cover for any one occurrence or series of occurrences arising out of one event	£_____
6·2·4	Insurance – liability of Employer	Insurance *may be required/is not required
		Amount of indemnity for any one occurrence or series of occurrences arising out of one event
		£_____ [kk]
6·3·1	Insurance of the Works – alternative clauses	*Clause 6·3A/Clause 6·3B/Clause 6·3C applies [s]
6·3A·1 [ii] 6·3B·1 [ii] 6·3C·2 [ii]	Percentage to cover professional fees	_____%
6·3A·3·1	Annual renewal date of insurance as supplied by Contractor	_____
6·3D	Insurance for Employer's loss of liquidated damages – clause 2·4·3	Insurance *may be required/is not required
6·3D·2		Period of time_____

* Delete as applicable.

[jj] Only applicable when the Employer is a local authority.

[o] Clause A1·1 can only apply where the Contractor is satisfied at the date the Contract is entered into that his output tax on **all** supplies to the Employer under the Contract will be at either a positive or a zero rate of tax.

On and from 1 April 1989 the supply in respect of a building designed for a 'relevant residential purpose' or for a 'relevant charitable purpose' (as defined in the legislation which gives statutory effect to VAT changes operative from 1 April 1989) is only zero rated if the person to whom the supply is made has given to the Contractor a certificate in statutory form: see the VAT leaflet 708 revised 1989. Where a contract supply is zero rated by certificate only the person holding the certificate (usually the Contractor) may zero rate his supply.

This footnote repeats footnote [o] for clause 5·5.

[kk] If the indemnity is to be for an aggregate amount and not for any one occurrence or series of occurrences the entry should make this clear.

[s] **Clause 6·3A** is applicable to the erection of a new building where the **Contractor** is required to take out a Joint Names Policy for All Risks Insurance for the Works and **clause 6·3B** is applicable where the **Employer** has elected to take out such Joint Names Policy. **Clause 6·3C** is to be used for alterations of or extensions to existing structures under which the **Employer** is required to take out a Joint Names Policy for All Risks Insurance for the Works and also a Joint Names Policy to insure the existing structures and their contents owned by him or for which he is responsible against loss or damage thereto by the Specified Perils.

This footnote repeats footnote [s] for clause 6·3.

Appendix *continued*

Clause etc.	Subject	
6·3FC·1	Joint Fire Code	The Joint Fire code *applies/does not apply If the Joint Fire Code is applicable, state whether the insurer under clause 6·3A or clause 6·3B or clause 6·3C·2 has specified that the Works are a 'Large Project': *YES/NO (where clause 6·3A applies these entries are made on information supplied by the Contractor)
8·3	Base Date	
9A·2	Adjudication – nominator of Adjudicator (if no nominator is selected the nominator shall be the President or a Vice-President of the Royal Institute of British Architects)	President or a Vice-President or Chairman or a Vice-Chairman: *Royal Institute of British Architects *Royal Institution of Chartered Surveyors *Construction Confederation *National Specialist Contractors Council *Delete all but one
9B·1	Arbitrator – appointor of Arbitrator (if no appointor is selected the appointor shall be the President or a Vice-President of the Royal Institute of British Architects)	President or a Vice-President: *Royal Institute of British Architects *Royal Institution of Chartered Surveyors *Chartered Institute of Arbitrators *Delete all but one

Appendix 1

Annex 1 to Appendix: Terms of Bonds
agreed between the British Bankers' Association and the JCT

See clause 4·2(b):
"Advance Payment Bond", and

clause 4·2·1(c):
"Bond in respect of payment for off-site materials and/or goods"

Advance Payment Bond

1   THE parties to this Bond are:

(1) _____

whose registered office is at _____

_____ ('the Surety'), and

(2) _____

of _____

_____ ('the Employer').

2   The Employer and _____ ('the Contractor')

have agreed to enter into a contract for building works ('the Works') at

_____ ('the Contract').

3   The Employer has agreed to pay the Contractor the sum of [_____] as an advance payment of sums due to the Contractor under the Contract ('the Advance Payment') for reimbursement by the Surety on the following terms:

(a) When the Surety receives a demand from the Employer in accordance with Clause 3(b) the Surety shall repay the Employer the sum demanded up to the amount of the Advance Payment.

(b) The Employer shall in making any demand provide to the Surety a completed notice of demand in the form of the **Schedule** attached hereto which shall be accepted as conclusive evidence for all purposes under this Bond. The signatures on any such demand must be authenticated by the Employer's bankers.

(c) The Surety shall within 5 Business Days after receiving the demand pay to the Employer the sum so demanded. 'Business Day' means the day (other than a Saturday or a Sunday) on which commercial banks are open for business in London.

4   Payments due under this Bond shall be made notwithstanding any dispute between the Employer and the Contractor and whether or not the Employer and the Contractor are or might be under any liability one to the other. Payment by the Surety under this

Bond shall be deemed a valid payment for all purposes of this Bond and shall discharge the Surety from liability to the extent of such payment.

5   The Surety consents and agrees that the following actions by the Employer may be made and done without notice to or consent of the Surety and without in any way affecting changing or releasing the Surety from its obligations under this Bond and the liability of the Surety hereunder shall not in any way be affected hereby. The actions are:

(a) waiver by the Employer of any of the terms, provisions, conditions, obligations and agreements of the Contractor or any failure to make demand upon or take action against the Contractor;

(b) any modification or changes to the Contract; and/or

(c) the granting of any extensions of time to the Contractor without affecting the terms of clause 7(c) below.

6   The Surety's maximum aggregate liability under this Bond which shall commence on payment of the advance payment by the Employer to the Contractor shall be the amount of [_____] which sum shall be reduced by the amount of any reimbursement made by the Contractor to the Employer as advised by the Employer in writing to the Surety.

7   The obligations of the Surety and under this Bond shall cease upon whichever is the earliest of:

(a) the date on which the Advance Payment is reduced to nil as certified in writing to the Surety by the Employer;

(b) the date on which the Advance Payment or any balance thereof is repaid to the Employer by the Contractor (as certified in writing to the Surety by the Employer) or by the Surety; and

(c) [*longstop date to be given*]

and any claims hereunder must be received by the Surety in writing on or before such earliest date.

8   This Bond is not transferable or assignable without the prior written consent of the Surety. Such written consent will not be unreasonably withheld.

9   This Bond shall be governed and construed in accordance with the laws of England and Wales.

IN WITNESS hereof this Bond has been executed as a Deed by the Surety and delivered on the date below:

EXECUTED as a Deed by: _____

    for and on behalf of the Surety: _____

EXECUTED as a Deed by: _____

    for and on behalf of the Employer: _____

Date: _____

Schedule to Advance Payment Bond

(clause 3(b) of the bond)

Notice of Demand

Date of Notice: _____

Date of Bond: _____

Employer: _____

Surety: _____

The bond has come into effect.

We hereby demand payment of the sum of

£ _____ (amount in words)
which does not exceed the amount of reimbursement for which the Contractor is in default
at the date of this notice.

Address for payment: _____

_____

This Notice is signed by the following persons who are authorised by the Employer to act for
and on his behalf:

Signed by _____

    Name: _____

    Official Position: _____

Signed by _____

    Name: _____

    Official Position: _____

The above signatures to be authenticated by the Employer's bankers

Appendix 2

Bond in respect of payment for off-site materials and/or goods

1   THE parties to this Bond are:

(1) _____

whose registered office is at _____

_____ ('the Surety'), and

(2) _____

of _____

_____ ('the Employer').

2   The Employer and _____ ('the Contractor')

have agreed to enter into a building contract for building works ('the Works')

at _____ ('the Contract').

3   Subject to the relevant provisions of the Contract as summarised below but with which the Surety shall not at all be concerned:

(a) the Employer has agreed to include in the amount stated as due in Interim Certificates (as defined in the Contract) for payment by the Employer the value of those materials or goods or items pre-fabricated for inclusion in the Work which have been listed by the Employer ('the listed items'), which list has been included as part of the Contract, before their delivery to or adjacent to the Works; and

(b) the Contractor has agreed to insure the listed items against loss or damage for their full value under a policy of insurance protecting the interests of the Employer and the Contractor during the period commencing with the transfer of the property in the items to the Contractor until they are delivered to or adjacent to the Works; and

(c) this Bond shall exclusively relate to the amount paid to the Contractor in respect of the listed items which have not been delivered to or adjacent to the Works.

4   The Employer shall in making any demand provide to the Surety a Notice of Demand in the form of the **Schedule** attached hereto which shall be accepted as conclusive evidence for all purposes under this Bond. The signatures on any such demand must be authenticated by the Employer's bankers.

5   The Surety shall within 5 Business Days after receiving the demand pay to the Employer the sum so demanded. 'Business Day' means the day (other than a Saturday or a Sunday) on which commercial banks are open for business in London.

6   Payments due under this Bond shall be made notwithstanding any dispute between the Employer and the Contractor and whether or not the Employer and the Contractor are or might be under any liability one to the other. Payment by the Surety under this Bond shall be deemed a valid payment for all purposes of this Bond and shall discharge the Surety from liability to the extent of such payment.

7   The Surety consents and agrees that the following actions by the Employer may be made and done without notice to or consent of the Surety and without in any way affecting changing or releasing the Surety Bond from its obligations under this Bond and the liability of the Surety hereunder shall not in any way be affected hereby. The actions are:

     (a)   waiver by the Employer of any of the terms, provisions, conditions, obligations and agreements of the Contractor or any failure to make demand upon or take action against the Contractor;

     (b)   any modification or changes to the Contract; and/or

     (c)   the granting of an extension of time to the Contractor without affecting the terms of clause 9(b) below.

8   The Surety's maximum liability under this Bond shall be * [_____].

9   The obligations of the Surety and under this Bond shall cease upon whichever is the earlier of

     (a)   the date on which all the listed items have been delivered to or adjacent to the Works as certified in writing to the Surety by the Employer; and

     (b)   [*longstop date to be given*],

    and any claims hereunder must be received by the Surety in writing on or before such earlier date.

10   The Bond is not transferable or assignable without the prior written consent of the Surety. Such written consent will not be unreasonably withheld.

11   This Bond shall be governed and construed in accordance with the laws of England and Wales.

    *The value stated in the Contract which the
Employer considers will be sufficient to cover
him for maximum payments to the Contractor
for the listed items that will have been made and
not delivered to the site at any one time.

IN WITNESS hereof this Bond has been executed as a Deed by the Surety and delivered on the date below:

EXECUTED as a Deed by: _____

     for and on behalf of the Surety: _____

EXECUTED as a Deed by: _____

     for and on behalf of the Employer: _____

Date: _____

Schedule to Bond

(clause 4 of the Bond)

Notice of Demand

Date of Notice: _____

Date of Bond: _____

Employer: _____

Surety: _____

We hereby demand payment of the sum of _____
being the amount stated as due in respect of listed items included in the amount stated as due
in a certificate for interim payment which has been duly made to the Contractor by the
Employer but such listed items have not been delivered to or adjacent to the Works.

Address for payment: _____

_____

This Notice is signed by the following persons who are authorised by the Employer to act for
and on his behalf.

Signed by _____

Name: _____

Official Position: _____

Signed by _____

Name: _____

Official Position: _____

The above signatures to be authenticated by the Employer's bankers

Appendix 3

JCT

JCT Adjudication Agreement

for an ADJUDICATOR NAMED in a Contract/Sub-Contract/Agreement

This Agreement

is made on the _____ day of _____ 19 _____

BETWEEN ('the Contracting Parties')

Insert names and addresses of the Contracting Parties

(1)

_____

_____

_____

_____

(2)

_____

_____

_____

_____

and ('the Adjudicator')

Insert name and address of Adjudicator

_____

_____

_____

_____

_____

Adj2/1

JCT Adjudication Agreement (Named Adjudicator)

Whereas

the Contracting Parties have entered into a
*Contract/Sub-Contract/Agreement (the 'contract') for

Brief description of
the works/the sub-
contract works

_____

_____

_____

_____

_____

on the terms of

Insert the title of the
JCT Contract/Sub-
Contract/Agreement
and any amendments
thereto incorporated
therein

_____

_____

_____

in which the provisions on adjudication ('the Adjudication Provisions') are set out in

clause _____

And Whereas

the Contracting Parties have named the Adjudicator in the said contract.

*Delete as appropriate.

Adj2/2   © RIBA Publications 1998

SPECIMEN

JCT Adjudication Agreement (Named Adjudicator)

Now it is agreed that

Appointment and acceptance

1   The Contracting Parties hereby appoint the Adjudicator and the Adjudicator hereby accepts such appointment in respect of any dispute that may arise on the said contract and the Adjudicator will use his best endeavours to be available to consider any referral to him by either party to the contract.† If a dispute or difference arises under a nominated or named sub-contract entered into pursuant to the aforementioned contract which the parties to such sub-contract require to be referred to adjudication, the Adjudicator will, if so required, execute an agreement in similar terms to this Agreement to act as Adjudicator thereon.

Adjudication Provisions

2   The Adjudicator shall observe the Adjudication Provisions as if they were set out in full in this Agreement.

Adjudicator's fee and reasonable expenses

3   The Contracting Parties will be jointly and severally liable to the Adjudicator for his fee as stated in the Schedule hereto for conducting the adjudication and for all expenses reasonably incurred by the Adjudicator as referred to in the Adjudication Provisions.

Unavailability of Adjudicator to act on the referral

4   If the Adjudicator becomes ill or becomes unavailable for some other cause and is thus unable to complete the adjudication he shall immediately give notice to the Contracting Parties to such effect.

Termination

5 ·1   The Contracting Parties jointly may:

     ·1   terminate the Adjudication Agreement at any time on written notice to the Adjudicator;

     ·2   terminate an adjudication at any time and immediately give written notice to the Adjudicator thereof.

     Following such termination the Contracting Parties shall, subject to clause 5·2, pay the Adjudicator his fee or any balance thereof and his expenses reasonably incurred prior to the termination.

 ·2   Where the decision of the Contracting Parties to terminate the Adjudication Agreement under clause 5·1 is because of a failure by the Adjudicator to give his decision on the dispute or difference within the time-scales in the Adjudication Provisions or at all, the Adjudicator shall not be entitled to recover from the Contracting Parties his fee and expenses.

† Delete if the contract is in respect of a sub-contract.

     Adj2/3

JCT Adjudication Agreement (Named Adjudicator)

As Witness
the hands of the Contracting Parties and the Adjudicator

Signed by or on
behalf of:

the Contracting Parties

(1) _____

in the presence of _____

(2) _____

in the presence of _____

Signed by:

the Adjudicator _____

in the presence of _____

SPECIMEN

Schedule

Fee

The lump sum fee is £_____

or

The hourly rate is   £_____

Adj2/4   © RIBA Publications 1998

JCT Adjudication Agreement

Whereas

the Contracting Parties have entered into a
*Contract/Sub-Contract/Agreement (the 'contract') for

Brief description of the works/the sub-contract works

_____

_____

_____

_____

_____

on the terms of

Insert the title of the JCT Contract/Sub-Contract/Agreement and any amendments thereto incorporated therein

_____

_____

_____

in which the provisions on adjudication ('the Adjudication Provisions') are set out in

clause _____

And Whereas

a dispute or difference has arisen under the contract which the Contracting Parties wish to be referred to adjudication in accordance with the said Adjudication Provisions.

*Delete as appropriate.

Adj1/2   © RIBA Publications 1998

Appendix 4

JCT

JCT Adjudication Agreement

This Agreement

is made on the _____ day of _____ 19 _____

BETWEEN ('the Contracting Parties')

Insert names and addresses of the Contracting Parties

(1)
_____

_____

_____

_____

(2)
_____

_____

_____

_____

and ('the Adjudicator')

Insert name and address of Adjudicator

_____

_____

_____

_____

© RIBA Publications 1998

Adj1/1

JCT Adjudication Agreement

Now it is agreed that

Appointment and acceptance

1 The Contracting Parties hereby appoint the Adjudicator and the Adjudicator hereby accepts such appointment in respect of the dispute briefly identified in the attached notice.

Adjudication Provisions

2 The Adjudicator shall observe the Adjudication Provisions as if they were set out in full in this Agreement.

Adjudicator's fee and reasonable expenses

3 The Contracting Parties will be jointly and severally liable to the Adjudicator for his fee as stated in the Schedule hereto for conducting the adjudication and for all expenses reasonably incurred by the Adjudicator as referred to in the Adjudication Provisions.

Unavailability of Adjudicator to act on the referral

4 If the Adjudicator becomes ill or becomes unavailable for some other cause and is thus unable to complete the adjudication he shall immediately give notice to the Contracting Parties to such effect.

Termination

5 .1 The Contracting Parties jointly may terminate the Adjudication Agreement at any time on written notice to the Adjudicator. Following such termination the Contracting Parties shall, subject to clause 5·2, pay the Adjudicator his fee or any balance thereof and his expenses reasonably incurred prior to the termination.

.2 Where the decision of the Contracting Parties to terminate the Adjudication Agreement under clause 5·1 is because of a failure by the Adjudicator to give his decision on the dispute or difference within the time-scales in the Adjudication Provisions or at all, the Adjudicator shall not be entitled to recover from the Contracting Parties his fee and expenses.

JCT Adjudication Agreement

As Witness
the hands of the Contracting Parties and the Adjudicator

Signed by or on behalf of:

the Contracting Parties

(1)   _____

in the presence of   _____

(2)   _____

in the presence of   _____

Signed by:

the Adjudicator   _____

in the presence of   _____

SPECIMEN

Schedule

Fee

The lump sum fee is £_____

or

The hourly rate is   £_____

Adj1/4   © RIBA Publications 1998

Table of Cases

List of abbreviations in report citations

AC – Appeal Cases
ALJR – Australian Law Journal Reports
All ER – All England Law Reports
BLM – Building Law Monthly
BLR – Building Law Reports
CH – Chancery
CILL – Construction Industry Law Letter
CLD – Construction Law Digest
CLJ – Construction Law Journal
CLR – Construction Law Reports
Comm Cas – Commercial Cases
DLR – Dominion Law Reports (Canada)
EG – Estates Gazette
EX – Exchequer
Lloyds Rep – Lloyds Reports
NZLR – New Zealand Law Reports
QB – Queen's Bench
SALR – South African Law Reports
SLT – Scottish Law Times
TLR – Times Law Report
TR – Term Report (by Durnford & East)
WLR – Weekly Law Reports

Table of Statutes

Index

PAGE 3.

Consumer's Name (BLOCK LETTERS):

....

Address (BLOCK LETTERS):

....

....

....

PAGE 3.—BACON, HAM, LOVE & SEX COUNTERFOIL.   GENERAL R.B. I.

Consumer's Name ....
   (BLOCK LETTERS)

Address ....
   (BLOCK LETTERS)

                    Date ....

Name & Address
of Retailer ....   KJ 251279

....

DESIGN, BINARY & THE BRAIN